New Physico-Chemical Techniques for the Characterization of Complex Food Systems

edited by

ERIC DICKINSON

Procter Department of Food Science
University of Leeds, UK

BLACKIE ACADEMIC & PROFESSIONAL
An Imprint of Chapman & Hall
London · Glasgow · Weinheim · New York · Tokyo · Melbourne · Madras

Published by
Blackie Academic & Professional, an imprint of Chapman & Hall
Wester Cleddens Road, Bishopbriggs, Glasgow G64 2NZ

Chapman & Hall, 2–6 Boundary Row, London SE1 8HN, UK

Blackie Academic & Professional, Wester Cleddens Road, Bishopbriggs, Glasgow G64 2NZ, UK

Chapman & Hall GmbH, Pappelallee 3, 69469 Weinheim, Germany

Chapman & Hall USA, 115 Fifth Avenue, Fourth Floor, New York NY 10003, USA

Chapman & Hall Japan, ITP-Japan, Kyowa Building, 3F, 2-2-1 Hirakawacho, Chiyoda-ku, Tokyo 102, Japan

DA Book (Aust.) Pty Ltd, 648 Whitehorse Road, Mitcham 3132, Victoria, Australia

Chapman & Hall India, R. Seshadri, 32 Second Main Road, CIT East, Madras 600 035, India

First edition 1995

Typeset in 10/12pt Times by Pure Tech India Ltd. Pondicherry, India
Printed in Great Britain by Cambridge University Press, Cambridge

ISBN 0 7514 0252 4

A catalogue record for this book is available from the British Library
Library of Congress Catalog Card Number: 95-75905

∞ Printed on acid-free text paper, manufactured in accordance with ANSI/NISO Z39.48-1992 (Permanence of Paper)

Preface

'There is no higher or lower knowledge, but only one, flowing out of experimentation.'
(Leonardo da Vinci, 1452–1519)

Food materials are complex in terms of composition, structure and mechanical properties. In order to understand the relationship between these different kinds of complexity, the experimental food scientist has a wide range of physico-chemical techniques at his or her disposal. But, in practice, of course, there are often severe limitations on the techniques which are available for any particular investigation. Apart from obvious constraints associated with instrument cost and accessibility, one common problem is a lack of knowledge by the non-expert about the capabilities and limitations of every new advance in instrumentation. No individual worker in the field of food science can become expert in more than a very small number of experimental techniques. On the other hand, most of us wish to know enough about the major emerging experimental technologies to enable us to make a realistic assessment of what they may have to contribute towards any new problems that we may meet.

This book collects together in a single volume an up-to-date set of introductory articles describing a range of new physico-chemical techniques which can be used to probe food structure at the molecular, colloidal and microscopic levels. Each individual chapter is written by an acknowledged expert in his field. The whole set of 14 chapters is based on the invited lectures delivered at a symposium on *New Physico-Chemical Techniques for the Characterization of Complex Food Systems* organized by the Department of Continuing Professional Education at the University of Leeds on 27–29th September 1994 at the Weetwood Hall Conference and Training Centre, Leeds, UK. Strongly represented amongst the contributions collected here are descriptions of different types of microscopy and spectroscopy, as well as various scattering and rheological techniques. The authors were each asked to do quite a difficult job—critically to describe their 'state-of-the-art' technique using a 'down-to-earth' style of approach. Each chapter introduces the essential background principles of the technique, discusses its main advantages and disadvantages, and illustrates its application with a number of well-chosen food-related examples. The book is meant to be appropriate for use as the main text in a graduate level course on 'physico-chemical techniques in food science'. It is therefore aimed towards those who teach and learn about food in universities and colleges, as well as active researchers and technologists in industry, research institutes, and academic institutions.

I am extremely grateful to the chapter authors for their hard work on this project and for their general promptness in delivering their manuscripts to me and in responding to my editorial queries.

E. Dickinson
University of Leeds

Contributors

S. Ablett	Unilever Research Laboratory, Colworth House, Sharnbrook, Bedford MK44 1LQ, UK
B.E. Brooker	Institute of Food Research, Earley Gate, Whiteknights Road, Reading, Berks RG6 2EF, UK
A.H. Darke	Unilever Research Laboratory, Colworth House, Sharnbrook, Bedford MK44 1LQ, UK
E. Dickinson	Procter Department of Food Science, The University of Leeds, Leeds LS2 9JT, UK
J. Eastoe	School of Chemistry, University of Bristol, Bristol BS8 1TS, UK
M.J. Gidley	Unilever Research Laboratory, Colworth House, Sharnbrook, Bedford MK44 1LQ, UK
R. Gray	The Department of Agriculture for Northern Ireland, Food Science Division, Newforge Lane, Belfast BT9 5PX, UK
S.E. Harding	Department of Applied Biochemistry and Food Science, University of Nottingham, Sutton Bonington LE12 5RD, UK
I. Heertje	Unilever Research Laboratorium, Vlaardingen, The Netherlands
B.P. Hills	Institute of Food Research, Norwich Research Park, Colney, Norwich NR4 7UA, UK
D.S. Horne	Hannah Research Institute, Ayr KA6 5HL, UK
H. Luyten	Department of Food Science, Wageningen Agricultural University, Wageningen, The Netherlands
A.J. McArthur	Unilever Research Laboratory, Colworth House, Sharnbrook, Bedford MK44 1LQ, UK
T.J. McMaster	H.H. Wills Physics Laboratory, Tyndall Avenue, Royal Fort, University of Bristol, Bristol BS8 1TL, UK

M.J. Miles H.H. Wills Physics Laboratory, Tyndall Avenue, Royal Fort, University of Bristol, Bristol BS8 1TL, UK

M. Pâques Unilever Research Laboratorium, Vlaardingen, The Netherlands

M.J.W. Povey Procter Department of Food Science, The University of Leeds, Leeds LS2 9JT, UK

A. Prins Department of Food Science, Wageningen Agricultural University, Wageningen, The Netherlands

S.B. Ross-Murphy Division of Life Sciences, King's College London, Campden Hill Road, Kensington, London W8 7AH, UK

T. van Vliet Department of Food Science, Wageningen Agricultural University, Wageningen, The Netherlands

R.H. Wilson Institute of Food Research, Norwich Laboratory, Norwich Research Park, Colney, Norwich NR4 7UA, UK

Contents

Chapter 1

Advances in Electron Microscopy

I. Heertje and M. Pâques
Unilever Research Laboratorium,
Vlaardingen, The Netherlands

1.1 Introduction

Most food products are composed of rather a limited variety of structural elements, such as droplets, air cells, granules, crystals, strands, micelles and interfaces. A proper understanding of the behaviour of a food product requires knowledge of its structure, i.e. the spatial arrangement of the structural elements and their interactions (Fig. 1.1). Interaction forces determine the consistency and the physical stability of food products.

The spatial arrangement of structural elements can be studied by visual observation techniques, such as light and electron microscopy. Also, the result of the interaction between the various components in a system can be studied in this manner. Visual observation techniques are therefore an important aid in the analysis of food structure (Vaughan, 1979; Aguilera and Stanley, 1990). Also the journal *Food Structure* (SMI, Chicago, formerly *Food Microstructure*) is an extensive source for microstructural studies, with special emphasis on relations between processing, molecular properties, microstructure and textural behaviour. In general, the information from image-forming systems is of a qualitative nature. However, by applying image processing and analysis techniques, quantitative information can also be extracted.

Among the image-forming techniques, electron microscopy (EM) plays an important role. It should, however, be realized that EM is only one

Ingredients ⟶ Process ⟶ Structure ⟶ Properties and Function

Conformation	Hardness
Denaturation	Gel strength
Molecular mass	Spreadability
Interaction	Mouthfeel
Crystal modification	Emulsion stability
Liquid crystallinity	Water binding

Fig. 1.1 Relation between composition, processing, structure and behaviour.

of many possible techniques for studying microstructure. It is certainly not a simple technique and it requires, in general, complicated preparation procedures, which can give rise to artefacts. Its use is justified when high resolution information, above the resolution limit of light microscopy (LM), is needed. In addition, for a complete description of a food system, integration of information at all levels of organization (Davis and Gordon, 1982) is always required.

Electron microscopy in itself is a vast expertise area (Hawkes and Valdrè, 1990). The field includes different types of electron microscopes for morphological observations, specific localization techniques, an impressive variety of preparation techniques, and techniques for image processing and analysis.

Two possibilities for image formation can be distinguished:

(i) *a fixed beam illumination and the simultaneous recording of all image points* This principle is applied in conventional transmission electron microscopy (CTEM) and also in classical LM.
(ii) *scanning the specimen with a small electron beam with sequential recording of the produced signal* This principle is applied in scanning transmission EM (STEM) as well as in the scanning EM (SEM). The latter technique is used for the observation of bulk surfaces, in contrast to transmission instruments (CTEM as well as STEM), which are used for the investigation of thin (electron transparent) specimens.

Apart from morphological observations by transmission and scanning EM it is also possible to perform compositional analysis of micro-environments. In this review attention will be given to X-ray microanalysis (XRMA), electron energy loss spectroscopy (EELS) and specific labelling techniques involving EM. The first two of these techniques are element-specific localization techniques. XRMA (Morgan, 1985) is based on the phenomenon of formation of X-rays when electrons impinge onto a sample. As these X-rays are element-specific, a compositional map of excitable elements can be obtained by sequential recording. XRMA can be applied in TEM and SEM. EELS is based on a characteristic energy loss of electrons transmitted through a sample. Compositional information is obtained by recording the specific energy loss using an electron spectrometer. By the nature of the technique, its application is restricted to thin specimens.

Specific labelling is based on the general principle of a specific affinity between ligand and receptor molecules, e.g. antigens and antibodies, sugars and lectins, enzymes and their substrates. In immuno-microscopy (Polak and Varndell, 1984; Polak and van Noorden, 1983, 1984), various compounds (antigens A_g) can be specifically identified, localized and quantified in a sample by the use of labelled (M) antibodies (A_b):

$$A_g + A_b \cdot M \rightarrow A_g \cdot A_b \cdot M.$$

The antigen is detected in the microscope by the localization of the marker from the $A_g \cdot A_b \cdot M$ complex. The technique is not only applicable to biological macromolecules, but in some cases also to small molecules. A special case is offered by lectins (L). These are proteins of a non-immunoglobulin nature which can bind specifically to carbohydrate moieties. They occur in a wide variety of plant materials. The use of lectins allows the specific localization of sugar (S) residues by a similar procedure as with antigens:

$$S + L \cdot M \rightarrow S \cdot L \cdot M.$$

Many types of markers can be used for specific labelling, but nowadays the most popular is colloidal gold (Beesley, 1989). Colloidal gold particles are very 'electron dense' and they can be prepared in the practical size range 1–20 nm.

Modern EM has much to offer in terms of ease of operation, technological breakthroughs, automation and data acquisition. Nevertheless, a good knowledge and much skill are still required to take high quality, high resolution micrographs. However, the real art in the EM of food products is specimen preparation, and in this respect there exists a close resemblance with biological samples. This is really not surprising because most food products, as well as biological samples, are mainly composed of water. This situation is not compatible with the grim EM conditions of high vacuum and high radiation dose. Therefore, extensive drying and/or stabilization regimes are required (Robards and Wilson, 1993) before samples can be introduced in the apparatus. In addition, for TEM, samples should be prepared very thin in order to make them transparent to the electron beam. This explains why SEM preparation techniques are in principle simpler than TEM preparation techniques.

It goes without saying that the applied extensive preparation procedures may give rise to artefactual observations in the final image. In order to avoid wrong interpretations, results from various preparation techniques and from other image-forming techniques should be compared. In addition, other structural information should be included in the final interpretation, and the EM of food products should not be pursued in isolation.

In what follows, various techniques are described and they are illustrated by examples taken from the foods area.

1.2 General techniques

Many developments in the design and performance of electron microscopes have taken place in the last decade. Also numerous observation techniques and procedures exist to obtain the best morphological and micro-compositional information. However, it is our contention that the real art in food microscopy is specimen preparation. Therefore, the emphasis is on some general principles of the main preparation techniques and their application, as illustrated by examples showing a varied number of structural elements and mutual interactions (Heertje, 1993a, b).

Elaborate preparation techniques are, in general, needed to allow proper EM observation of food products. Two important main routes for sample preparation can be distinguished: chemical methods and physical methods. In the first approach, chemicals are used to dehydrate (and solidify) liquid or semi-solid samples. These chemical agents may induce changes in molecules and structures and give rise to artefacts. In the second approach the samples are solidified by freezing, and the frozen aqueous samples are subsequently treated in a number of different ways (Echlin, 1992). The cryo-fixation is a very critical step because of the formation of ice crystals, which may disturb the structure. Very high freezing rates are therefore required. Many ways of achieving this have been described (Echlin, 1992).

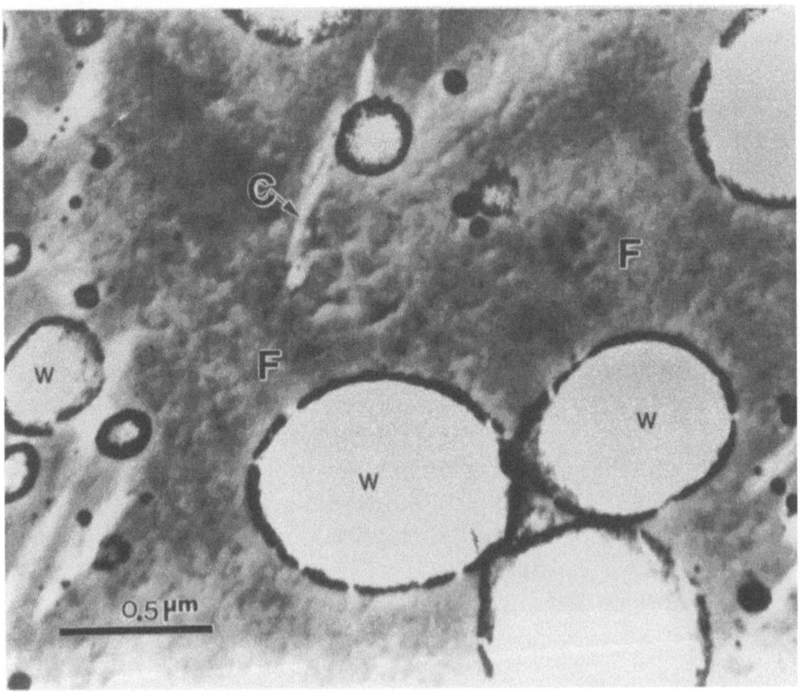

Fig. 1.2 Water droplets (W) in a margarine. Linear structures (C), indicative of fat crystals, can be observed in the continuous fat matrix (F). Dark structures at the water–fat matrix interface indicate emulsifier. Technique: ultramicrotomy.

1.2.1 Techniques in transmission electron microscopy

(a) Ultramicrotomy. Due to the limited penetration of the electron beam, it is necessary to prepare ultra-thin sections of food specimens which do not exceed 100 nm and are preferably thinner. In preparing these thin specimens complicated preparation schemes are required involving fixation, dehydration, infiltration and embedding, thin sectioning and staining. These steps have been amply described (Glauert and Reid, 1974; Robards and Wilson, 1993) and they require much skill and expertise. Attainable resolution with this technique is about 2 nm. Consequently, this type of TEM is a powerful preparation technique with a high information content. However, it is an elaborate technique, requiring much effort, and by its very nature the occurrence of artefacts should always be suspected.

Examples of water-in-oil and oil-in-water emulsion structure and foam structure illustrate the application to food products. Water droplets are important structural elements in emulsion-type food products such as margarine, butter and low-fat spreads. The droplets are stabilized by soluble emulsifiers, such as monoacylglycerols and lecithins and/or by solid particles. Figure 1.2 shows a water droplet in margarine and the

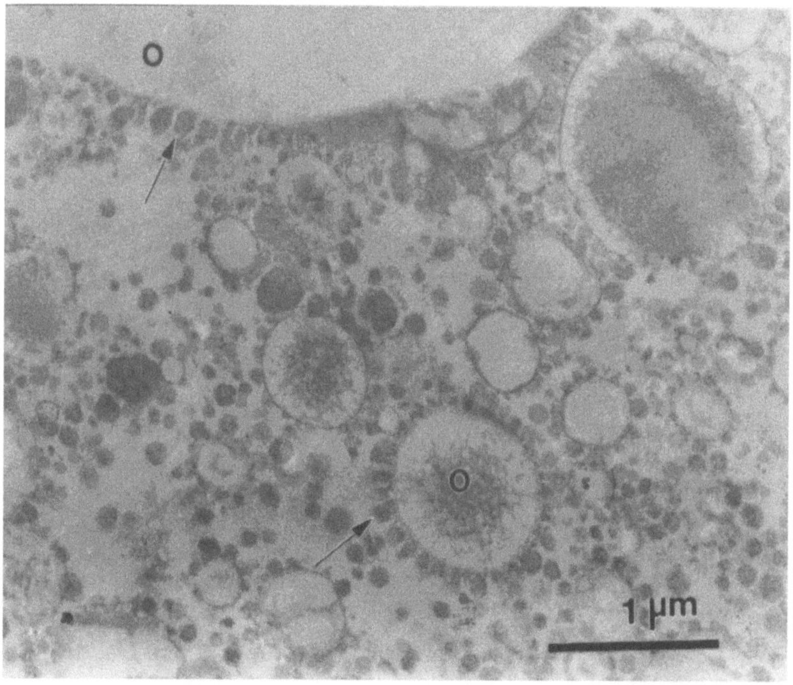

Fig. 1.3 Oil droplets (O) in homogenized milk, covered with casein micelles (arrows). Technique: ultramicrotomy.

presence of emulsifier at the interface between the water and the oil. Oil or fat droplets occur as structural elements in such diverse products as milk, cream, cheese, mayonnaise, dressings, meat products and ice-cream. Figure 1.3 shows milk fat droplets covered with casein micelles in a homogenized milk. In many food products such as bread, cake, whipped cream, ice-cream, beer and chocolate mousse, gas cells play an important role. The gas cells are often immobilized in a solid matrix, whereas in other cases stabilization is achieved by protein films around the air cells. Air cells in whipped cream are stabilized by the combined action of protein and fat globules (Brooker *et al.*, 1986). Figure 1.4 clearly demonstrates the localization of fat globules at the air–water interface and remnants of the protein film. The examples in Figs 1.2–1.4 indicate the versatility of ultra-microtomy in the analysis of food microstructure.

(b) Negative staining. In negative staining the object is immersed in a solution of a heavy metal salt such as ammonium molybdate, sodium phosphotungstate or uranyl acetate. After drying, the specimen is embedded in the electron-dense metal stain; the specimen itself appears

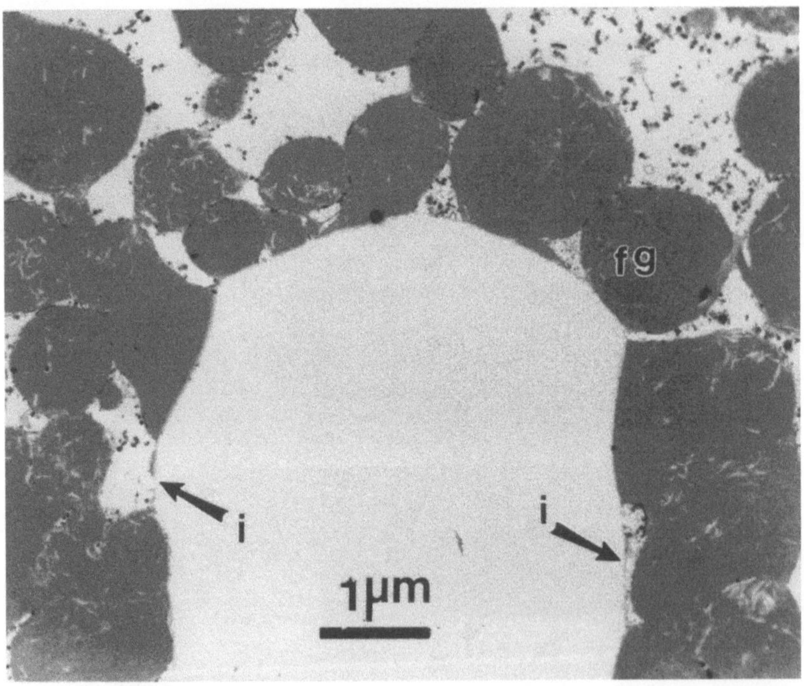

Fig. 1.4 Air cell in whipped cream. The air cell is stabilized by fat globules (fg) adsorbed at the interface and remnants of a protein film (i). Technique: ultramicrotomy. (Courtesy B. Brooker, Institute of Food Research, Reading, UK.)

translucent which allows for the observation of the internal structure of many objects. In its simplest form a solution or suspension of particles or macromolecules is mixed with the stain, sprayed onto a substrate, air-dried and observed in the TEM. Many variations exist (Hayat and Miller, 1990; Kiselev *et al.*, 1990; Robards and Wilson, 1993). Attainable resolution is about 2 nm, although under special conditions, avoiding radiation damage, higher resolutions can be obtained (Kiselev *et al.*, 1990; Bremer *et al.*, 1992). Besides being a simple and quick procedure, this method often yields images of regular arrays of macromolecules that show a well preserved order. This allows averaging of images of particles in order to improve the signal-to-noise ratio and the application of digital image processing techniques to elucidate the three-dimensional structure of macromolecules and macromolecular assemblies.

There are numerous pitfalls and some limitations in the application of negative staining (Hayat and Miller, 1990). High salt concentrations and pH changes induced by drying, dehydration, flattening of particles, and aspecificity of staining, may give rise to artefacts, wrong interpretations and irreproducible results.

Fig. 1.5 Network of glutenin polymers, showing elongated strands and spherical particles. Technique: negative staining.

In particular, biopolymer strands can be properly visualized by this technique. Strands may be formed by aggregation of macromolecules or macromolecular assemblies in globular form (particularly proteins) which usually produce particle gels (Dickinson, 1990). Alternatively, strands may consist of extended random coil molecules, which form association networks with entanglements or junction zones. A study on wheat flour proteins (Graveland *et al.*, 1993) has demonstrated that the glutenin polymers form large aggregates. Network formation at a molecular level is observed, the structure of which depends on the type of flour. Figure 1.5 shows a network of elongated strands with spherical globules alongside. It is suggested that the strands consist of high molecular weight glutenin sub-units, whereas the spherical globules consist of low molecular weight sub-units. A schematic drawing of glutenin aggregation during processing, based on TEM observations using the negative stain technique, is presented in Fig. 1.6.

The morphology of strands of polysaccharides has also been amply investigated. Strands greatly varying in size, shape, flexibility and branching can be observed. A typical example is offered by a recent study

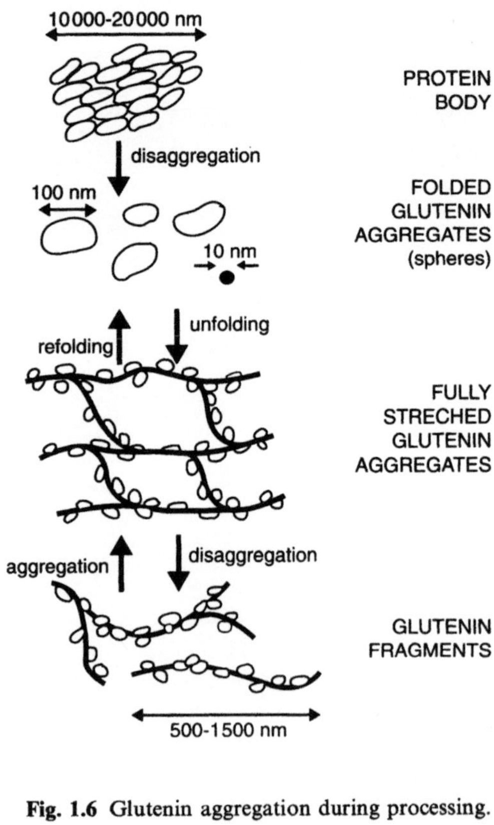

Fig. 1.6 Glutenin aggregation during processing.

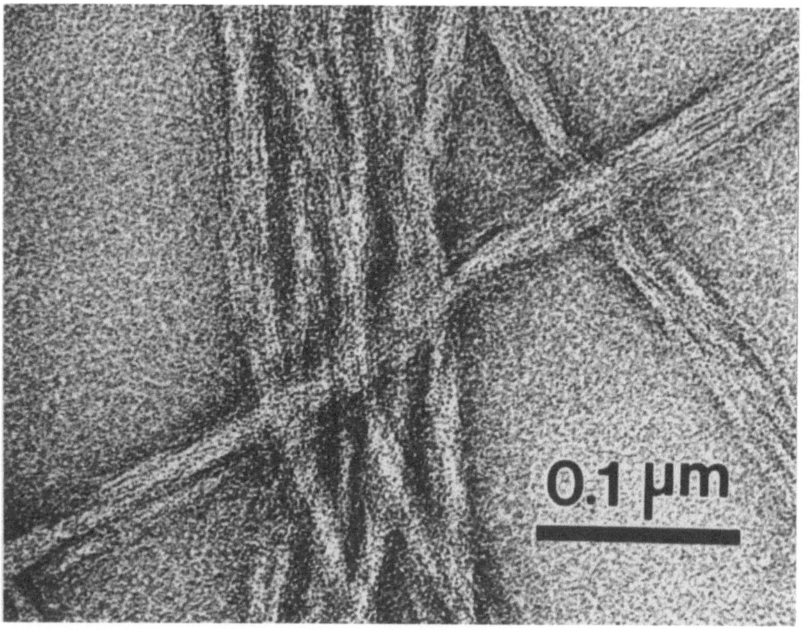

Fig. 1.7 Stiff polysaccharide strands observed in a curdlan gel. Technique: negative staining. (Courtesy T. Harada.)

on curdlan, a bacterial polysaccharide, discovered by Harada *et al.* (1991). Different types of molecular association in curdlan can be distinguished depending on the processing conditions and the chemical treatment. An example showing stiff curdlan strands composed of microfibrils is presented in Fig. 1.7.

Negative staining has also been applied to study the heat-induced association between the whey protein β-lactoglobulin and casein micelles or κ-casein. This association appears to have important consequences for the final properties of dairy products. Minor variations in pH have been shown to have large effects on the strength of adherence of protein aggregates to casein micelles (Creamer *et al.*, 1978). In this context, model experiments with β-lactoglobulin and casein micelles have been performed as a function of pH (Figs 1.8 and 1.9). At pH 7.0, separate micelles and thread-like truncated protein structures are observed and the granular micelle structure is clearly visible. At pH 6.5, β-lactoglobulin is concentrated on the micelles which now have a woolly fuzzy appearence, apparently due to a reaction of the whey protein with the micelle surface. Apparently, a minor change in pH has a very substantial effect on the reactions taking place at the micellar surface. To get a better insight into the nature of the free protein particles at pH 7.0,

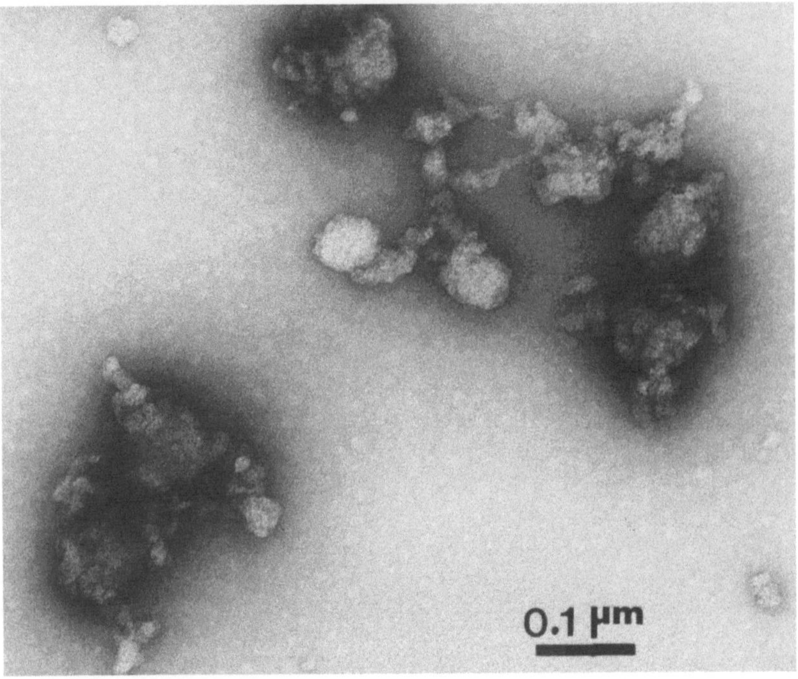

Fig. 1.8 Heat-induced association of β-lactoglobulin and casein micelles at pH 6.5. Strong interaction. Technique: negative staining.

micrographs of heated solutions of κ-casein, β-lactoglobulin and of a 1:1 mixture of both proteins have been prepared. Heated κ-casein often exhibits spherical particles (Fig. 1.10). Heated β-lactoglobulin shows irregular thread-like particles which appear to be connected into a network (Fig. 1.11). The micrograph of the tested mixture of the two proteins exhibits noodle-like truncated particles (Fig. 1.12) similar to those observed for the interaction between β-lactoglobulin and casein micelles at pH 7.0 (Fig. 1.9). Apparently, the free protein particles represent the product of interaction between β-lactoglobulin and κ-casein. This has been further substantiated by chemical evidence.

(c) Replication and shadowing. In TEM the specimen must be sufficiently thin to allow the transmission of electrons. In the case of free surfaces, or surfaces obtained after cleaving the material in some way, replica techniques can be applied. In these techniques, a thin electron-transparent layer which faithfully moulds the surface contours is prepared and then stripped off from the specimen (Willison and Rowe, 1980). The contrast in the replicas is obtained by differences in thickness in the direction of the electron beam which can be further enhanced by

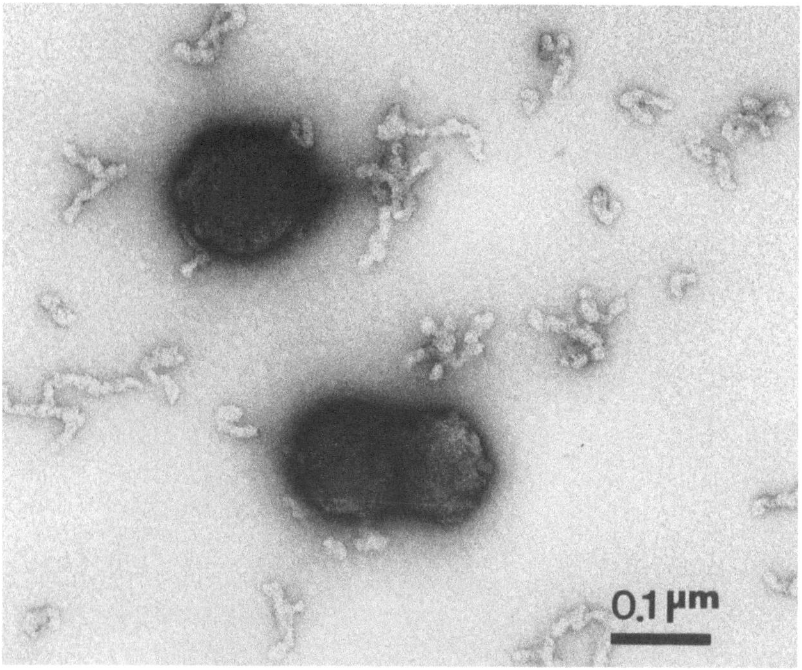

Fig. 1.9 Heat-induced association of β-lactoglobulin and casein micelles at pH 7.0. No interaction. Technique: negative staining.

shadowing, i.e. the evaporation of metal from a source set at an oblique angle relative to the specimen surface. Many experimental approaches and many replicating materials, such as plastics, tapes and solutions, have been described (Willison and Rowe, 1980). However, in biology and food science, vacuum evaporation from a carbon source has become the most popular because it allows the best resolution. The attainable lateral resolution is of the order of 5 nm, but also the other types of resolution should be distinguished. Particle height resolution of the order of 2 nm can be reached when specimens are deposited on very smooth surfaces (e.g. freshly cleaved mica) and with very low shadowing angles. This should be considered in the imaging of extended bio-macromolecules such as DNA, xanthan and carrageenan. Step resolution, which is also of the order of 2 nm and very similar to particle height resolution, applies when steps between parallel planes are present. An example showing 4 nm growth steps on a triacylglycerol fat crystal is presented in Fig. 1.13.

In general, replica techniques have lost a lot of their original popularity due to the development of scanning EM which is now the preferred technique for the examination of surfaces. This trend will be strengthened by the development of field emission SEM, which offers a much better

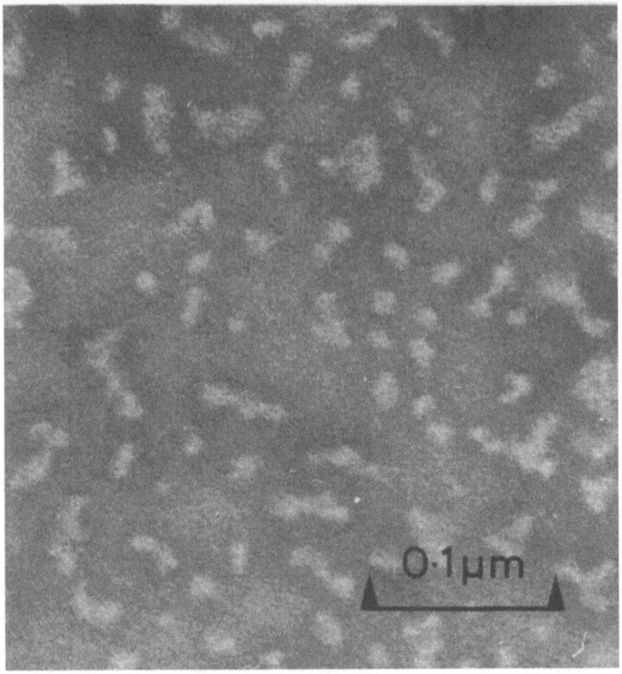

Fig. 1.10 Heated κ-casein, showing mostly spherical and some extended particles. Technique: negative staining.

resolution than conventional SEM, especially in low voltage applications. Nevertheless, in a number of cases replication is still the technique of choice for: (i) flexible elongated macromolecules or macromolecular assemblies; (ii) beam sensitive specimens, such as fat crystals at room temperature; (iii) freeze-fractured or freeze-etched samples. This latter technique has become extremely popular for investigating food products containing large amounts of water (Heertje *et al.*, 1985; Humphreys, 1989; Buchheim and Dejmek, 1990). Applications in these three areas are presented below.

Flexible elongated macromolecules. Biopolymers such as proteins and polysaccharides are present in many food systems, often in the form of gels. These gels are formed by networks of polymer strands which often consist of long, flexible macromolecules with a characteristic contour length, extension and flexibility. Replica techniques are well suited to image this type of molecule. Freshly cleaved mica is used as the substrate to deposit the strands. Various options to do this have been discussed and critically reviewed: spraying (Snoeren *et al.*, 1976), spraying with glycerol

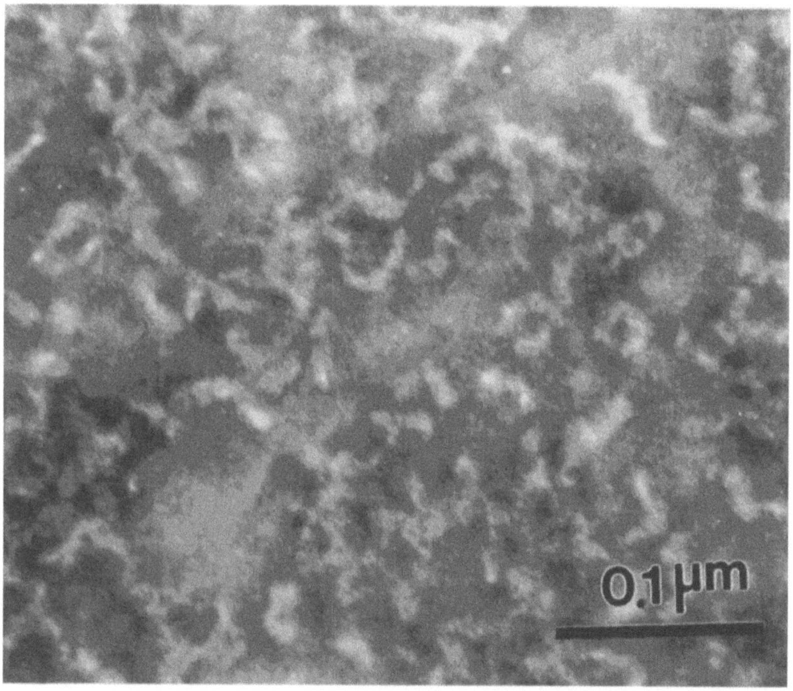

Fig. 1.11 Heated β-lactoglobulin, showing thread-like particles connected into a network. Technique: negative staining.

(Stokke and Brandt, 1990), the use of the mica sandwich (Hermansson, 1989), and micro-deposition (Robards and Abeysekera, 1992).

The application of the spray method is illustrated by a study of the interaction between κ-carrageenan and κ-casein or casein micelles, which is of importance in the structuring of many manufactured dairy products. Micrographs of κ-carrageenan, κ-casein and a mixture of both components, are presented in Fig. 1.14 which shows very fine carrageenan threads with a diameter of about 3 nm (Fig. 1.14(a)), spherical κ-casein particles with a diameter of about 15 nm (Fig. 1.14(b)), and the micro-structure of the mixture (Fig. 1.14(c)). Some particles of κ-casein appear to be adhering to the carrageenan threads, sometimes accompanied by chain formation. Similar observations have been reported by Snoeren *et al.* (1976).

A further study deals with the interaction between κ-carrageenan and casein micelles. Mixtures of κ-carrageenan and casein micelles heated for 10 min to temperatures of 60°C, 80°C or 90°C at pH 6.6 have been observed. A typical micrograph showing the interaction, after heating at 80°C, is presented in Fig. 1.15. It shows that, in the presence of carrageenan, the micelles partly disintegrate into smaller units that partly

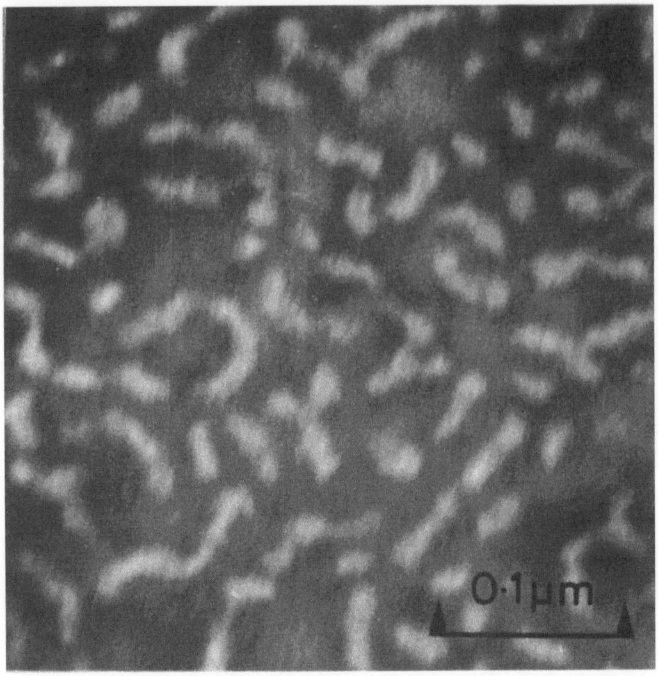

Fig. 1.12 Heated mixture of κ-casein and β-lactoglobulin, showing noodle-like particles. Technique: negative staining.

adhere to the carrageenan threads and form chains. The degree of disintegration appears to be dependent on the temperature. A schematic drawing of these observations is shown in Fig. 1.16 (Heertje, 1993b). The interactions are considered to be important for final product properties.

Another study deals with the influence of cation type and concentration on the gel formation of κ-carrageenan. A glycerol-spray method has been used. Figure 1.17 presents micrographs of heated (80°C) solutions of κ-carrageenan in 0.1 M KCl and 0.01 M NaCl. A strong influence of the electrolyte type and its concentration on the size, thickness and flexibility (persistence length) of the strands is indicated (Hermansson *et al.*, 1991). This is in agreement with the well-known influence of potassium ions on the gelation behaviour of κ-carrageenan. Firm cohesive gels of κ-carrageenan are obtained in the presence of potassium ions, whereas under identical conditions of concentration and ionic strength, no gels are formed in the presence of sodium ions.

Beam-sensitive specimens. The morphology of fat crystals plays an important role in the final product properties of various fat products such as margarine, butter and shortening. Product attributes such as

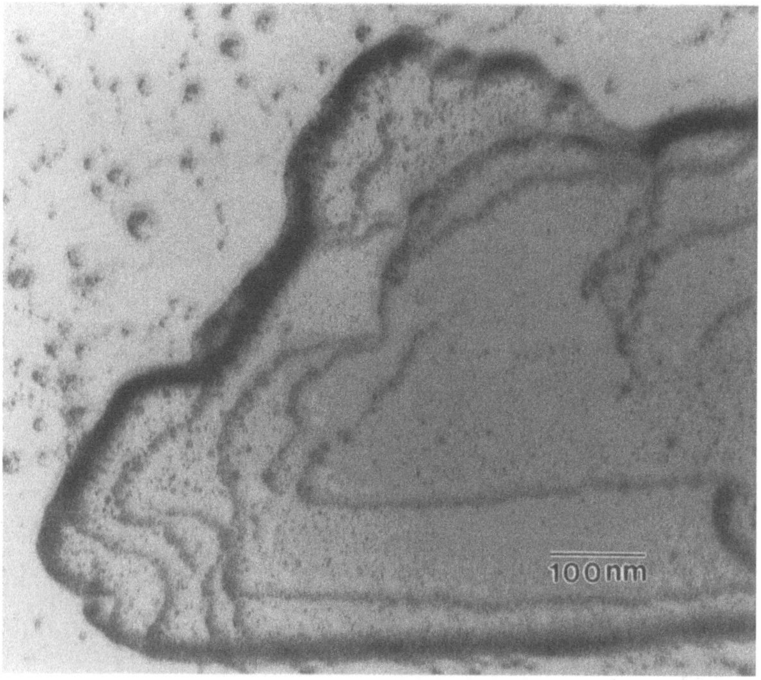

Fig. 1.13 Growth steps on a triacylglycerol fat crystal. Step height: 4.5 nm approximately. Technique: replica.

spreadability, hardness and work softening are determined at least partly by the shape and the size of the individual fat crystals and the way in which these crystals interact to form clusters, agglomerates and networks. Replica techniques are well suited to provide quantitative information on the size and shape of the individual fat crystals. However, in most fat products, large amounts of oil are also present, and this has to be removed for proper analysis. This can be achieved by washing with an appropriate detergent solution (Poot *et al.*, 1975) after which the crystal suspension is applied to a glass microscope slide and replicated after drying. An alternative technique has been developed recently which applies the fixation of the fat crystals to a carbon film with replication (Heertje and Leunis, 1994). Examples of fat crystal populations obtained by the carbon fixation technique are presented in Fig. 1.18. The large crystals (Fig. 1.18(a)) and the small crystals (Fig. 1.18(b)) were obtained respectively by slow and rapid cooling of a liquid fat blend below the melting temperature of the solid phase. Surface area distributions obtained from this type of picture are presented in Fig. 1.19. It appears that crystals in slowly cooled samples are approximately a factor 20 times larger in surface area than those in rapidly cooled samples.

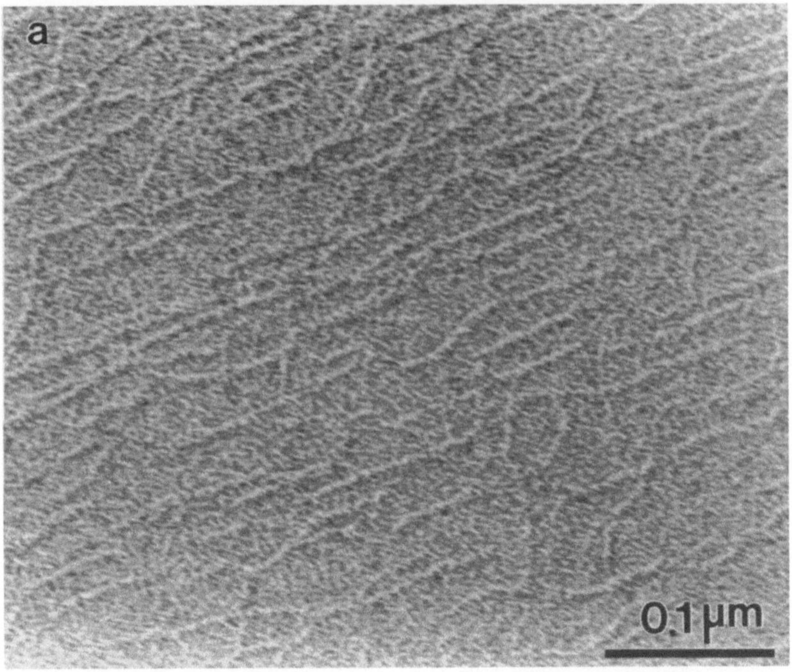

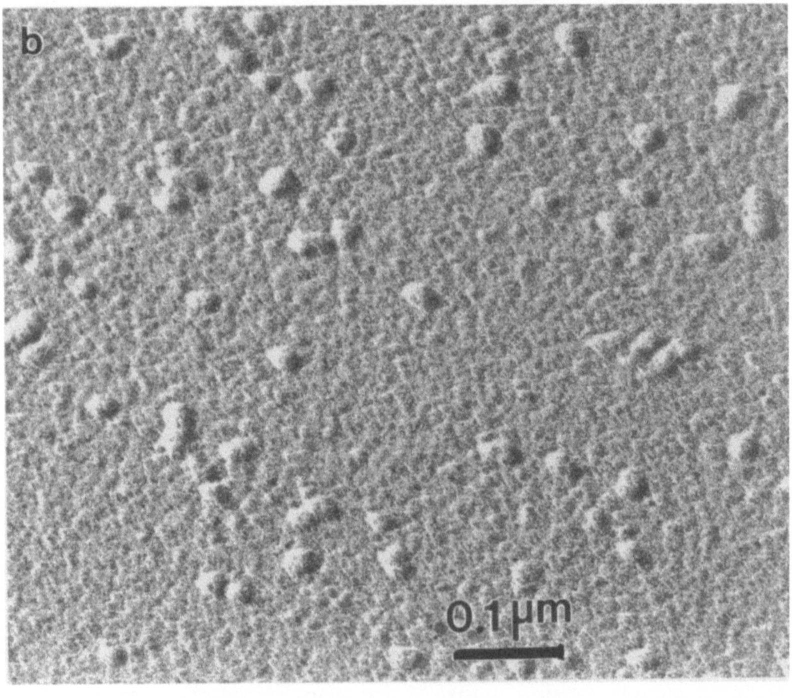

Fig. 1.14 (a) κ-carrageenan, (b) κ-casein,

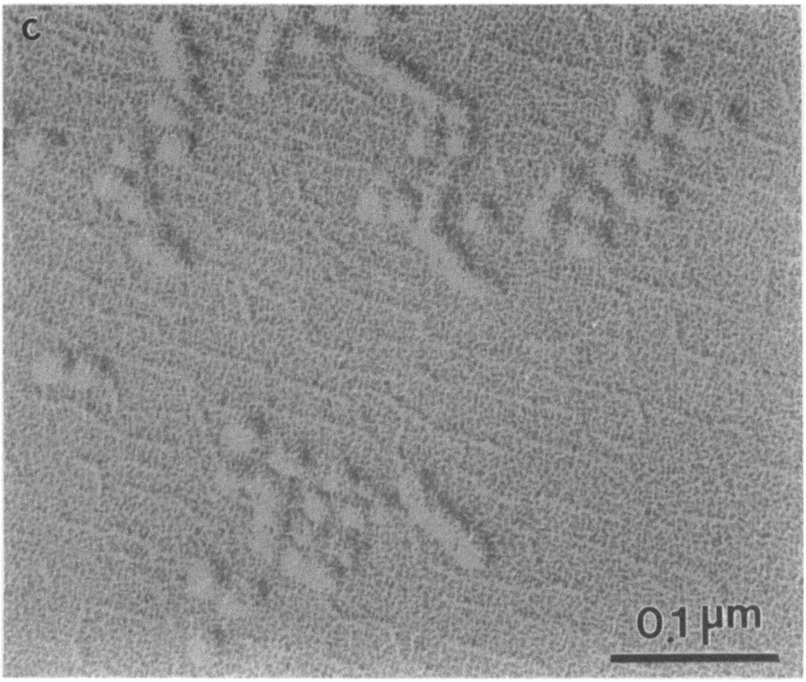

Fig. 1.14 (c) a mixture of both components; heated to 60°C for 10 min. Technique: replica of material sprayed on mica.

Freeze-fractured samples. Freeze-fracturing and etching, with replication, is a widely applied technique in the biosciences in general and in food science in particular (Heertje *et al.*, 1985; Humphreys, 1989; Buchheim and Dejmek, 1990). In particular, the Buchheim group has vastly contributed to the popularity of the technique of freeze-fracturing in the foods area. Extensive reviews and experimental details have been published (Willison and Rowe, 1980; Robards and Sleytr, 1985; Echlin, 1992). A schematic drawing of the basic principle, showing the sequence of steps from cooling the sample to the final cleaning of the replica, is depicted in Fig. 1.20. The most critical step in this technique and in all cryopreservation techniques is the cooling step. Very high cooling rates ($\sim 10^4$ K s^{-1}) and/or cryoprotectants are needed to avoid damage to the delicate structures by ice crystals. Numerous sample cooling procedures have been described (Robards and Sleytr, 1985; Echlin, 1992) to limit the ice crystal size as much as possible. Ideally, complete vitrification of water should be obtained, but in practice this can only be achieved for very thin specimens under special conditions (as for the thin film technique) or by high pressure freezing (Echlin, 1992).

The big advantage of this approach is that, ideally, no chemicals are used in the preparation of the samples and so the samples can be studied

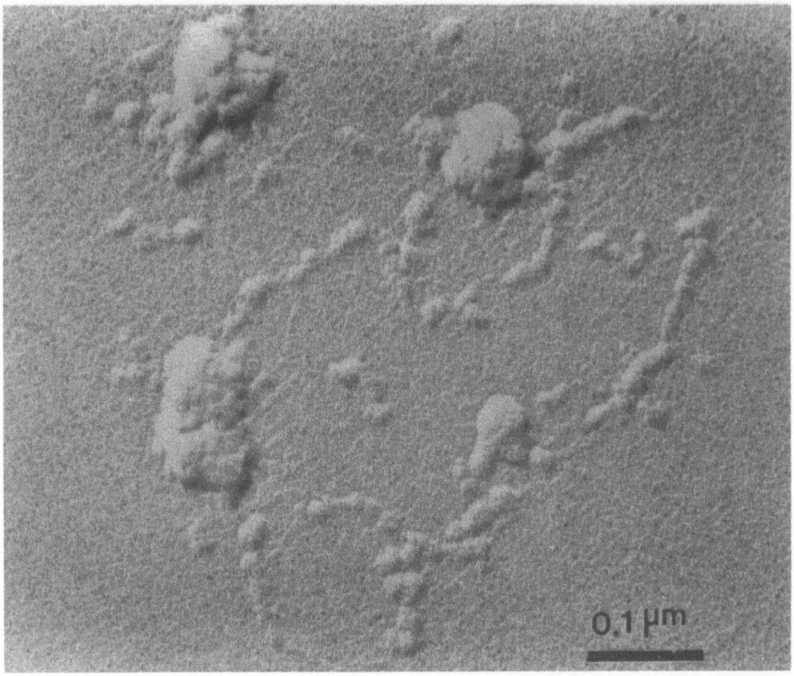

Fig. 1.15 A mixture of κ-carrageenan and casein micelles heated to 80°C for 10 min. Note the partial break-up of casein micelles and the adherence of their remnants to carrageenan threads accompanied by chain formation. Technique: replica of material sprayed on mica.

in their 'natural' aqueous environment. However, this technique also has its artefacts. It is very difficult to ascertain that small ice crystal damage has been completely avoided (e.g. in the case of gel-type structures). In addition, plastic deformation, dislocation and collapse may occur during fracturing and subsequent etching. The extent of damage depends on the nature of the specimen; systems containing biopolymers such as proteins and carbohydrates are quite susceptible to damage of this type.

Freeze fracture replication has *inter alia* been applied to water-in-oil and oil-in-water emulsions, protein particles in milk, gels of biopolymers, and starch-containing foods. Figure 1.21 shows water droplets in a margarine sample. The mean droplet size influences rheological, sensorial and microbial properties. Fracture occurs either over the surface of the water droplets or directly through the droplets. The droplets are seen to be stabilized by fat crystals covering the surface.

In Fig. 1.22 the emulsion structure of butter is presented. At the start of butter production water is the continuous phase, but during churning and subsequent working, the fat phase becomes the continuous phase. Through more intensive working, the water becomes more finely dis-

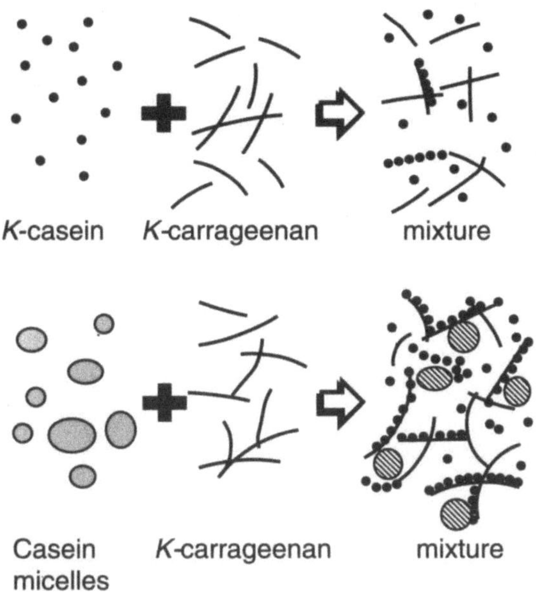

Fig. 1.16 Schematic drawing of the interactions of κ-carrageenan with κ-casein and casein micelles. At low temperatures (< 60°C), no appreciable disintegration of casein micelles occurs, whereas at high temperatures (90°C) almost complete disintegration occurs accompanied by the appearance of loose casein particles and the aggegation of casein particles in threads.

persed in the product. Nevertheless, in contrast to margarine, water continuity in the butter microstructure is still observed. It has been proposed that the water-containing membranes of many fat globules in butter lie so close to water droplets that a pathway is offered for macroscopic transport of water molecules (water channels). This type of structural arrangement can indeed be observed in Fig. 1.22, although water, as such, cannot be detected.

(d) Other cryotechniques

Cryosectioning. Cryosectioning or cryo-ultramicrotomy describes the process whereby samples are sectioned below 273 K to produce thin slices of material which can be viewed by TEM (Echlin, 1992). Sections can remain in and be observed in the frozen hydrated state, or they can be observed after freeze-drying. The latter process gives a marked increase in contrast. In general, contrast is a weak point in micrographs of cryosectioned material. The most fruitful applications of cryosections deal with specific localization techniques, in particular XRMA, EELS and ImmunoEM. In the of case ImmunoEM, in contrast to the chemical

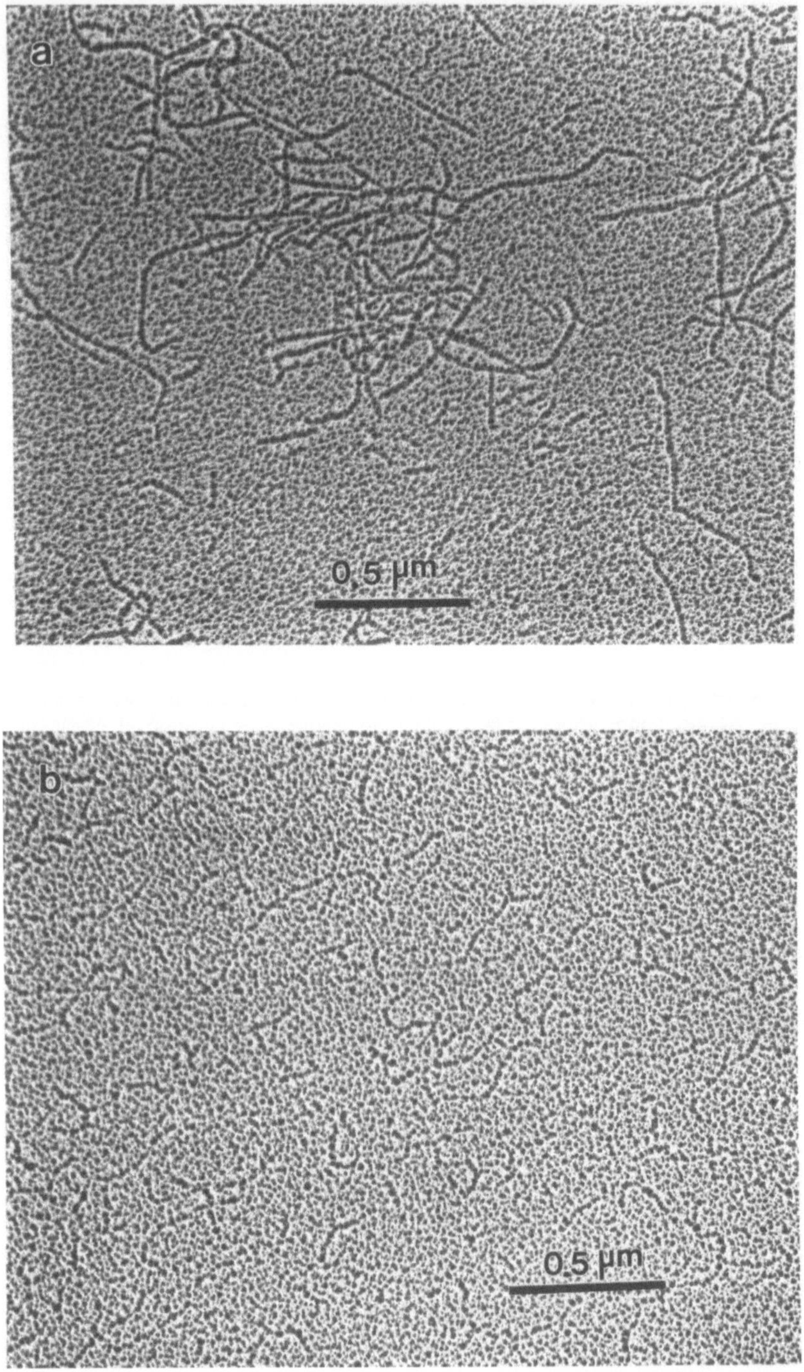

Fig. 1.17 Strands in κ-carrageenan: (a) strands in 0.1 M KCl have a large persistence length, while (b) strands in 0.01 M NaCl have a small persistence length. Technique: replica of glycerol containing aqueous solutions after vacuum-drying on mica.

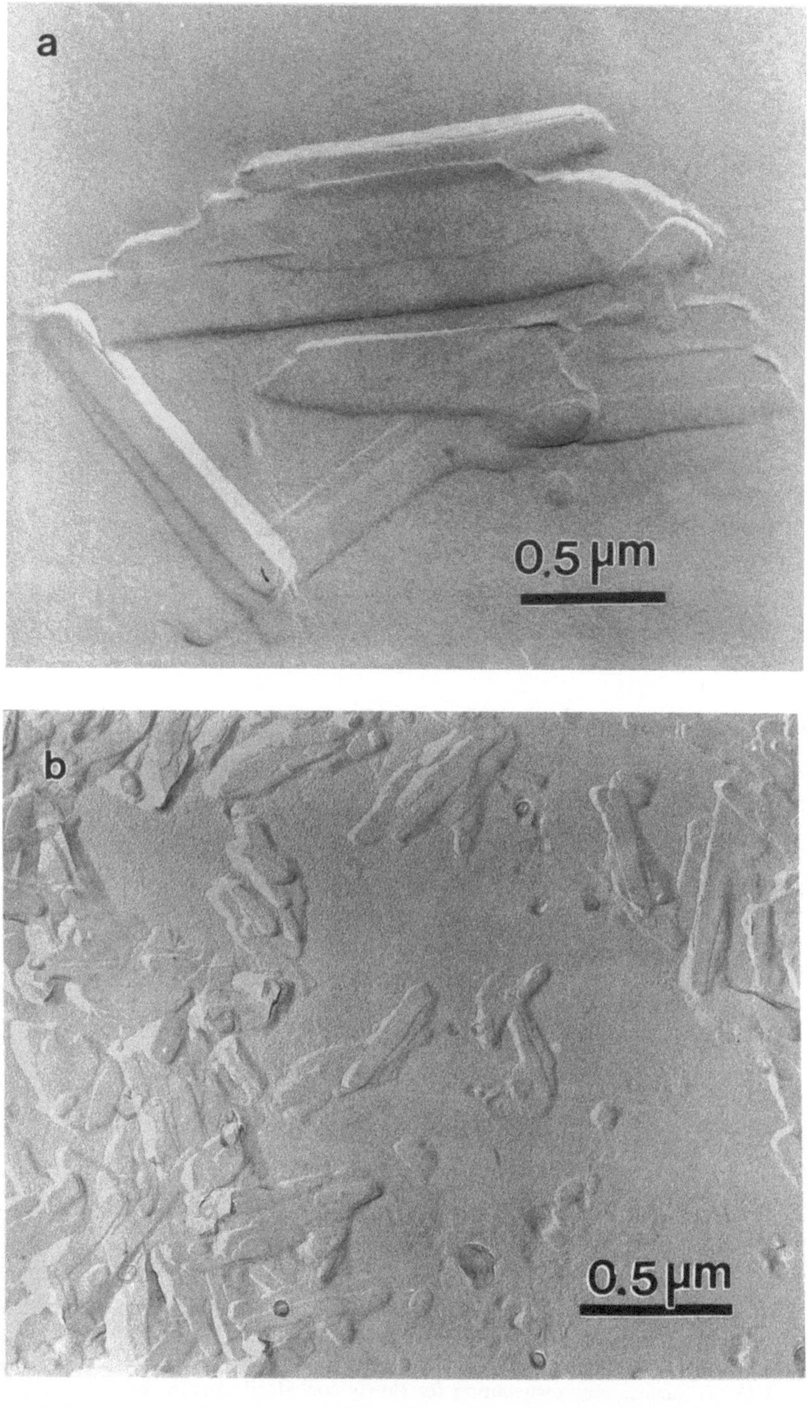

Fig. 1.18 Fat crystals after (a) slow cooling and (b) rapid cooling of a liquid fat blend. Technique: replica after fixation to a carbon film.

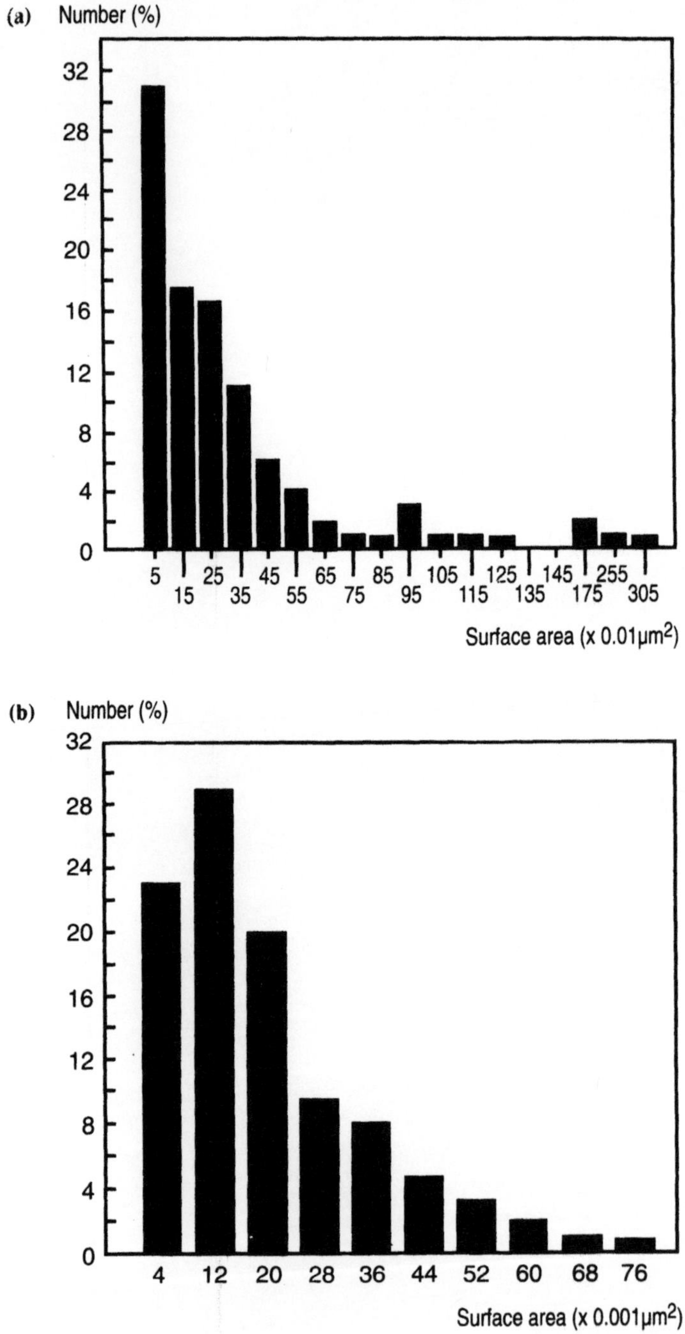

Fig. 1.19 (a) Surface area distribution for slowly cooled oil sample. Average surface area: $0.34\,\mu m^2$. (b) Surface area distribution for rapidly cooled oil sample. Average surface area: $0.018\,\mu m^2$.

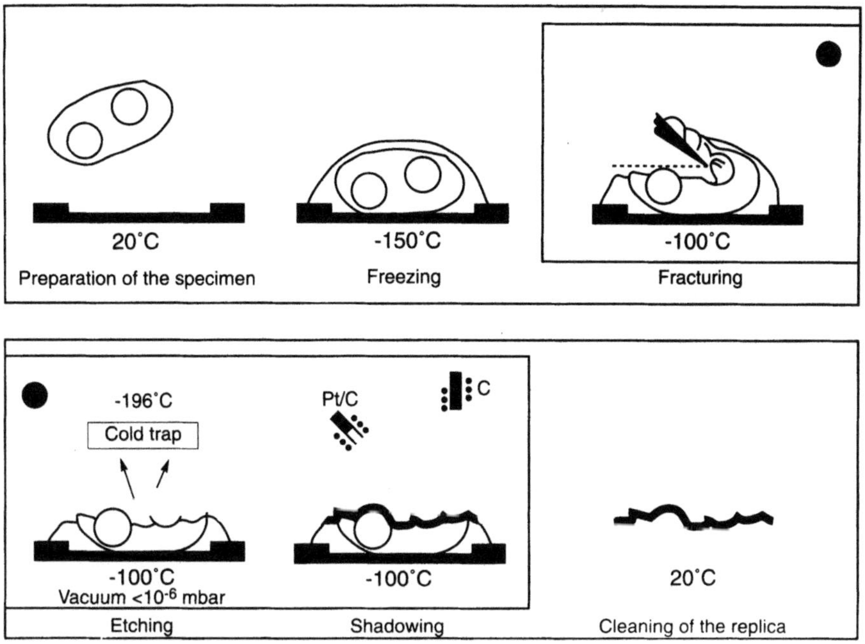

Fig. 1.20 Sequence of steps in freeze-fracture replication.

fixation techniques, the freeze techniques allow better retention of antigenicity. In the other cases, frozen material provides the best option to fix soluble and labile substances in order to allow proper microanalysis.

Figure 1.23 shows the fat distribution and the dense protein matrix in a micrograph of a low-fat cheese obtained by means of cryo-sectioning. The protein distribution in a processed cheese is shown in Fig. 1.24. Protein strands are observed as in earlier work (Heertje *et al.*, 1981) involving different preparative techniques. Such strands influence sensorial and rheological properties of the cheese.

Attempts to cryo-section water-in-oil emulsion products, such as fatty spreads, appear not to be very successful. In general, the cryo-sectioning of food products for morphological observations only is of limited significance over other preparative techniques. Its application in localization studies is discussed later.

Low temperature embedding. In the standard embedding techniques used in ultramicrotomy, the samples are embedded at temperatures above 273 K. In the low temperature variant, embedding and polymerization in resin occur at temperatures considerably below 273 K. Different types of resins have been developed (Lowicryls, LR gold, Nanoplast). Polymerization is induced by UV radiation. Before embedding, the samples have

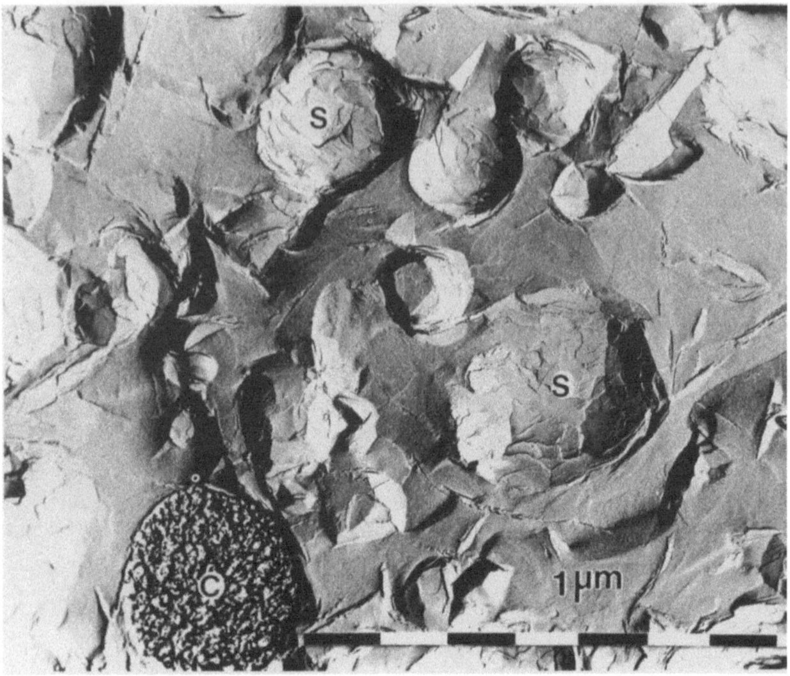

Fig. 1.21 Water droplets in a margarine. The sample breaks either around the surface (S) of the droplet or through (C) the droplet. Technique: freeze-fracture replication.

to be dehydrated. Two main routes have become popular: freeze substitution and progressive lowering of temperature (PLT). In freeze substitution, ice in the frozen hydrated specimen is removed and replaced by an appropriate organic solvent. In PLT, unfrozen fixed specimens are passed through increasing concentrations of dehydrating agents such as ethanol or acetone. By increasing stepwise the concentration of dehydrating agent, the dehydration can be applied at progressively lower temperatures without freezing the sample. The functionality of macromolecules and macromolecular assemblies and the antigenicity of proteins are much better preserved than in conventional embedding techniques. Low temperature embedding has therefore become very popular in combination with immunolocalization (see previous section on ImmunoEM).

Figure 1.25 shows part of the structure of ice-cream, obtained by freeze substitution of the frozen product and low temperature embedding. Ice-cream is a complex structure of large ice crystals, air cells, oil droplets, sugar and biopolymeric substances. Figure 1.25 indicates the presence of numerous air cells with a specific size distribution. With conventional embedding techniques a faithful analysis of the air cells would be impossible.

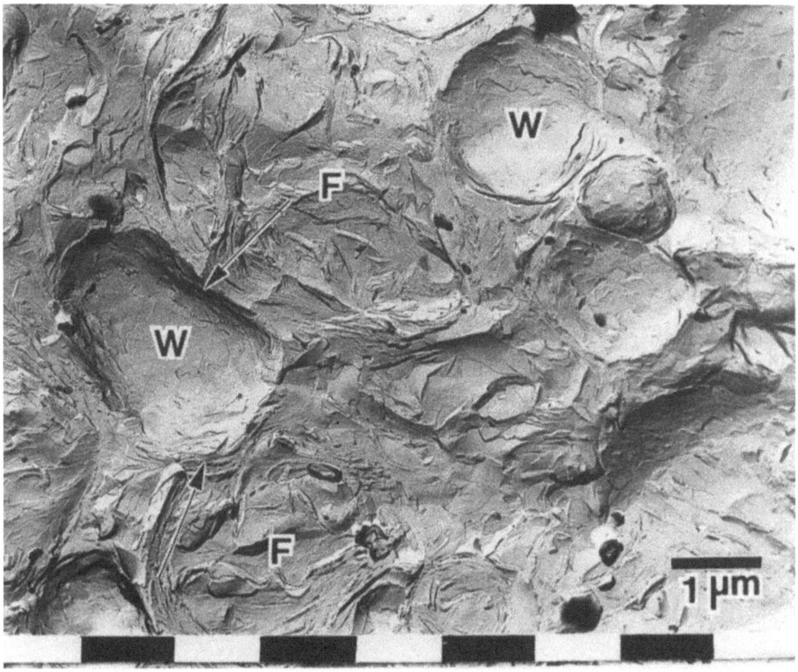

Fig. 1.22 Butter structure, showing membranes of fat globules in close contact with water droplets. Arrow indicates position of deformation of the globule surface. (W) Water droplet; (F) fat globule. Technique: freeze-fracture replication.

Thin film technique. All the electron microscopical preparation techniques described thus far lead to some damage of the specimen. Dehydration, embedding, adsorption, contrast media, ice crystal formation all give rise to structural changes, and so the observed material may be quite different from the original sample. Following the discovery that it is possible to vitrify water and thereby completely prevent ice crystal formation (Dubochet and McDowall, 1981), it has become feasible to study specimens in their native, hydrated, state. The method involves (a) the formation of a very thin film of material, (b) very rapid cooling to the vitreous state, (c) cryo-transfer into the microscope keeping the temperature below 140 K, and (d) low-dose observation of the cold specimen at the low temperature with the water in its vitreous state. The thin film method is rather simple to apply. With the development of vibration-free cold stages, cryo-transfer units (Echlin, 1992) and low-dose imaging in combination with computer-enhanced image analysis, high resolution images can be obtained. By virtue of the preparation technique (i.e. the necessity for thin films), the method is restricted to low viscosity materials such as suspensions or dispersions of viruses

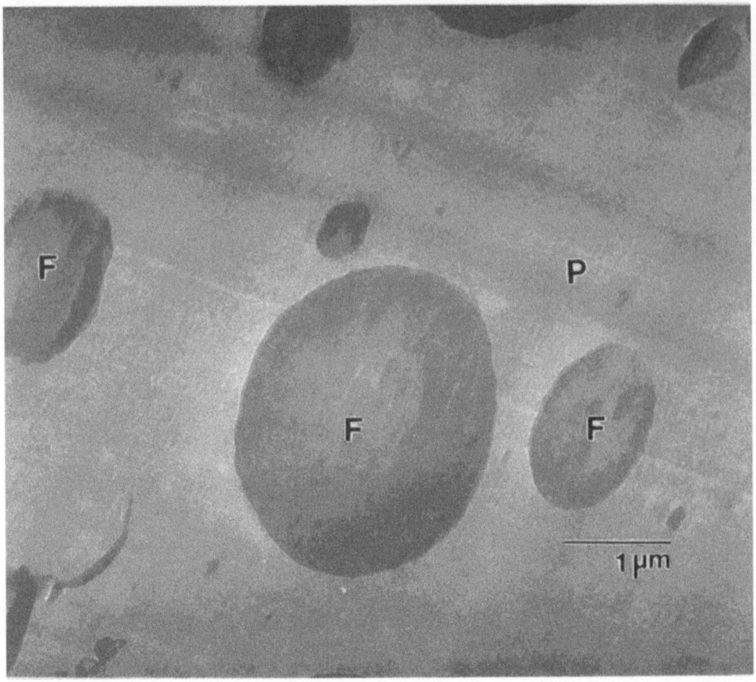

Fig. 1.23 Fat (F) and protein (P) distribution in a low-fat cheese. Technique: cryo-sectioning.

(Adrian *et al.*, 1984), unilamellar or multilamellar vesicles (Talmon, 1986; Frederik *et al.*, 1989), and solutions of macromolecules (Lepault and Erk, 1992; Sugiyama *et al.*, 1994). There is also growing interest in the study of microemulsions in the foods area. Figure 1.26 shows an example for a microemulsion system consisting of a triacylglycerol mixture, phosphatidyl choline, a co-surfactant and water. Spherical particles of size *ca.* 70 nm are observed. Similar structures are currently being observed by freeze-fracture replication.

1.2.2 Techniques in scanning electron microscopy

Scanning electron microscopy was introduced into food research in the late 1960s. The SEM instrument became very popular because of some special features. Three-dimensional pictures with a great depth of field are obtained. In addition, the technique allows the observation of bulk specimens, since, for image formation, use is made of secondary (or back-scattered) electrons emerging from the specimen surface. By applying appropriate fracturing techniques, not only the surface, but also the internal structure of a specimen, can be analysed. Sample preparation

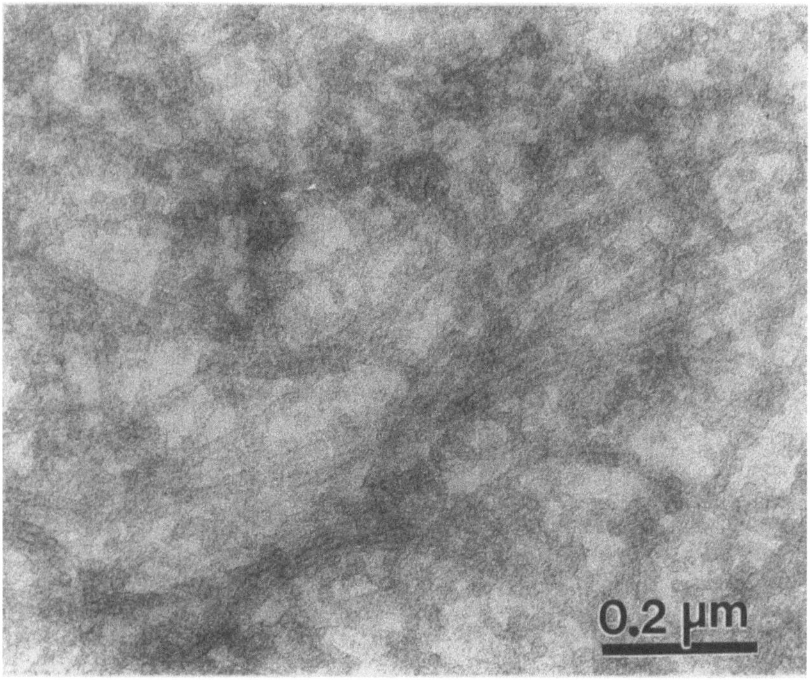

Fig. 1.24 Protein strands in a processed cheese. Technique: cryo-sectioning.

is, in general, much easier than for TEM because electron-transparent samples are not required. Of course, proper fixation and dehydration techniques are still needed. Special to SEM are provisions to prevent charging of the specimen. If not employed, severe image distortion and radiation damage may occur. The most popular way to prevent specimen charging is by coating with a thin film of conducting material, although other techniques can also be applied (Goldstein *et al.*, 1981). In this context, two developments need special mention: environmental SEM (ESEM) and field emission SEM (FESEM).

ESEM overcomes the limitations inherent in conventional SEM, i.e. the need for dehydration and charging of specimen. This is achieved by allowing a certain gas pressure (1–20 Torr) in the EM column. The gas can be any inert gas or water vapour; consequently samples can be observed at saturated water vapour pressure. In addition, the presence of the gas atmosphere prevents specimen charging (Danilatos, 1991, 1993). The applied pressures still allow a maximum resolution of about 4–5 nm. Conceptually, this method is a major breakthrough, but up until now it has not become popular in the foods area. It should be realized that water mobility and radiation damage in strongly hydrated food samples

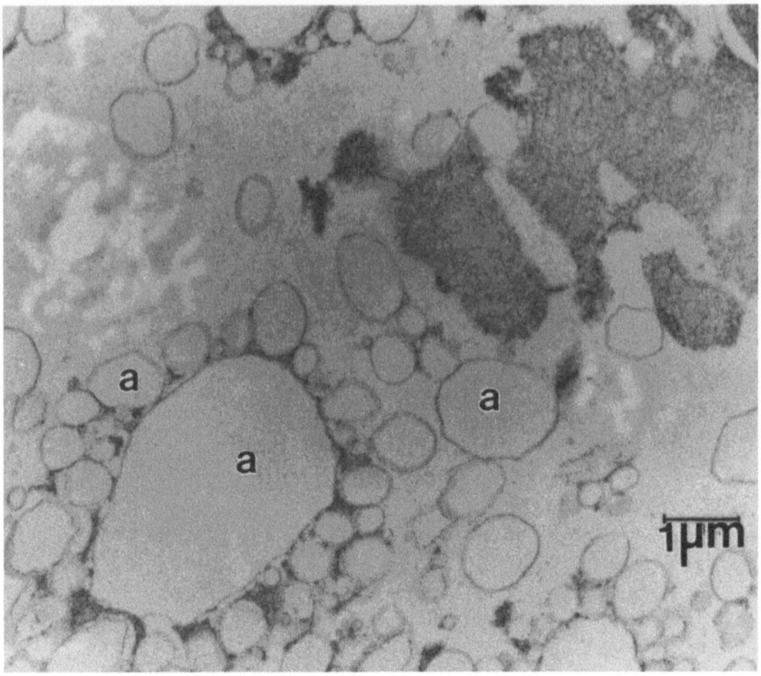

Fig. 1.25 Air cells (a) in ice-cream. Technique: freeze substitution and low temperature embedding. (Courtesy Mr Colin Smith, Unilever Research, Colworth, UK.)

often prevent proper image formation. In addition, the application requires much experimental skill and effort.

The development of FESEM has been stimulated by recent improvements in lens systems and field emission electron sources, which allow the formation of very small probe sizes, with enough current to produce high resolution information, even at low electron voltages (Joy and Pawley, 1992). It is, in particular, the latter option which expands the possibilities for SEM observations. Because charging artefacts can be avoided at low voltage, high resolution information of uncoated (or lightly coated) specimens can be obtained in the voltage range from 1 to 5 keV. It may be expected that low voltage FESEM, particularly in combination with cryo-techniques, will be exploited in the foods area in the near future.

The main preparation techniques for SEM are discussed and illustrated below by appropriate examples.

(a) Air-drying and vacuum-drying. In general, air-drying or vacuum-drying is not recommended because distortion and shrinkage may occur due to surface tension effects. Nevertheless, in case of low moisture

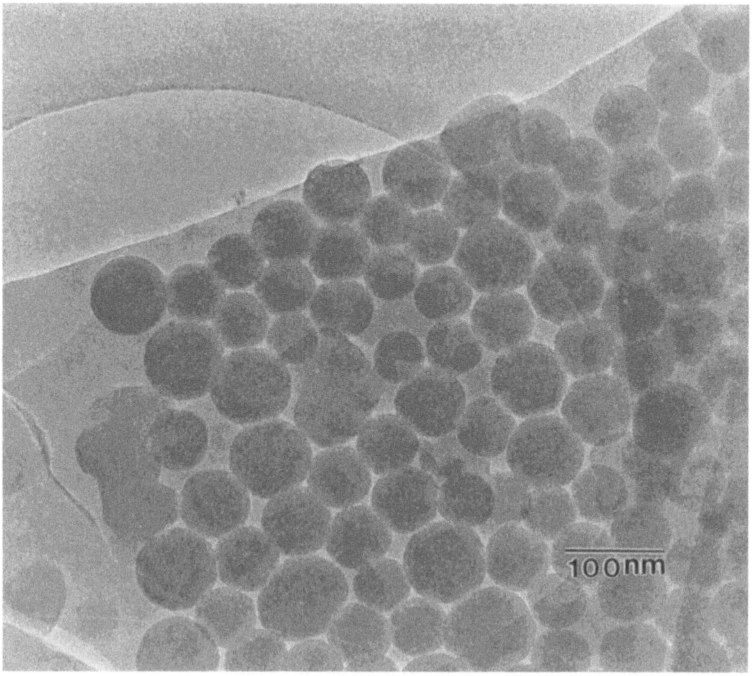

Fig. 1.26 Oil-in-water microemulsion system, showing a close-packed structure of microdroplets (70 nm approximately). Technique: freezing a thin film of material.

substances such as powders, acceptable structural preservation can be obtained (Varriano-Marston, 1977). This procedure is very simple and requires minimal sample preparation.

The method has been applied to investigations of dry fractionation of flour (Fig. 1.27). Various fractions were characterized through differences in the size and shape of the starch and protein particles and by their interactions. Relationships with functional properties can be established.

(b) Freeze-drying. Freeze-drying is a dehydration process in which the water in the sample is first converted to ice, which is then sublimed away at low temperatures and high vacuum (Echlin, 1992). As with all cryo-techniques, structural damage caused by ice crystallization during the freezing step may occur. Also shrinkage of the structure may take place. It is, however, difficult to predict beforehand how serious the structural alterations will be. In this respect there is a strong specimen–technique interaction and the results should therefore be critically evaluated. Depending on the specimen, special pre-treatments, such as chemical fixation and washing, before the freezing step may be required.

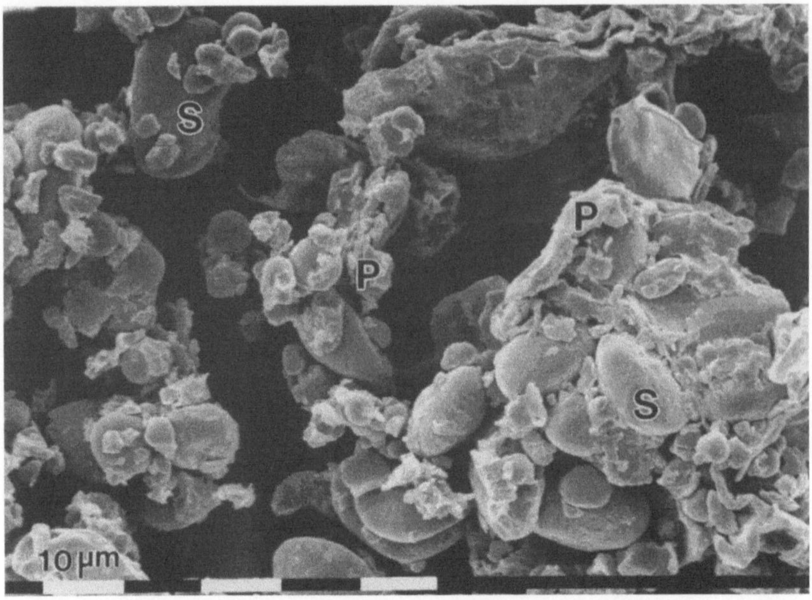

Fig. 1.27 Fractionated flour particles showing aggregates of starch granules (S) and protein (P). Technique: SEM, air-drying.

Properly applied it is a powerful method in the investigation of three-dimensional food structure as can be illustrated by a number of examples.

Yoghurt is a three-dimensional network structure of protein strands formed out of casein micelles aggregating in a liquid phase. Functional properties such as consistency and syneresis are dependent on the nature of the network structure. This structure can be influenced by various variables, such as the pre-treatment of the milk, the protein concentration, and the addition of whey proteins. Pioneering work in this area has been performed by Kalab (1993). Figure 1.28 shows an example in which an extra contribution to the consistency of the yoghurt is offered by an extracellular polysaccharide network.

Dough for breadmaking is prepared by mixing flour and water whilst adding yeast and salt. Starch and proteins become hydrated and, together with the flour lipids, form a visco-elastic matrix which is able to retain occluded air bubbles. In the rising stage, carbon dioxide produced by yeast from sugars diffuses to the air cells, thus causing expansion of the dough. It is assumed that the structure of the dough, and in particular the dough film surrounding the gas cells, is responsible for proper gas retention during the baking process. In this context, the dough structure has been investigated as a function of wheat variety, the presence of additives and the rise time (Fig. 1.29).

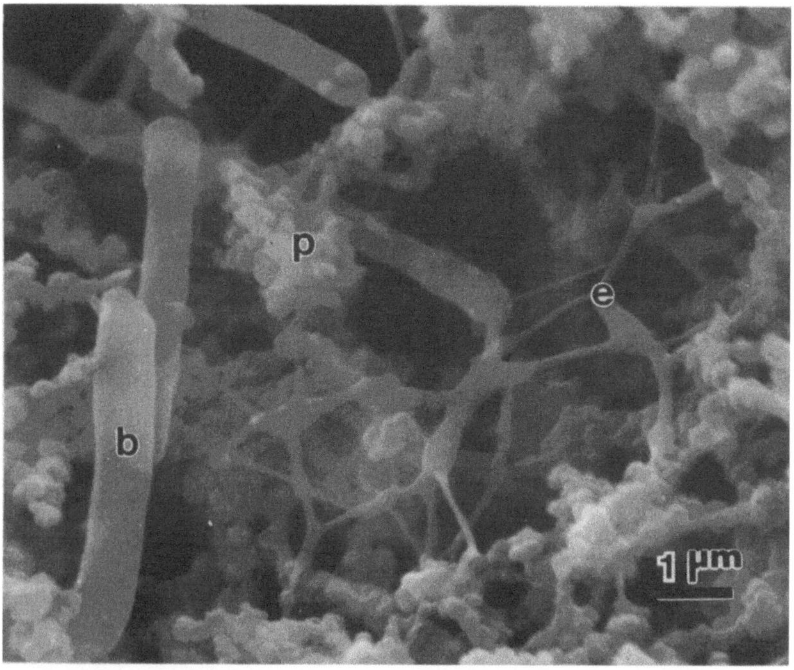

Fig. 1.28 Yoghurt structure showing bacteria (b), extracellular polysaccharide (e) and protein (p). Technique: SEM, freeze-drying.

(c) Critical point drying (CPD). In CPD, drying is accomplished at the gas–liquid critical point. At the critical point the vapour and liquid phase are in equilibrium without a phase boundary and consequently without surface tension effects. The technique is carried out using CO_2 which has a critical temperature of 304 K and a critical pressure of 7.39 MPa. The method involves dehydration of the sample with fluids such as ethanol and acetone, followed by immersion in liquid CO_2. Alternatively, an intermediate fluid may be used. The procedure is performed in a high pressure bomb. Proper structural information can, in general, be obtained, although material may be leached out by the CO_2 and thermal and pressure stresses may induce image distortion. Shrinkage can, at least partly, be avoided by special fixation and preparation regimes. The technique has been amply applied in the dairy field (Kalab, 1981, 1993). The structure of fresh cheese as a function of heat treatment of the starting milk has been investigated by this drying technique, as well as by solvent drying (see below).

(d) Solvent drying (SD). An alternative to CPD is offered by solvent drying. After immersion in suitable solvents, samples can be air-dried. There are three criteria for an ideal drying solvent (Robards and Wilson,

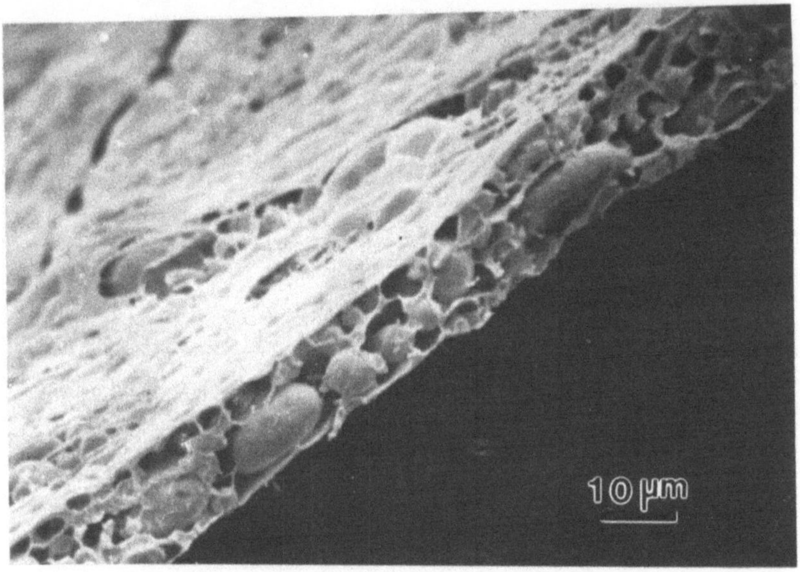

Fig. 1.29 Cell wall structure between gas cells in fermented dough showing starch granules, protein and a covering protein film. Technique: SEM, freeze-drying.

1993): (i) no reaction with the specimen; (ii) low boiling point and surface tension; (iii) solubility in the common dehydrating solvents. First the samples are treated with a dehydrating solvent and subsequently with the drying solvent. Various drying solvents have been applied: tetramethylsilane, hexamethyldisilazane, Freon 113, Peldri II (Nation, 1983; Reville and Cotter, 1991; Frau *et al.*, 1993). Hexamethyldisilazane in itself has a rather high boiling point, and may react with the dehydrating solvent ethanol with formation of volatile compounds.

In a number of cases SD gives results comparable to CPD, but in other cases considerable distortion of the structure is observed. When it can be applied successfully, SD is a very simple preparation technique and its use saves considerable time over CPD. Figure 1.30 shows the structure of a fresh cheese obtained by dehydration from ethanol and SD by hexamethyldisilazane. Results are equivalent to those obtained by CPD. Proper information about the three-dimensional structure (i.e. the state of protein aggregation, the dimensions of the casein protein strands and the void spaces) will help to elucidate the role of processing and composition in the formation of this type of dairy product.

(e) Low temperature scanning EM (LTSEM). The previously described techniques provide ways to free the sample from water. LTSEM is mainly

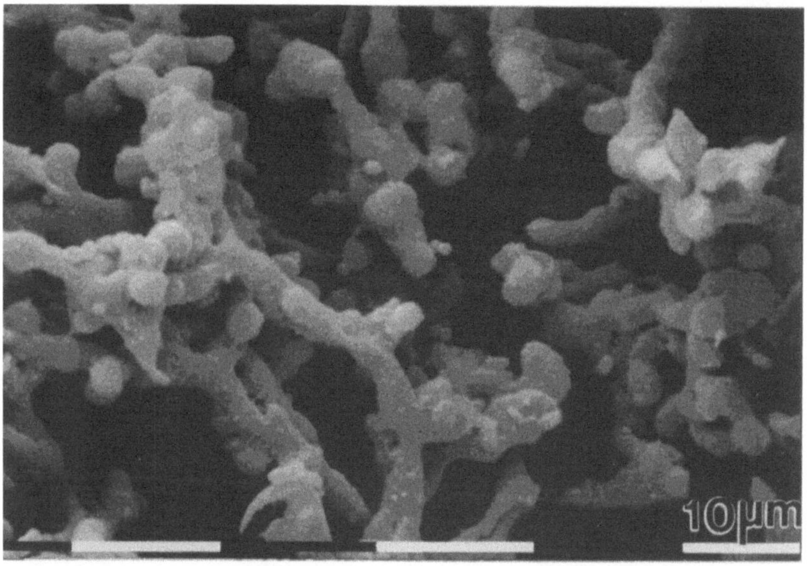

Fig. 1.30 Protein structure in a fresh cheese. Technique: SEM, solvent drying.

concerned with the observation of frozen specimens containing a substantial part of the original water. Conceptually it is a very simple technique. The sample is frozen and observed in the frozen state on the cold stage of a SEM. Artefacts associated with drying methods are avoided. In general, for traditional microscopes operating at a high voltage, a coating layer must be applied. Also fractured surfaces, which reveal the internal structure, can be imaged. In most cases fracturing and coating is performed in a special vacuum chamber attached to the microscope, enabling direct transport into the cold stage of the SEM.With a view to the importance of the cooling step in all cryo-techniques, small samples should be applied. However, most SEM specimens are bulk specimens which do not allow such high cooling rates as obtainable with TEM specimens. Consequently, ice crystal damage may present a problem. The spatial resolution of LTSEM is in general lower than for freeze-fracture replication. However, with the introduction of FESEM, it can be envisaged that this gap will be closed (Herman and Muller, 1993). A review, dealing with food and other industrial and biological applications, has been published (Sargent, 1988).

LTSEM can make a unique contribution to the study of foam structure. In many food products like bread, cake, whipped cream, ice-cream, beer and chocolate mousse, the distribution of gas cells plays an important role. Often we are dealing here with fragile systems which can properly be observed only in the well-stabilized frozen state. Air cells in a whipped

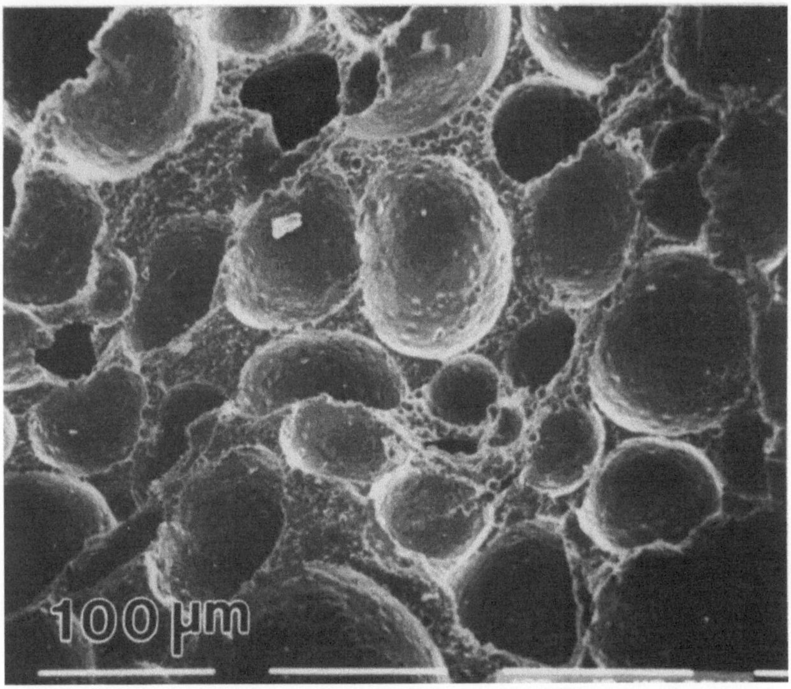

Fig. 1.31 Air cells in a whipped cream showing interface stabilized by fat globules and protein film. Technique: low temperature SEM.

cream are presented in Fig. 1.31. Stabilization is achieved by the combined action of fat globules and a protein film at the interface. Also ice-cream can profitably be examined at low temperatures. When the temperature is kept below −120°C, no evaporation of water occurs and the ice crystals are properly retained in the structure and can clearly be distinguished from the air cells (Fig. 1.32). Ice crystal size and air cell size are important attributes for ice-cream functionality. Figure 1.33(a) shows the structure of a concentrated oil-in-water cream before evaporation of the water by keeping the cold stage at −150°C. Figure 1.33(b) shows the same cream after the evaporation of the water by bringing the cold stage to a temperature of −80°C. The globular fat phase can be clearly distinguished. By comparing the morphologies, the original location of the water can be determined. This approach was used to study the phase inversion from an oil-in-water cream to a system having a dispersed water phase (Heertje, 1993a).

Low temperature SEM techniques are also used in the analysis of the three-dimensional structure of fatty spreads. To achieve this, a new methodology has had to be developed (Heertje *et al.*, 1987). An important step in the procedure is the removal of liquid oil from the fat sample by

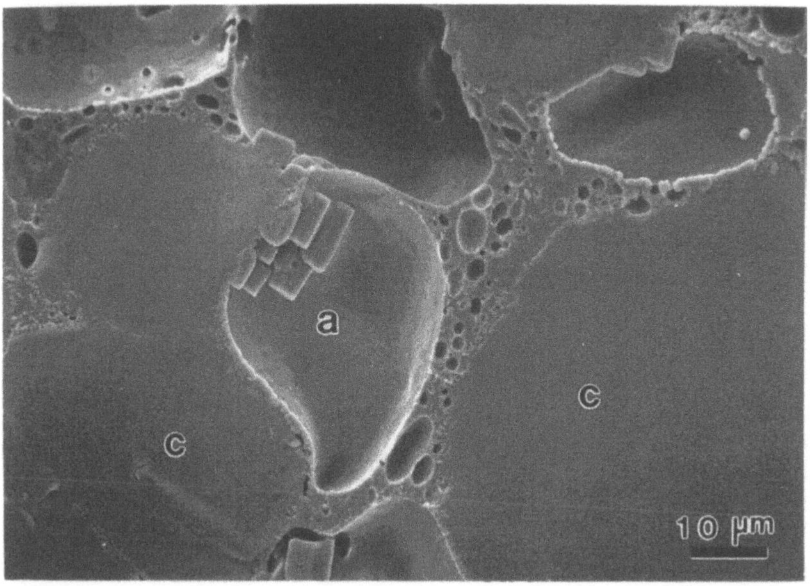

Fig. 1.32 Ice-cream containing air cells (a), ice crystals (c) and continuous matrix (m). Technique: low temperature SEM. (Courtesy Mr C. Smith, Unilever Research, Colworth, UK.)

an appropriate solvent, since liquid oil would completely obscure the observation of the solid fat crystalline matrix and it is this solid fat crystalline matrix which is mainly responsible for the final product properties. The de-oiled specimen is observed at low temperature to avoid beam damage and melting of the delicate porous structure.

Margarine is composed of a continuous network of fat crystals or fat crystal aggregates (Fig. 1.34), whereas butter has a much more discontinuous structure (Fig. 1.35), containing fat globules with no interaction or a limited interaction with the rest of the matrix (Juriaanse and Heertje, 1988). This is reflected in functional properties such as hardness, spreadability, mouthfeel, emulsion stability and salt release of both products. By parallel plate compression (Fig. 1.36), some of these properties can be determined. The maximum stress σ_{max} is a measure of the product hardness. The product softening or plasticity, which is determined by the ratio of the stress at infinity, σ_{∞}, to the maximum stress, appears to be much higher for margarine than for butter. On deformation, many more bonds are broken in the connected margarine structure than are broken in the discontinuous butter structure; this is in accordance with the observed microstructure (Heertje, 1993b).

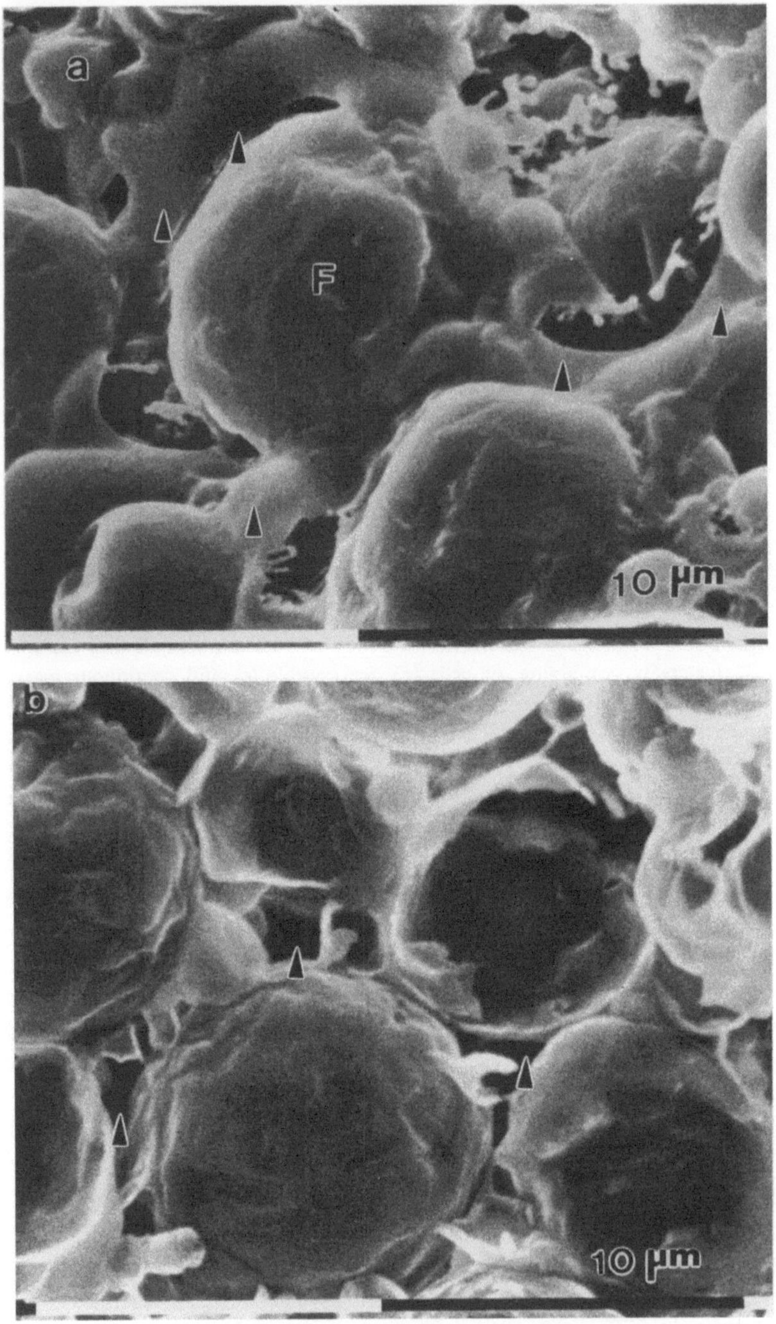

Fig. 1.33 An oil-in-water cream (a) before evaporation of water and (b) after evaporation of water. Arrows indicate position of water (a) and the original position of water (b). Technique: low temperature SEM.

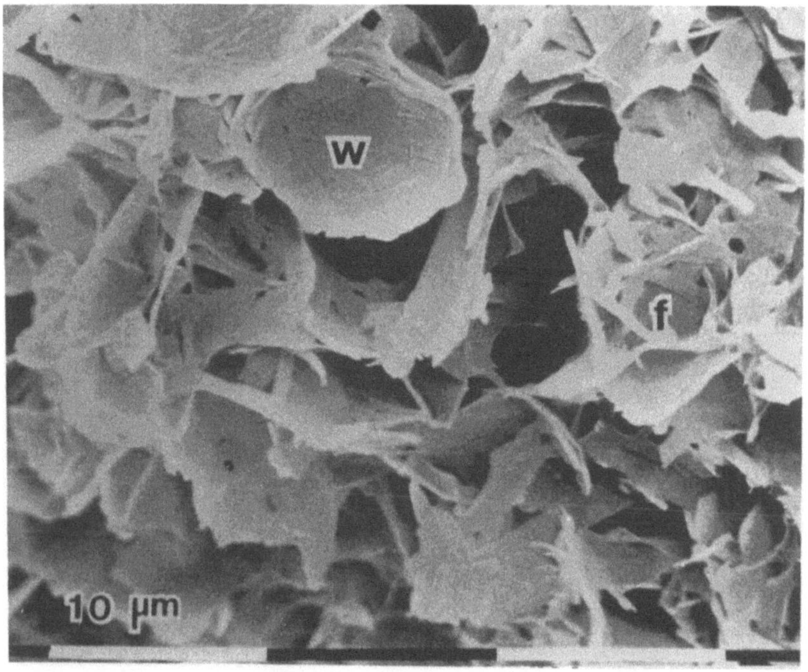

Fig. 1.34 Fat crystalline network structure (f) and water droplet structure (W) in margarine. Crystalline shell structure around water droplets is clearly discernible. Technique: low temperature SEM, after de-oiling.

1.2.3 Specific localization techniques

(a) X-Ray microanalysis (XRMA). XRMA is an element-specific localization technique which can be applied in both TEM and SEM. Because of beam broadening in the specimen, the resolution (of the order of 0.01 μm) for thin samples in TEM is much higher than that for bulk samples in SEM (~ 1 μm). Small amounts can be detected ($\geqslant 10^{-19}$ g), but not low concentrations ($\leqslant$ 250 ppm).

To perform XRMA, a probe-forming system and an X-ray detector system are required. To detect the element specific X-rays, wavelength-dispersive or energy-dispersive spectrometers are used. The latter are the most popular, consisting of a semi-conductor detector, an amplifier, a multi-channel analyser, a display system, and a computer. The detector is normally fitted with a beryllium window which absorbs low-energy X-rays. This limits the list of detectable elements to sodium and heavier ($Z \geqslant 11$). However, window-less detectors, and detectors with very thin windows, have been developed, enabling analysis of boron ($Z = 5$) (Morgan, 1985; Roomans, 1990).

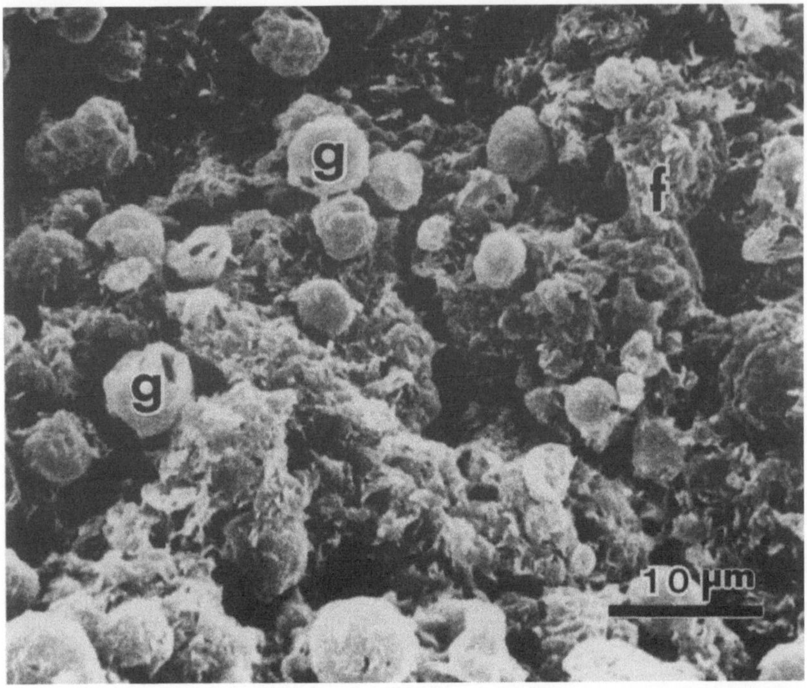

Fig. 1.35 Fat structure in butter, showing a discontinuous arrangement of fat globules (g) and fat crystals (f). Technique: low temperature SEM, after de-oiling.

Because most food products contain considerable amounts of water, specimen preparation is the most critical step in the XRMA of food samples. It has therefore to be safeguarded that elements are not removed or transported (or even introduced) during the preparation procedures. In this respect, low temperature techniques have much to offer. The XRMA method requires much experimental skill and is not without pitfalls (Roomans, 1990). Applications in the fields of dairy science, cereals and bakery, plant, fruits, vegetables and meat are reported. Special mention should be made to the work of Brooker (1987, 1990, 1991). The localization of phosphorus and calcium in cheese and of potassium in cocoa solids has been described. Frozen samples are usually flattened by means of a rotating diamond cutting tool (cryo-milling) to allow proper X-ray analysis. Also corrosion problems with containers for foods and drinks have successfully been tackled (Helwig and Biber, 1990; Charbonneau, 1991).

(b) Electron energy loss spectroscopy (EELS). EELS is an element specific localization technique, but it can only be applied to thin

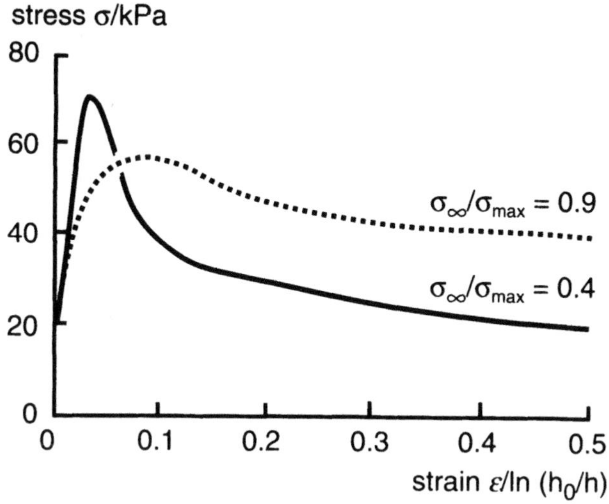

Fig. 1.36 Stress–strain curves for butter (—) and margarine (. . .), obtained from parallel plate compression, where h_0 and h are heights of sample before and after compression, respectively. The product softening, which is determined by the ratio $\sigma_{max}/\sigma_{\infty}$, is much more pronounced for margarine than for butter.

specimens in TEM. A resolution of 5 nm is clearly demonstrated for biological specimens. For non-biological specimens a resolution of better than 1.5 nm has been demonstrated. The sensitivity is high and consequently very small quantities of individual elements can be detected, certainly when present in high local concentrations. It is claimed that, in contrast to XRMA, light elements in particular (with the exception of hydrogen and helium) can be analysed, and that the use of EELS is complementary to XRMA. But heavier elements can also be analysed (Trebbia *et al.*, 1990).

To analyse the energy loss, an electron spectrometer, usually fitted below the EM column, and a detection unit are needed. In this way EELS spectra can be recorded. Alternatively (and this has become very popular in the biosciences), images (maps) can be produced from electrons that have suffered a specific energy loss (electron spectroscopic imaging). Many aspects of the quantification procedures are discussed in a special issue of the *Journal of Microscopy* (Knowles, 1994), containing a selection of papers from the *Fourth European Workshop on Electron Spectroscopic Imaging and Analysis Techniques*, Münster, 1993.

What has been said about specimen preparation for XRMA applies *mutatis mutandis* for EELS, with the complication that only very thin specimens ($\leqslant$ 50 nm) can be used. This probably explains why the application of EELS to food products has been rather limited. In addition, difficulties with past instrumentation and laborious data handling procedures did not stimulate its use. It may be expected, however, that technological improvements and computational advances will stimulate future applications.

Recent work by McManus and McMahon (1994) on the localization of calcium in dairy products has revealed that the distribution of this element over the casein fraction is far from uniform. This may have consequences for the aggregation behaviour of the constituting caseins.

(c) Specific labelling methods. Specific labelling methods are rapidly becoming very popular in the foods area for the unequivocal identification of structural elements with a specific composition. The specificity between target and identifier is based on the ligand–receptor principle (see Introduction) and should always be checked by performing control incubations (Polak and Varndell, 1984; Robards and Wilson, 1993). Examples of ligand–receptor combinations are antibodies–antigens, enzymes–substrates and lectins–sugars. The main markers are the reaction product of peroxidase, ferritine, latex spheres and colloidal gold particles (Graham and Karnovsky, 1966; Polak and van Noorden, 1983, 1984). Colloidal gold is a universal marker and the most popular nowadays. Because of their distinct spherical shape of any desired diameter (between 1 and 25 nm) colloidal gold particles can be applied in multiple labelling experiments (Bendayan, 1982) and quantitative studies (Polak and Varndell, 1984). The preparation of a colloidal gold solution is based on the controlled reduction of gold chloride ($HAuCl_4$). Various regimes have been developed (Frens, 1973; Slot and Geuze, 1981). The conjugation of ligand or receptor molecules in a molecular monolayer on the outer surface of the gold particles turns the marker into a probe (Roth, 1983).

The specific affinity between ligand and receptor is based on physicochemical, sequential and conformational characteristics of the molecules. Improper sample preparation (in particular, fixation and embedding) and handling (e.g. storage) of the receptor molecules can change physicochemical and conformational properties causing loss of specificity or introducing aspecific affinity for other domains. Specific labelling methods can be applied in conjunction with various TEM preparation methods such as ultramicrotomy, freeze-fracture replication, and negative staining and SEM preparation methods (Horrisberger and Rosset, 1977). In order to retain specificity, cryo-preparation techniques are preferred. Gold particles can be detected in TEM and SEM by normal imaging modes or by special imaging modes in the case of very small particles (Otten *et al.*, 1992). Various applications are now discussed.

Network-forming biopolymers are used in the texturization of food products. In mixed gels containing proteins and carbohydrates, phase

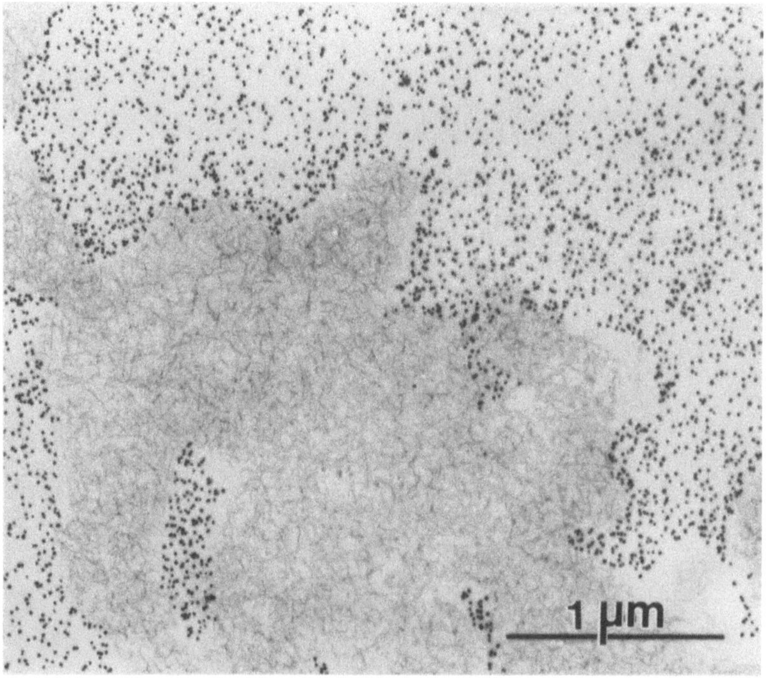

Fig. 1.37 Phase separation in a mixed polymer system. The protein phase (gelatin) is labelled with colloidal gold particles. The carbohydrate phase (pectin) is not labelled and shows strands. Technique: specific labelling based on electrostatic interaction on thin-sectioned material. (Courtesy Mr C. Smith, Unilever Research, Colworth, UK.)

separation of the two biopolymers may occur. This can be advantageously analysed by specific labelling methods. Figure 1.37 shows the specific labelling of the protein (gelatin) phase in a mixed system with pectin. The application is based on electrostatic interaction between positively charged protein and negatively charged gold particles at a pH value below the isoelectric point of the protein present in the sample. Phase separation is considered to be important for controlling the rheological properties of mixed biopolymer systems.

Immuno-localization techniques have been used (Ambruster and Desai, 1993) to identify β-lactoglobulin, casein, bovine whey protein, and egg albumin in low-fat frozen desserts, reduced-fat process cheese, and salad dressings. The enzymes cellulase and hemicellulase, conjugated with colloidal gold, have been used for the localization of cellulose material in dressings. Figure 1.38 shows β-lactoglobulin labelling of particles in processed cheddar cheese. Lectin-labelled gold has been used by Horrisberger *et al.* (1978) and Horrisberger and Vonlanthen (1980) for the localization of glycosylated κ-casein in bovine casein micelles and

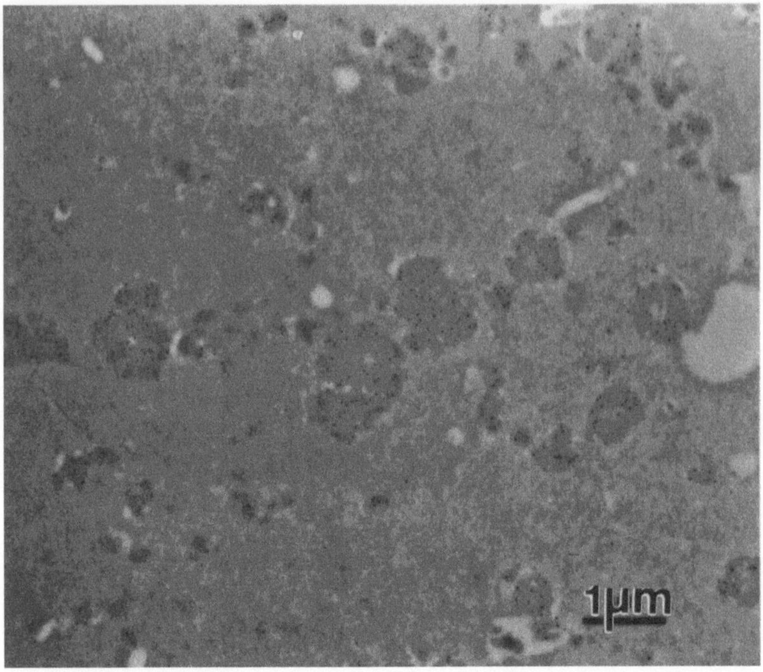

Fig. 1.38 Specific labelling of β-lactoglobulin in processed cheddar cheese. Technique: specific labelling with anti-β-lactoglobulin/gold on low temperature embedded, sectioned material. (Courtesy Dr B. Ambruster, Gatan Inc., USA.)

α-galactomannan in yeast cells. The identification of glycosylated κ-casein was done in TEM (ultramicrotomy) and was found to be mainly located in the bridging network interconnecting the casein micelles and it appeared to be loosely associated with the micelles. The case of α-galactomannan localization was studied in SEM and TEM, and it appeared to be present on the yeast cell surface at the grow-site (SEM) and in two layers in the cell wall (TEM). To study the function of emulsifiers in stabilizing emulsions, glycoproteins have been localized on milk fat globule membranes using lectin (wheat germ agglutinin) labelled gold (Fig. 1.39).

1.2.4 Image processing and analysis

In general, the information from image-forming experimental arrangements is of a qualitative nature. However, by applying image processing and analysis techniques, quantitative information can also be extracted.

All images generated by imaging systems contain imperfections. To improve the quality, image enhancement procedures such as contrast

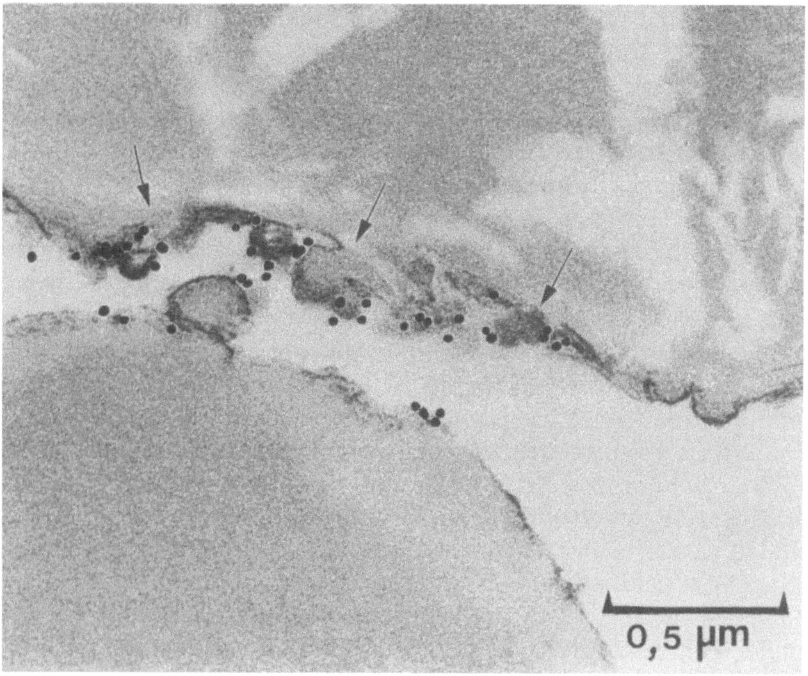

Fig. 1.39 Specific labelling of glyco-compounds on the milk fat globule membrane. Technique: specific labelling with lectin (wheat germ agglutinin)/gold on low temperature embedded, sectioned material.

manipulation and edge enhancement can be applied. This is often the first step for proper quantification by image analysis.

Traditionally EM images are in analogue form. Image processing is done in a computer environment. Processing software can only cope with digital image files. Most SEM instruments currently available on the market can generate digital images. In the last few years slow-scan CCD cameras have become available and are being introduced in TEM. In principle, they can replace photographic film in nearly all TEM applications. High-quality images can become available in a computer environment within seconds and can easily be transferred to other locations. A high resolution printer is needed for making hard copies. Procedures for the quantitative analysis of images have been amply discussed (Russ, 1990; Friel and Prestridge, 1991).

Image analysis can supply information on particle number, size distribution, surface area and shape (Friel and Prestridge, 1991). Conformational data for strand-like polymers can be used to deduce properties such as contour length distributions, end-to-end distances, and average persistence length. Three-dimensional information from images can be retrieved from thin serial sections of the specimen, using stereopairs, or by computerized reconstruction (Frank, 1989, 1992; Frank and Radermacher, 1992; Cookson, 1994). Procedures for the quantification of image data from strands of biopolymer molecules are described by Stokke *et al.* (1988, 1993) and Stokke and Brandt (1990). This can be used to study structure–function relationships of alginate, xylinan, xanthan, scleroglucan, biozan, welan, and so on.

A computer program to analyse the network structure of biopolymer networks and casein gels has been developed (Hermansson, 1993). Parameters considered to be of interest are the frequency of junction points, the pore-size distribution, the number of strands in a junction, and the angles between strands in a junction. Applying such an approach, if feasible, would significantly contribute to a better understanding of food structure.

The semi-crystalline structure at the molecular level of potato starch granules has been revealed by electron optical tomography on negatively stained potato starch granules and cryo-electron diffraction of frozen hydrated (unstained) granule fragments (Oostergetel and van Bruggen, 1993). The crystalline domains in amylopectin appear to be organized in left-handed helices, which act as a well-ordered skeleton for the starch granule. This particular type of super-helical lamellar structure has not been reported previously.

1.3 Microstructure and rheology of ovalbumin gels: example of an integrated study

Understanding the gelation behaviour of proteins is of importance in order to be able to control the textural properties of many food systems, and it may lead to proposals for product or process improvement. In this context the formation of heat-set ovalbumin gels, in different media, has been studied by microstructural, rheological and conformational observations (Heertje and van Kleef, 1986). Gels were prepared by heating at 100°C for 20 min.

1.3.1 Microstructural observations

An ovalbumin gel, prepared at pH 5, shows an inhomogeneous microstructure by TEM (Fig. 1.40) and a corresponding granular microstructure by SEM (Fig. 1.41). A similar structure is observed by critical point drying. An ovalbumin gel, prepared at pH 10, shows a homogeneous microstructure by TEM (Fig. 1.42) and a cellular micro-structure by SEM (Fig. 1.43). The cellular structure is induced by ice crystal formation during the freezing step. This is further substantiated by

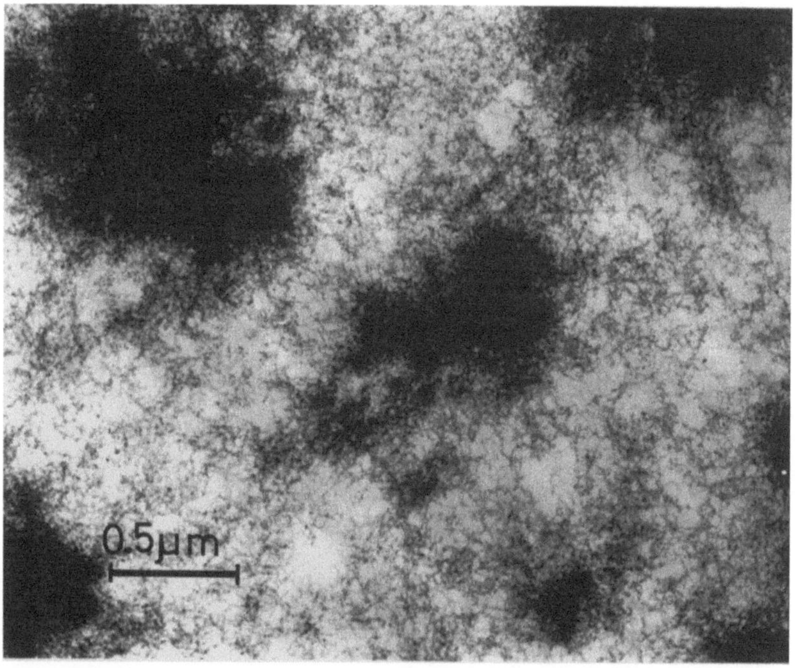

Fig. 1.40 Heat-set ovalbumin gel (20 wt%) at pH 5. Dark area represents the protein. Note the inhomogeneous structure. Technique: ultramicrotomy.

observations with critical point drying, again indicating a homogeneous microstructure. This example illustrates the importance of comparing several preparation techniques, in order to arrive at a reliable understanding of the microstructure.

1.3.2 Rheological observations

Tensile experiments were carried out on an Instron Universal Testing Instrument 1122. Tensile strips were prepared from heat-set ovalbumin sheets (van Kleef, 1986). Results of tensile experiments are shown in Fig. 1.44. A large difference in rheological behaviour is observed. The breaking stress (τ_b) for the pH 10 gel is considerably (about a factor of 30) larger than for the pH 5 gel. The deformation at the break (λ_b) differs by a factor of two.

1.3.3 Conformational observations

By ^{1}H NMR the influence of temperature on the unfolding of the protein molecules in D_2O is determined. Up to 70°C the solutions at pD 5 and

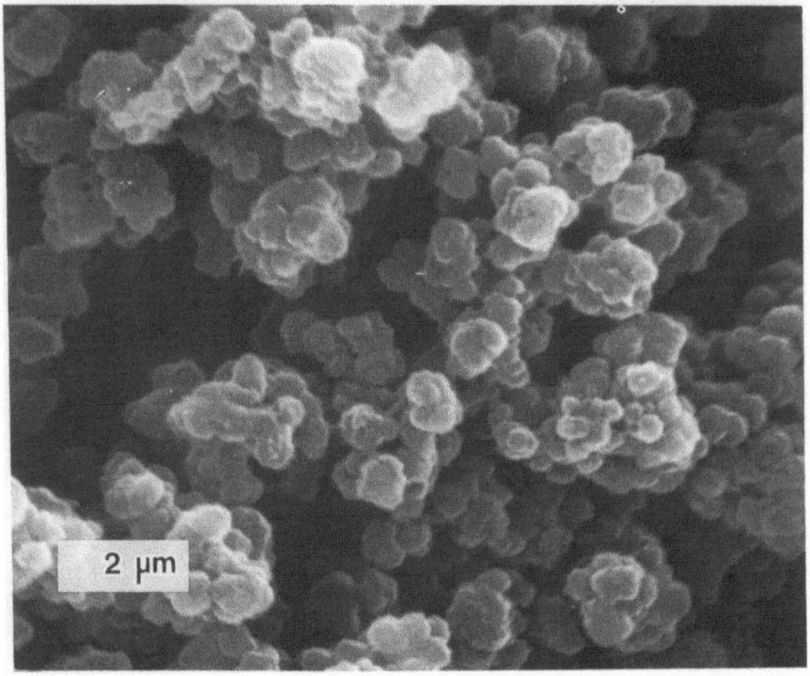

Fig. 1.41 Heat-set ovalbumin gel (20 wt%) at pH 5. Note the granular structure. Technique: SEM, freeze-drying.

pD 10 behave in a similar way. With increasing temperature ovalbumin molecules at pD 10 completely unfold, whereas at pD 5 only partial unfolding with aggregation occurs.

1.3.4 Discussion

In addition to the differences in microstructure and rheological properties, pH 5 gels are inhomogeneous, opaque and show water exudation, whereas pH 10 gels are homogeneous, transparent and do not show syneresis. It further follows from the NMR studies that network formation of the pH 10 gels occurs via flexible unfolded protein chains, whereas at pH 5 this formation occurs via protein aggregates. This is not surprising: at the low pH, near the isoelectric point of ovalbumin (pH 4.5), the protein conformation is strongly stabilized by hydrogen bonds and hydrophobic interactions and so gelation is expected to take place between these globular entities. At high pH, the protein molecule is negatively charged, which leads to electrostatic repulsions and at elevated temperatures to flexible unfolded protein chains without much interac-

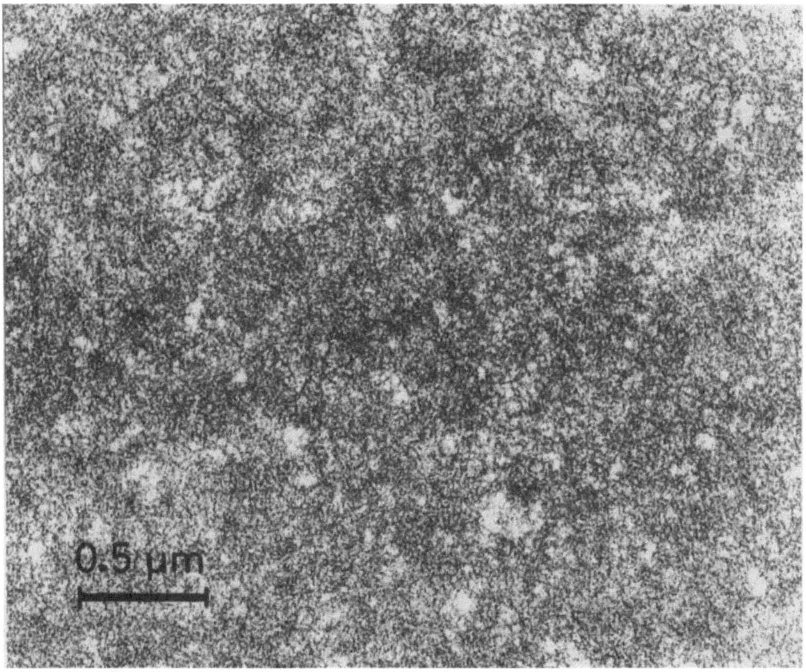

Fig. 1.42 Heat-set ovalbumin gel (20 wt%) at pH 10. Dark area represents the protein. Note the homogeneous distribution of protein filaments. Technique: ultramicrotomy.

tion. Indeed the microstructure observed is in good harmony with this picture; when there are strong inter- and intra-molecular interactions (pH 5), a granular inhomogeneous microstructure (Figs 1.40 and 1.41) is observed, whereas at high pH, with small interaction forces between the largely unfolded protein molecules, a homogeneous, uniform distribution of protein over the gel volume is found (Figs 1.42 and 1.43).

The rheological properties (Fig. 1.44) of the resulting gels are consistent with this difference in microstructure and with the character of the forces between the protein molecules. The structure which is composed of flexible, unfolded, but not strongly interacting, protein chains (pH 10) can be extended further without breaking than the aggregated structure composed of compact, strongly interacting protein molecules (pH 5). The regions of low protein concentration in the inhomogeneous gels (pH 5) will tend to act as weak points, thereby resulting in a low breaking stress. A summary of the properties of the different ovalbumin gels is given in Table 1.1.

This final example illustrates that information on all levels of structural organization is needed in order to arrive at a proper description of the behaviour of food products. It is clear that EM should not be

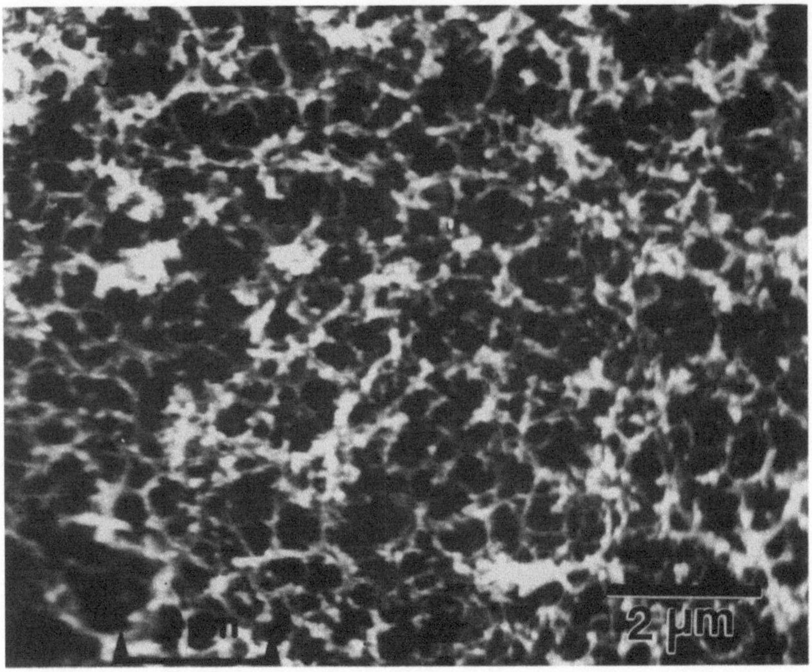

Fig. 1.43 Heat-set ovalbumin gel (20 wt%) at pH 10. Note the cellular structure, induced by ice crystal formation. Technique: SEM, freeze-drying.

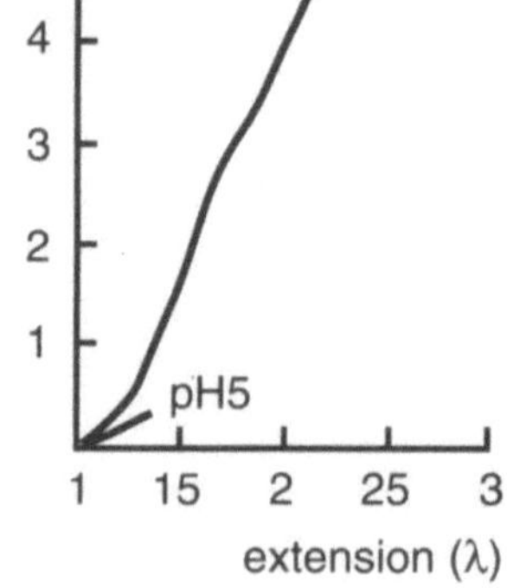

Fig. 1.44 Tensile stress–strain curves of ovalbumin gels (20 wt%) at pH 5 and pH 10. Cross-head speed 10 cm min^{-1}; 20°C.

Table 1.1 Summary of properties of ovalbumin gels

Property	pH 5	pH 10
Microstructure	Inhomogeneous, granular; network of aggregated protein	Homogeneous, uniform; network of protein filaments
Appearance	Opaque	Transparent
Rheology	Low breaking stress	High breaking stress
Conformation (NMR)	Rigid globular protein chain	Flexible unfolded protein chain
Protein–protein interaction (hydrophobic, hydrogen bond)	Strong	Weak
Protein–water interaction	Weak	Strong
Water release	Yes	No

pursued in isolation. Properly integrated with other characterization and image-forming techniques, however, it does offer a wealth of information on the relation between structure and function of traditional and new food products.

References

Adrian, M., Dubochet, J., Lepault, J. and McDowall, A. (1984) Cryo-electron microscopy of viruses. *Nature London*, **308**, 32–36.

Aguilera, J.M. and Stanley, D.W. (1990) *Microstructural Principles of Food Processing and Engineering*, Elsevier Applied Science, London.

Armbruster, B.L. and Desai, N. (1993) Identification of proteins and complex carbohydrates in some commercial low-fat dairy products by means of immunolocalization techniques. *Food Struct.*, **12**, 289–299.

Beesley, J.E. (1989) *Colloidal Gold: A New Perspective for Cytochemical Marking*, Oxford University Press–Royal Microscopical Society.

Bremer, A., Henn, C., Engel, A. and Baumeister, W. (1992) Has negative staining still a place in biomacromolecular electron microscopy? *Ultramicroscopy*, **46**, 25–111.

Brooker, B.E. (1987) The crystallization of calcium phosphate at the surface of mould-ripened cheese. *Food Microstruct.*, **6**, 25–33.

Brooker, B.E. (1990) Identification and characterization of cocoa solids and milk proteins in chocolate using X-ray microanalysis. *Food Struct.*, **9**, 9–21.

Brooker, B.E. (1991) Recent developments in the application of X-ray microanalysis to the study of food systems. *Food Struct.*, **10**, 145–152.

Brooker, B.E., Anderson, M. and Andrews, A.T. (1986) The development of structure in whipped cream. *Food Microstruct.*, **5**, 277–285.

Buchheim, W. and Dejmek P. (1990) Milk and dairy-type emulsions, in *Food Emulsions* (eds. K. Larson and S.E. Friberg), Marcel Dekker, New York, pp. 203–246.

Budd, P.M. and Goodhew, P.J. (1988) *Light-Element Analysis in the Transmission Electron Microscope: WEDX and EELS*, Oxford University Press–Royal Microscopical Society.

Charbonneau, J.E. (1991) Application of scanning electron microscopy and X-ray microanalysis to investigate corrosion problems in plain and enamelled three piece welded food cans. *Food Struct.*, **10**, 171–184.

Cookson, J. (1994) Three-dimensional reconstruction in microscopy. *Proc. R. Microsc. Soc.*, **29**, 3–10.

Creamer, L.K., Berry, G.P. and Matheson, A.R. (1978). The effect of pH on protein aggregation in heated skim milk. *N.Z. J. Dairy Sci. Technol.*, **13**, 9–15.

Danilatos, G.D. (1991) Review and outline of environmental SEM at present. *J. Microsc.*, **162**, 391–402.

Danilatos, G.D. (1993) A bibliography of environmental scanning electron microscopy. *Microsc. Res. Tech.*, **25**, 529–534.

Davis, E.A. and Gordon, J. (1982) Food microstructure: an integrative approach, *Food Microstruct.*, **1**, 1–12.

Dickinson, E. (1990) Particle gels. *Chem. Ind.*, **October**, 595–599.

Dubochet, J. and McDowall A.W. (1981) Vitrification of pure water for electron microscopy. *J. Microsc.*, **124**, 3–4.

Echlin, P. (1992) *Low-Temperature Microscopy and Analysis*, Plenum Press, New York.

Frank, J. (1989) Advances in microscopy—part I: three-dimensional imaging techniques in electron microscopy. *Biotechniques*, **7**, 164–173.

Frank, J. (1992) *Electron Tomography: Three-Dimensional Imaging with the Transmission Electron Microscope*, Plenum Press, New York.

Frank, J. and Radermacher, M. (1992) Three-dimensional reconstruction of single particles negatively stained or in vitreous ice. *Ultramicroscopy*, **46**, 241–262.

Frau, M., Blanco, A., Canellas, J. and Rossello, C. (1993) Use of Peldri II (a fluorocarbon compound) in the preparation of cheese samples for examination by scanning electron microscopy. *J. Dairy Res.*, **60**, 129–132.

Frederik, P.M., Stuart, P.H., Bomans, P.H.H. and Busing, W.M. (1989) Phopholipid: nature's own slide and coverslip for cryo-electron microscopy. *J. Microsc.*, **153**, 81–92.

Frens, G. (1973) Controlled nucleation for the regulation of the particle size in monodisperse gold suspensions. *Nature (London) Phys. Sci.*, **241**, 20–22.

Friel, J.J. and Prestridge, E.B. (1991) Grain sizing by image analysis. *Adv. Mater. Proc.*, **139**, 33–37.

Glauert, A.M. and Reid N. (1974) Fixation, dehydration and embedding of biological specimens and ultramicrotomy, in *Practical Methods in Electron Microscopy* (ed. A.M. Glauert), North-Holland, Amsterdam, vol. 3, pp. 1–207.

Goldstein, J.I., Newbury, D.E., Echlin, P., Joy, D.C., Fiori, C. and Lifshin, E. (1981) *Scanning Electron Microscopy and X-ray Microanalysis*, Plenum Press, New York.

Graham, R.C. and Karnovsky, M.J. (1966) The early stages of absorption of injected horseradish peroxidase in the proximal tubulus of mouse kidney: ultrastucture cytochemistry by a new technique. *Histochem. Cytochem.*, **14**, 291–307.

Graveland, A., Henderson, M.H., Pâques, M. and Veerman, S.M. (1993) Practical applications of molecular glutenin structures, breakdown and reassembly in a wheat flour dough, in *Gluten Proteins*, Association of Cereal Research, Detmold, pp. 393–403.

Harada, T., Kanzawa, Y., Kanenaga, K., Koreeda, A. and Harada, A. (1991) Electron microscopic studies on the ultrastructure of curdlan and other polysaccharides in gels used in foods. *Food Struct.*, **10**, 1–18.

Hawkes, P.W. and Valdrè, U. (eds.) (1990) *Biophysical Electron Microscopy*, Academic Press, London.

Hayat, M.A. and Miller, S.E. (1990) *Negative Staining*, McGraw-Hill, New York.

Heertje, I. (1993a). Microstructural studies in fat research. *Food Struct.*, **12**, 77–94.

Heertje, I. (1993b) Structure and function of food products: a review. *Food Struct.*, 12, 343–364.

Heertje, I. and Leunis, M. (1994) Measurement of shape and size of fat crystals by electron microscopy. *Food Struct.*, **13**.

Heertje, I. and van Kleef F. (1986) Observations on the microstructure and rheology of ovalbumin gels. *Food Microstruct.*, **5**, 91–98.

Heertje, I., Boskamp, M.J., van Kleef, F. and Gortemaker, F.H. (1981) The microstructure of processed cheese. *Neth. Milk. Dairy J.*, **35**, 177–179.

Heertje, I., Visser, J. and Smits, P. (1985) Structure formation in acid milk gels. *Food Microstruct.*, **4**, 267–277.

Heertje, I., Leunis, M., van Zeijl, W.J.M. and Berends, E. (1987) Product morphology of fatty products. *Food Microstruct.*, **6**, 1–8.

Helwig, E.J. and Biber, H.E. (1990) The use of the scanning electron microscope in investigating container corrosion by canned food and beverages. *Food Struct.*, **9**, 195–202.

Herman, R. and Muller, M. (1993) Progress in scanning electron microscopy of frozen-hydrated biological specimens. *Scanning Microsc.*, **7**, 343–350.

Hermansson, A.M. (1989) Rheological and microstructural evidence for transient states during gelation of kappa-carrageenan in the presence of potassium. *Carbohydr. Polym.*, **10**, 163–181.

Hermansson, A.M. (1993) Microstructure of protein gels related to fuctionality, in *Food Proteins, Structure–Function Relationships.* IUFoST Monograph Series, Elsevier Applied Science, Barking, UK.

Hermansson, A.M., Eriksson E. and Jordansson, E. (1991) Effects of potassium, sodium and calcium on the microstructure and rheological behaviour of kappa-carrageenan gels. *Carbohydr. Polym.*, **16**, 297–320.

Horrisberger, M. and Rosset, J. (1977) Colloidal gold, a useful marker for transmission and scanning electron microscopy. *J. Histochem. Cytochem.*, **25**, 295–305.

Horrisberger, M. and Vonlanthen, M. (1979) Multiple marking of cell surface receptors by gold granules. Simultaneous localization of three lectin receptors on human erythrocytes. *J. Microsc.*, **115**, 97–102.

Horrisberger, M. and Vonlanthen, M. (1980) Localization of glycosylated κ-casein in bovine casein micelles by lectin-labelled gold granules. *J. Dairy Res.*, **47**, 185–191.

Horrisberger, M., Vonlanthen, M. and Rosset, J. (1978) Localization of α-galactomannan and of wheat germ agglutinin receptors in *Schizosaccharomyces pombe. Arch. Microbiol.*, **119**, 107–111.

Humphreys, S.H. (1989) Freeze-fracture of manufactured foods. *J. Electr. Microsc. Technol.*, **13**, 300–308.

Joy, D.C. and Pawley, J.B. (1992) High-resolution scanning electron microscopy. *Ultramicroscopy*, **47**, 80–100.

Juriaanse, A.C. and Heertje, I. (1988) Microstructure of shortenings, margarine and butter—a review. *Food Microstruct.*, **7**, 181–188.

Kalab, M. (1981) Electron microscopy of milk products: a review of techniques. *Scanning Electron Microsc.*, **III**, 453–472.

Kalab, M. (1993) Practical aspects of electron microscopy in dairy research. *Food Struct.*, **12**, 95–114.

Kiselev, N.A., Sherman, M.B. and Tsuprun, V.L. (1990) Negative staining of proteins. *Electron Microsc. Rev.*, **3**, 43–72.

Knowles, K.M. (1994) Editorial. *J. Microsc.*, **174**, 131–132.

Lepault, J. and Erk, I. (1992) Time-resolved cryo-electron microscopy. *Proc. 10th Eur. Congress on Electron Microscopy*, Granada, vol. III, p. 91.

McManus, W.R. and McMahon, D.J. (1994) Calcium elemental mapping of freeze-dried milk and yoghurt through electron spectroscopic imaging (ESI). *Abstracts, Food Structure Meeting, Toronto.* Scanning Microscopy Intl., Chicago.

Morgan, A.J. (1985) *X-Ray Microanalysis in Electron Microscopy for Biologists*, Oxford University Press—Royal Microscopical Society.

Mulder, M. (1957) The presence of moisture in butter. *Neth. Milk Dairy J.*, **11**, 25–42.

Nation, J.L. (1983) A method using hexamethyldisilazane for preparation of soft insect tissues for scanning electron microscopy. *Stain Technol.*, **58**, 347–351.

Oostergetel, G.T. and van Bruggen, E.F.J. (1993) The crystalline domains in potato starch granules are arranged in a helical fashion. *Carbohydr. Polym.*, **21**, 7–21.

Otten, M.T., Stenzel, D.J., Cousens, D.R., Humbel, B.M., Leunissen, J.L.M., Stierhof, Y.D. and Busing, W.M. (1992) High-angle annular dark-field STEM imaging of immunogold labels. *Scanning*, **14**, 282–289.

Polak, J.M. and van Noorden, S. (eds.) (1983) *Immunocytochemistry*, John Wright, Bristol.

Polak, J.M. and van Noorden, S. (1984) *An Introduction to Immunocytochemistry: Current Techniques and Problems*, Oxford University Press—Royal Microscopical Society.

Polak, J.M. and Varndell, I.M. (eds.) (1984) *Immunolabelling for Electron Microscopy*, Elsevier, Amsterdam.

Poot, C., Dijkshoorn, W., Haighton, A.J. and Verburg, C.C. (1975) Laboratory separation of crystals from plastic fats using detergent solution. *J. Am. Oil Chem. Soc.*, **70**, 69–72.

Reville, W.J. and Cotter, M.P. (1991) An evaluation of the usefulness of air drying biological samples from tetramethylsilane in preparation for scanning electron microscopy. *J. Electron Microsc.*, **40**, 198–202.

Robards, A.W. and Abeysekera R.M. (1992) Improved freeze preparation of biopolymer molecules for transmission electron microscopy. *Proc. 10th Eur. Congress EM*, Granada, Vol. III, pp. 77–78.

Robards, A.W. and Sleytr U.B. (1985) Low temperature methods in biological electron microscopy, in *Practical Methods in Electron Microscopy* (ed. A.M. Glauert), Elsevier, Amsterdam, vol. 10.

Robards, A.W. and Wilson, A.J. (1993) *Procedures in Electron Microscopy*, Wiley, Chichester.

Roomans, G.M. (1990) X-ray microanalysis, in *Biophysical Electron Microscopy*, (eds. P.W. Hawkes and U. Valdrè), Academic Press, London, pp. 347–406.

Roth, J. (1983) The colloidal gold marker system for light and electron microscopic cytochemistry, in *Techniques in Immunocytochemistry 2*, (eds. G.R. Bullock and P. Petrusz), Academic Press, London, pp. 217–284.

Russ, J.C. (ed.) (1990) *Computer-Assisted Microscopy: The Measurement and Analysis of Images*, Plenum Press, New York.

Sargent, J.A. (1988) Low temperature scanning electron microscopy: advantages and applications. *Scanning Microsc.*, **2**, 835–849.

Slot, J.W. and Geuze, H.H. (1981) Sizing of protein A colloidal gold probes for immunoelectron microscopy. *J. Cell Biol.*, **90**, 533–536.

Snoeren, T.H.M., Both, P. and Schmidt, D.H. (1976) An electron microscopic study of carrageenan and its interaction with κ-casein. *Neth. Milk Dairy J.*, **30**, 132–141.

Stokke, B.T. and Brant, D.A. (1990) The reliability of wormlike polysaccharide chain dimensions estimated from electron micrographs. *Biopolymers*, **30**, 1161–1181.

Stokke, B.T., Elsgsaeter, A. and Kitamura, S. (1993) Macrocyclization of polysaccharides visualized by electron microscopy. *Int. J. Biol. Macromol.*, **15**, 63–68.

Stokke, B.T., Smidsrød, O., Marthinsen, A.B.L. and Elsgsaeter, A. (1988) Conformational analysis of xanthan and welan using electron microscopy, in *Water-Soluble Polymers for Petroleum Recovery* (eds. G.A. Stahl and D.N. Schulz), Plenum Publishing Corp., New York, pp. 243–252.

Sugiyama, J., Rochas, C., Turquois, T., Taravel, F. and Chanzy, H. (1994) Direct imaging of polysaccharide aggregates in frozen aqueous dilute systems. *Carbohydr. Polym.*, **23**, 261–264.

Talmon, Y. (1986) Imaging surfactant dispersions by electron microscopy of vitrified specimen. *Colloids Surf.*, **19**, 237–248.

Trebbia, P., Jeanguillaume, C. and Walls, M. (1990) EELS: instrumentation, theory, practice and applications, in *Biophysical Electron Microscopy*, (eds. P.W. Hawkes and U. Valdrè), Academic Press, London, pp. 413–448

van Kleef, F.S.M. (1986) Thermally induced protein gelation. *Biopolymers*, **25**, 31–39.

Varriano-Marston, E. (1977) A comparison of dough preparation procedures for scanning electron microscopy. *Food Technol.*, **October**, 32–36.

Vaughan, J.G. (ed.) (1979) *Food Microscopy*, Academic Press, London.

Willison J.H.M. and Rowe A.J. (1980) Replica, shadowing and freeze-etching techniques, in *Practical Methods in Electron Microscopy*, (ed. A.M. Glauert), North-Holland, Amsterdam, vol. 8.

Chapter 2
Imaging Food Systems by Confocal Laser Scanning Microscopy

B.E. Brooker
Institute of Food Research,
Reading, UK

2.1 Introduction

Light microscopy is a well developed and increasingly widely used technique for studying the microstructure and composition of food systems in relation to their physical properties and processing behaviour (Vaughan, 1979). However, because the large depth of field of the light microscope produces images containing information from different focal planes in thick specimens, good quality and high resolution images of the internal structure of foods can only be obtained from smears, squashes or thin sections of the sample. Procedures which apply substantial shear and compressive forces may destroy or damage structural elements, and sectioning is time consuming and involves chemical processing steps that may introduce artefacts and make image interpretation difficult.

2.2 Principles of operation

Confocal laser scanning microscopy (CLSM) overcomes these problems because in this instrument image formation does not depend on transmitting light through the specimen, and therefore bulk specimens can be used for the first time in light microscopy. The instrument uses a focused, scanning laser to illuminate a sub-surface layer of the specimen in such a way that information from this focal plane passes back through the specimen and is projected onto a pinhole (confocal aperture) in front of a detector (Fig. 2.1). Only the light from a defined focal plane in the specimen is able to pass through the confocal aperture, reach the detector, and produce an image which is effectively an optical slice. By moving the specimen up and down relative to the focused laser light, large numbers of consecutive optical sections with improved lateral resolution (compared with conventional light microscopy) can be obtained with a

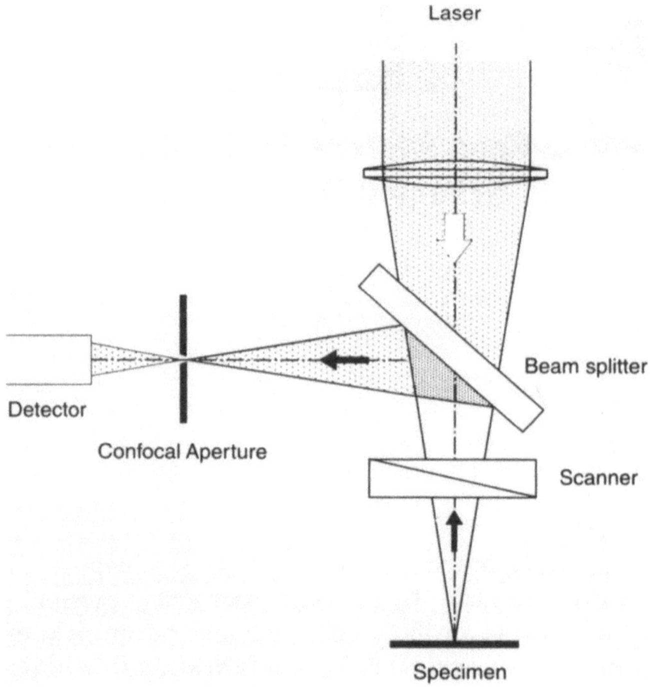

Fig. 2.1 Diagrammatic representation of an incident light confocal system showing the excitation beam and the reflected fluorescent light.

minimum of sample preparation. The thickness of the optical slices can be controlled in two ways: (i) by increasing the numerical aperture (NA) of the objective lens which reduces section thickness for any given diameter of the confocal aperture, and (ii) by varying the diameter of the confocal aperture, a larger aperture producing thicker slices, and *vice versa*. The thinnest slice that can be obtained with a high NA lens is less than 1 μm. This principle of confocality was first described by Minsky (1957) and has been adapted for use in a number of designs for commercial instruments.

By collecting a number of consecutive images in different planes of the specimen and then superimposing them electronically, three-dimensional information can be obtained by reconstruction of a solid model or by the construction of an anaglyph (Carlsson *et al.*, 1985). Alternatively, a series of time-resolved images of some dynamic process can be re-assembled to produce a moving image showing the changes in structure with time. These manipulations can be achieved using the 'on-board' CLSM software, or by first transferring the data to an image analyser or some such peripheral data processor. For an in-depth treatment of the optics and principles of operation of CLSM, the reader is referred to Pawley (1990) and Wilson (1990).

In the biological sciences, the term confocal microscopy is frequently assumed to be synonymous with confocal fluorescence microscopy. However, it should be pointed out there exists a wide range of alternative confocal techniques such as reflected bright field (Cogswell and Sheppard, 1990a), and differential phase contrast and differential interference contrast microscopy (Cogswell and Sheppard, 1990b), which are based on the analogous methods of conventional light microscopy and which each have their place in characterizing the microstructure of complex food systems. Reflected bright field confocal microscopy, in particular, is very useful for imaging reflective structures in foods such as sugar crystals. Another adaptation of the confocal principle for use in a microscope resulted in the tandem scanning microscope (TSM) which has certain distinct advantages compared with CLSM (Petráň *et al.*, 1968, 1985). The most significant of these is that it is possible to use real-time imaging to observe rapid dynamic changes in a specimen; in CLSM, the scan time is commonly 1 or 2 s, which requires a very stable specimen, and can only successfully record relatively slow processes. Furthermore, a conventional light source can be used to obtain TSM images of the real colour in a specimen by direct viewing rather than by the indirect viewing that is necessary in the case of CLSM (Boyde *et al.*, 1990).

2.3 Fluorescent dyes – practical aspects

Most commercially available CLSM instruments are used as confocal fluorescence microscopes, although it is common for them to also have facilities for transmitted laser light and reflected light microscopy. In the fluorescence mode of operation, images of various chemical components, such as proteins, carbohydrates, lipids and ions, are produced by using the laser light to excite a selective fluorescent dye that has already been introduced or allowed to diffuse into the food system. Alternatively, specific localization of chemical components can be achieved using antibodies, enzymes and ligands labelled with appropriate fluorescent dyes. If the laser(s) in the CLSM instrument produces light of two or more wavelengths, and dyes are introduced into the sample which excite at these differing wavelengths, it is possible to obtain multiple images of the same field showing the spatial distribution of different chemical components, such as fat and protein, which can then be superimposed to show their relative distributions. The Ar/Kr laser, for example, produces three lines at 488, 568 and 647 nm.

It should be noted that a large number of highly specialized fluorochromes are commercially available which were originally developed principally for biomedical applications (physiological indicators and tracers, pH indicators, and membrane probes), but which can be used in foods to image selectively some of the chemical components (Haugland, 1992). The dyes referred to below are the most suitable for general use but information on more specialized applications is given by Tsien and Waggoner (1990) and Haugland (1992).

The determination of the dye that produces the best images, and the conditions under which the dye is best applied, is a matter of experiment when unfamiliar food products are encountered. However, the diffusion of fluorescent dyes into food materials can only occur if both are brought into molecular contact, either by direct application of the powdered dye to the surface or by applying a solution of the dye, although in the latter case, care must be taken to ensure that this treatment does not affect the integrity or structure of the specimen (Brooker, 1991; Blonk and van Aalst, 1993). Examples of stains that have been found to produce satisfactory results in a variety of food materials are discussed further below. However, it should be pointed out that dyes which have long proved to be satisfactory for conventional fluorescence microscopy of relatively thick samples of foods may not produce good results with confocal microscopy. In the confocal mode, the excited fluorochrome must emit sufficient light from the thin optical plane being studied to produce an image with high signal/noise ratio, although this also depends on the characteristics and design of the microscope optics and its efficiency of light detection.

Many fluorescent dyes are available for studying the distribution of protein in foods such as dairy products, emulsions and spreads, batters and doughs, and confectionery products. These include fluorescein isothiocyanate (FITC) and acridine orange, which excite at about 500 nm, rhodamine 123 and Texas Red, which excite at about 560 nm, or Cy5 and the phycocyanins which excite in the region of 630 nm. When an oil phase is present in a food, it can be imaged using Nile Red, an oil-soluble dye that fluoresces strongly in hydrophobic environments but has weak fluorescence under slightly hydrophobic and hydrophilic conditions (Greenspan *et al.*, 1985). When oil is a disperse phase in water, the lipid is best stained with Nile Blue, which is non-fluorescent at the wavelengths used in confocal microscopy, but which contains small amounts of Nile Red which readily partition into the oil droplets, thus allowing them to be imaged. However, fat crystals, which are of considerable functional importance in foods and occur together with oil in butter, shortenings and confectionery products, cannot be positively stained because the dye is unable to penetrate their crystal lattice (Cain, 1947). Instead, crystals may only be seen by negative contrast so long as there is a strongly fluorescing and dominant oil phase in the background (Fig. 2.2).

In situations when the oil phase is continuous (e.g. butter), Nile Red is always the preferred dye, and it can be applied directly to the surface

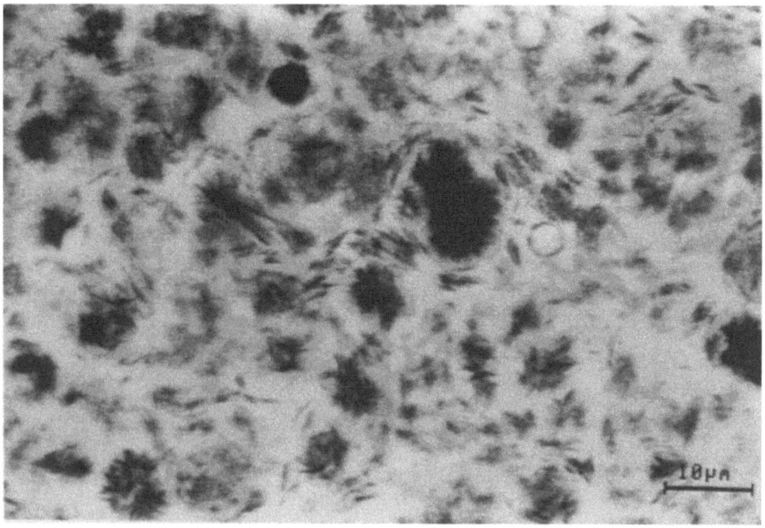

Fig. 2.2 A shortening stained with Nile Red. The dominant continuous oil phase is fluorescent and the non-staining fat crystals and crystal aggregates are visible by negative contrast.

of the specimen as a fine dusting of powder or as a solution in oil or in a mixture of polyethylene glycol and glycerol (Brundrett *et al.*, 1991) followed by the application of a coverslip. For solid foods, such preparations need to be left for some hours before being examined to allow the dye enough time to diffuse into the matrix. If, however, the sample is liquid (e.g. molten chocolate), the dye can be stirred in until completely dissolved and then examined immediately, either whilst still in the molten state if a hot sub-stage is attached to the microscope stage, or after allowing it to solidify.

In conventional light microscopy, the ability to control the temperature of a specimen by the use of a hot or cold stage has always offered a number of advantages because it makes it possible to maintain the examined food system in a desired state for a prolonged period. In confocal microscopy where bulk specimens are routinely used, the temperature controlled stage is especially important, and it offers the ability to study the three-dimensional structure of fragile foods such as frozen products, fermenting cereal products, and crystalline fats and foams, without them being adversely affected by fluctuating changes in the ambient temperature.

Carbohydrates are frequently important components of food systems that can be specifically localized and distinguished from related carbohydrate molecules using one of the many carbohydrate-binding proteins (lectins) conjugated with a fluorescent dye (Yiu, 1993). The exact choice of lectin to be used depends on the composition and structure of the carbohydrate to be imaged (Liener *et al.*, 1986). It is possible to identify at least two carbohydrates with different chemical composition in any given food using lectins with different specificities conjugated to fluorochromes which excite at different laser wavelengths, e.g. FITC-labelled concanavilin A (terminal α-D-mannosyl and α-D-glucosyl residues) at 490 nm and tetramethylrhodamine isothiocyanate-labelled *Helix aspersa* agglutinin (terminal *N*-acetyl-α-D-galactosaminyl residues) at 540 nm.

When the use of a fluorescent dye has made it possible to obtain an image of the internal structure of a food material, quantitative information can then be extracted from it with the use of image analysis. This is often essential if the microstructure is to be related to some physical property of the bulk specimen and a relationship between the two properly established. Confocal microscopy lends itself to image analysis because the pictures are already digitized, and they can therefore be easily transferred from one instrument to another. Moreover, the range of grey levels in the image can be reduced electronically using brightness and contrast controls, and therefore much of the lengthy image processing that is normal with conventional images can be avoided.

The diffusion rate constants of fluorochrome-labelled molecules through liquids have been determined by Blonk *et al.* (1993) using an established technique, fluorescence recovery after photobleaching (FRAP), in conjuction with confocal microscopy. The bringing together of these two techniques makes it possible to obtain detailed structural information and therefore to select the area of the specimen in which diffusion phenomena are to be studied. Thus, FRAP provides information about rates of diffusion of molecules such as flavour compounds, colouring agents or functional ingredients (e.g. antioxidants) through food matrices in relation to food structure.

2.4 Applications to food systems

The type of information that can be obtained from food systems in general is best illustrated by reference to a number of the most common foods and the most difficult to examine. For animal and plant cells, the reader is referred to White *et al.* (1987) and Pawley (1990). Comparatively few papers have yet been published dealing with food-related topics.

2.4.1 Chocolate

The particulate components of chocolate, namely milk protein, sugar and cocoa, have been imaged by light microscopy using sections (8 μm thick) that have been de-fatted and stained (Heathcock, 1985). The thinner optical sections and improved resolution produced by confocal microscopy results in high quality images that reveal features that have not been described previously for chocolate.

In order to image the particulate components of chocolate, the simplest approach is to stain the continuous fat phase with Nile Red so that the particles are visible by negative contrast (Fig. 2.3). In milk crumb chocolate, the milk protein can be seen as grey material closely associated with particles of cocoa and sugar (Fig. 2.3), whereas in milk powder chocolate, the particles of full cream milk powder may remain intact or produce fragments that can be easily identified (Heathcock, 1985).

Cocoa and sugar particles are angular structures that are very similar in appearance and cannot be distinguished on the basis of their morphology. However, cocoa particles can be recognized because they are weakly auto-fluorescent when excited at about 630 nm, and sometimes they contain droplets of residual cocoa butter that have not been extracted during pressing.

In all milk and plain chocolates that have been examined, large numbers of both types of particle contain a number of fine cracks which are produced during the particle size reduction step (refining) in the early stages of processing. Three-dimensional imaging of individual particles has shown that many cracks run from one side of a particle to another

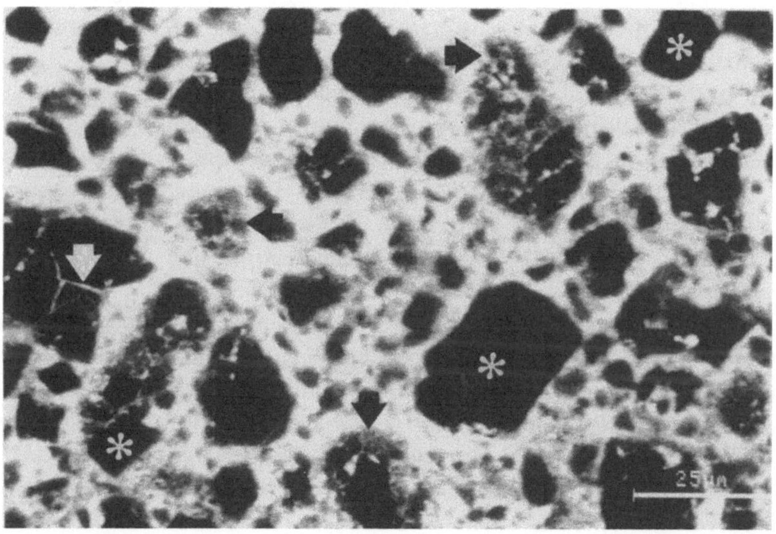

Fig. 2.3 Crumb milk chocolate stained with Nile Red whilst in the molten state and then allowed to solidify. Particles of cocoa, sugar (asterisks) and milk protein (black arrows) are visible by negative contrast. Note the cracks in one of the cocoa particles (white arrow).

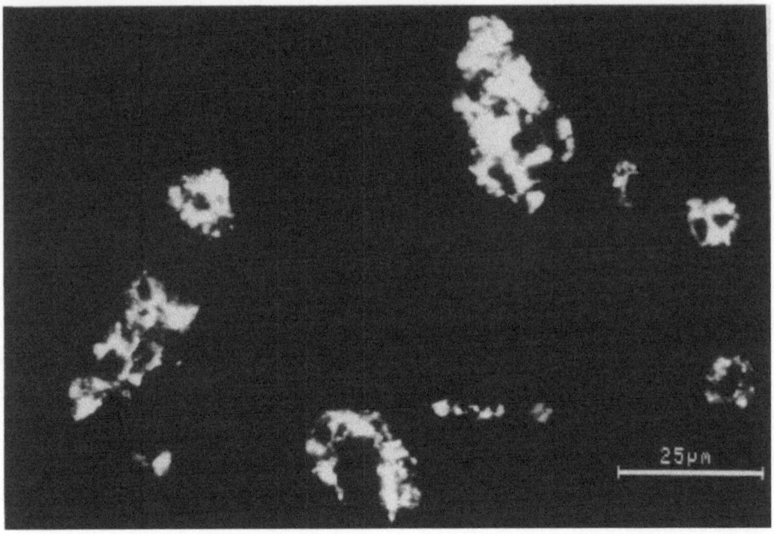

Fig. 2.4 Same field as the crumb milk chocolate in Fig. 2.3 but excited at 630 nm. A Cy5-labelled polysaccharide was added to the chocolate whilst it was still molten. After 5 min, the chocolate was allowed to solidify. The polysaccharide is localized in the areas of milk protein.

and are open at both ends. Many particles can therefore be imagined to be composed of several sub-particles which are held closely together by the presence of the cocoa butter in the fine cracks. The separation of the sub-particles would be possible only if the cocoa butter in the cracks was liquid and if continuous shear was applied in order to encourage the movement of additional oil from the continuous phase into the cracks. These conditions are encountered during 'conching', and it may well be that the separation of sub-particles and the concomitant change in particle-size distribution contribute to the changes in the rheological properties of chocolate that take place during this conching process (Chevalley, 1975).

The detection and localization of a functional ingredient added to chocolate (or to any other food system) may be achieved using double-labelling techniques provided an appropriate fluorochrome-labelled compound is available. Thus, in Fig. 2.4 the distribution of a Cy5-labelled hydrophilic polysaccharide in relation to the population of particles can be seen 5 min after it has been added to molten chocolate. Similarly, the fate of lecithin which is normally added to chocolate to reduce viscosity, can be followed using one of a range of Texas Red or tetramethylrhodamine-labelled phospholipids. In both cases, the distribution of the ingredient relative to the particles of cocoa and sugar can only be determined if Nile Red has also been incorporated.

2.4.2 Emulsions

Many foods are emulsions at the microstructural level. They include products such as low-fat spreads, margarines, salad dressings and toffee. Their bulk rheological and organoleptic properties depend critically on the nature and amount of the disperse phase and the composition and concentration of the continuous phase, together with the strength of interaction between the two. Characterization of emulsions by CLSM is relatively simple compared with conventional microscopy because there is no pre-treatment, as in the case of toffee (Lewis, 1986), or compression of a thin layer of the emulsion which may lead to destabilization and redistribution of the disperse phase. This is especially true of some low-fat spreads and whippable emulsions.

Whether the emulsion is of the oil-in-water or water-in-oil type, a fluorescent dye may be introduced into the oil phase, into the aqueous phase, or into both if two exciting wavelengths are available from the laser. If only one dye is used, e.g. a water soluble dye to stain the aqueous phase (and any hydrophilic particles dispersed in it), it is tempting to assume that all of the particles seen by negative contrast represent oil droplets. This is often the case, but not always so. For example, Figs 2.5 and 2.6 are of a low-fat oil-in-water salad dressing to which particles of soya protein and modified starch have been added to compensate for the changes in the rheology produced by the reduction in oil content. Using

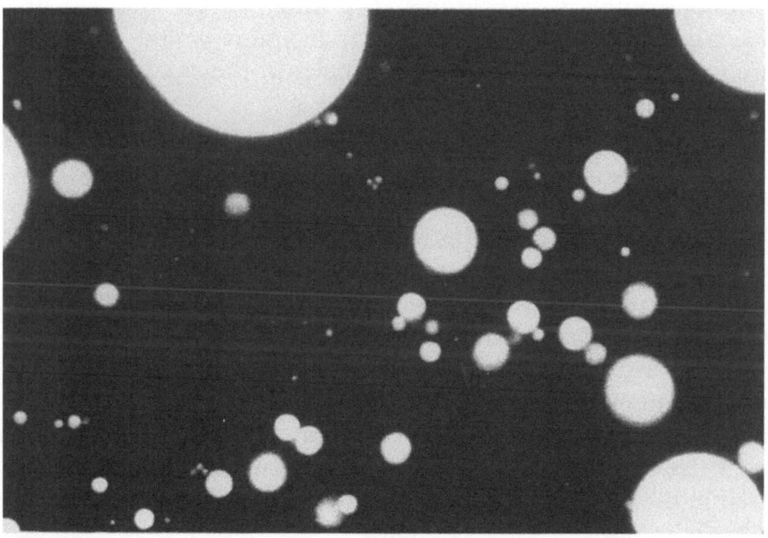

Fig. 2.5 Image of a double-labelled low-fat salad dressing stained with Nile Red and Fast Green FCF. The oil phase consists of many droplets. Excited at 488 nm.

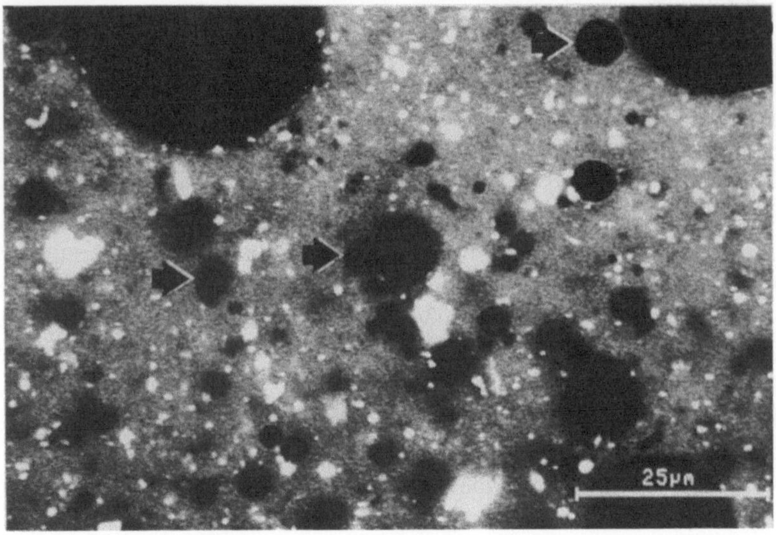

Fig. 2.6 As Fig. 2.5 except excited at 630 nm. The continuous aqueous phase is slightly fluorescent because of soluble proteins and contains strongly fluorescent particles of soya protein. Large, circular black areas represent the positions of oil droplets. Some particles (arrows) are non-fluorescent.

double fluorochrome labelling, it is possible to find particles of fluorescing soya protein and other particles of starch which do not fluoresce. If only the oil-soluble fluorochrome (Nile Blue) or the water-soluble dye had been used, the population of starch grains would not have been apparent, and the total number of particles in the aqueous phase would have been considerably underestimated.

In cases where the emulsion is a solid, as in margarines, shortenings and low-fat spreads, the microstructure of very large bulk samples can be examined *in situ*, provided they are stable when placed on the microscope stage (Brooker, 1993a). One of the features of interest in these products is the size and distribution of water droplets since these influence microbiological stability and texture, and contain many of the important flavour compounds that will be released during mastication. Fat crystals also are of major functional importance, but, except for some shortenings, these are difficult to image in detail because of their small size and the relatively small amount of oil present, as discussed above. In such cases, the masses of crystals appear as patches of grey material in the background when Nile Red or Nile Blue has been used to stain the oil phase.

The principles for staining different phases in these spreadable emulsions follow the general principles given above, but Heertje (1993) and Blonk and van Aalst (1993) have shown that it is also possible to obtain detailed information of particulate proteins, such as gelatin and casein aggregates, in the aqueous phase with the use of selective protein stains. Air bubbles incorporated by accident during processing, or by design as an integral part of the food structure, as in shortenings, whipped creams and toppings, cannot be imaged directly and appear as black areas by negative contrast in optical sections.

When the disperse phase of an emulsion is stabilized by the adsorption of proteins from an aqueous phase that has been stained using a selective protein dye (such as fluorescein isothiocyanate), the protein oil–water interface is visible as a bright line in optical sections. The thickness of the fluorescent layer is considerably thicker than would be expected from physical measurements, principally because of optical effects arising from the difference in refractive index between the oil and water phases. This effect has been put to good use by Heertje *et al.* (1990) to study competition between proteins and emulsifiers at oil–water interfaces. Using an inverted microscope, they imaged the interface between a drop of oil and a drop of water that had been brought into contact. They studied the displacement of FITC-labelled sodium caseinate from the interface by a number of purified but unlabelled monoacylglycerols and obtained quantitative information on the impact of emulsifier type and concentration in this process. The effect of the protein displacement was that the fluorescent interface reduced in intensity progressively as the unlabelled emulsifier adsorbed. Similarly, Blonk and van Aalst (1993) used two fluorescent agents to demonstrate the displacement of FITC-labelled sodium caseinate from an oil–water interface by the more surface active and oil-soluble phospholipid *N*-(Texas Red sulfonyl) dipalmitoyl-L-α-phosphatidyl-ethanolamine.

2.4.3 Cereal products – batters and doughs

The main structural elements of batters and doughs are starch granules, fat droplets, water-soluble and water-insoluble proteins, air, and, in the case of bread, yeast cells. Starch grains can be imaged using Nile Blue by exciting at 630 nm (Fig. 2.7) or with FITC-labelled concanavilin A (Yiu, 1993). The size distributions of air bubbles in cake batters with and without added fat were obtained by Brooker (1993c) using projections of the bubbles in micrographs produced by transmission laser miscroscopy. The gluten network and protein associated with the surface of starch grains respond well to the protein stains mentioned above (Heertje *et al.*, 1987; Blonk and van Aalst, 1993), especially to Fast Green FCF, and to safranine O.

The importance of confocal microscopy with these materials is that it is possible to examine bulk samples under conditions of controlled temperature (with an appropriate stage) and to observe structural

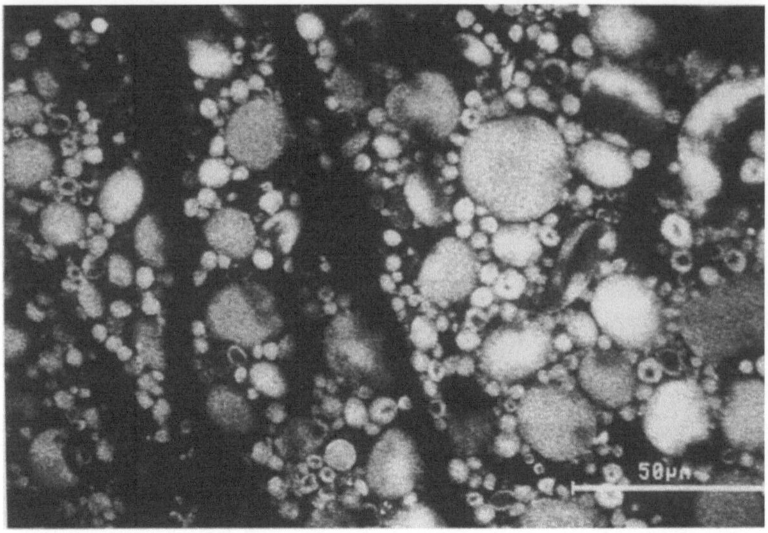

Fig. 2.7 Bread dough at the end of mixing produced by the Chorleywood baking process showing the distribution of large and small starch grains after staining with Nile Blue (630 nm excitation and 680 long-pass filter). Unstained dark strands between the grains represent the position of gluten fibres.

changes accompanying the proofing of doughs and the baking process itself (Heertje *et al.*, 1987). To do this, it is usually necessary to limit the vertical movement of the batter or dough using a cell with a fixed height which allows expansion to take place unimpeded only in the horizontal plane.

The continuous matrix of confectionery wafers produced from complex batters and cooked at high temperature for a short time is auto-fluorescent. Therefore, the structure can be examined by confocal microscopy, without the addition of fluorochromes, by exciting at 488 nm (Beckett *et al.*, 1994). It is interesting to note that, in the case of fermented wafer batters, the yeast cells are also rendered auto-fluorescent, probably by interaction with other (but unknown) components in the batter during cooking, since the yeasts do not fluoresce in uncooked batter or when they have been heated for the same time in the absence of cereal components.

2.4.4 Dairy products

Although milk and dairy products are diverse in appearance and in their physical properties, they are composed of the same chemical and structural components. These can be imaged by confocal microscopy using the same general methods of preparation and staining described

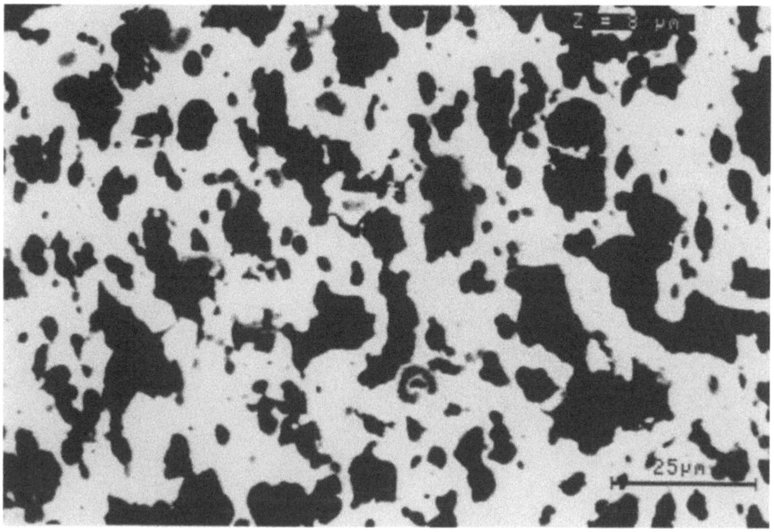

Fig. 2.8 Mature cheddar cheese stained with Fast Green FCF (630 excitation) showing the continuous protein phase. The continuous fat phase is visible by negative contrast.

above for use with emulsions and protein-containing foods. Fat globules, whether in cream (Brooker, 1993b) or in a solid matrix, such as cheese (Fig. 2.8), can be identified with the use of Nile Blue and Nile Red, respectively (Brooker, 1991). Cream quality is affected by the conditions under which it is stored, separated, pasteurized and packaged (Foley *et al.*, 1971), and the quality can be determined from the free fat content and the degree of globule coalescence. Apart from methods that involve chemical extraction procedures, the stability of the emulsion can be measured by CLSM because non-globular fat can be identified and quantified using image analysis (Brooker, 1991). Whereas extraction may cause disruption of fragile globules and lead to artificially high values, confocal microscopy is non-invasive and as a consequence produces more accurate results.

Casein micelles and the continuous matrices they form in products such as cheese and yoghurt require the use of protein dyes such as FITC, rhodamine 123, Fast Green FCF or acridine orange. Thus, Bremer *et al.* (1991) studied gel formation in sodium caseinate dispersions that had been labelled with FITC or rhodamine B and obtained quantitative structural information which was used to test ideas about the fractal properties of the casein gels. The use of acridine orange in milk that is being fermented to produce cheese curd, yoghurt (Fig. 2.9), kefir or *Streptococcus acidophilus* milk, permits changes in the structure and distribution of micellar casein to be monitored continuously and at the

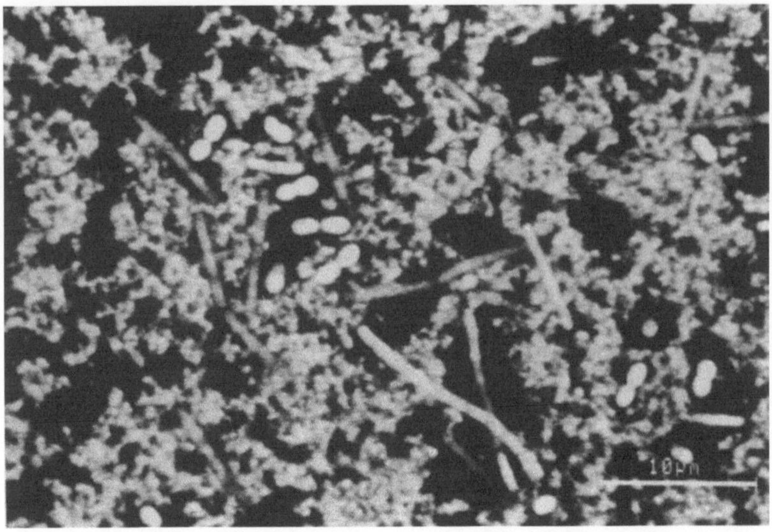

Fig. 2.9 Yoghurt prepared from skimmed milk showing the continuous, finely divided protein matrix produced by the aggregation of casein micelles. The fermenting bacteria are visible as rods and cocci (*Streptococcus thermophilus* and *Lactobacillus bulgaricus*). Stained with acridine orange and excited at 488 nm.

same time allows the growth of bacterial colonies to be followed in one or more optical planes.

The disperse aqueous phase of butters and modified butters (low-fat or containing fractionated fats) can be examined in negative contrast as described above for margarines using Nile Red or by the use of water-soluble dyes, even in cases where there is no continuous aqueous phase (Blonk and van Aalst, 1993). However, it is much more difficult to obtain useful information from the crystalline fat phase, as with other fat-containing foods.

2.5 Concluding remarks

It is hoped that this brief survey of confocal microscopy as applied to food materials has given a good impression of the range and depth of information that can be obtained using the technique. Although the CLSM technique already has wide application in the bio-medical area, and has started to be exploited in industrial food research, its use to probe the structure of foods is still in its infancy. The availability of a wider choice of lasers to provide even more wavelengths of light with which to image even more chemical components than is presently possible will increase the power of confocal microscopy, especially if it is used in conjunction with other physical techniques such as Raman spectroscopy and FRAP.

References

Beckett, S., Livings, S. and Schroeder, P. (1994) Edible wafers need physics. *Phys. World*, **7**, 49–53.

Blonk, J.C.G. and van Aalst, H. (1993) Confocal scanning light microscopy in food research. *Food Res. Int.*, **26**, 297–311.

Blonk, J.C.G., Don, A., van Aalst, H. and Birmingham, J.J. (1993) Fluorescence photobleaching recovery in the confocal scanning light microscope. *J. Microsc.*, **117**, 363–374.

Boyde, A., Jones, S.J., Taylor, M.L., Wolfe, L.A. and Watson, T.F. (1990) Fluorescence in the tandem scanning microscope. *J. Microsc.*, **157**, 39–49.

Bremer, L.G.B., van Vliet, T. and Walstra, P. (1991) Confocal scanning laser microscopy as a tool to study the fractal nature of casein gels, in *Past, Present and Future of Electron Microscopy in Agricultural Research* (ed. A. Boekestein), Dienst Landbouwkundig Onderzoek-Technische en Fysische Dienst voor de Landbouw, Wageningen, Netherlands, pp. 41–46.

Brooker, B.E. (1991) The study of food systems using confocal laser scanning microscopy. *Microsc. Anal.*, **26**, 13–15.

Brooker, B.E. (1993a) Modern methods of imaging. *Biologist*, **40**, 42–45.

Brooker, B.E. (1993b) The stabilization of air in foods containing fat—a review. *Food Struct.*, **12**, 115–122.

Brooker, B.E. (1993c) The stabilization of air in cake batters—the role of fat. *Food Struct.*, **12**, 285–296.

Brundrett, M.C., Kendrick, B. and Peterson, C.A. (1991) Efficient lipid staining in plant material with Sudan 7B or Fluoral Yellow 088 in polyethylene glycol-glycerol. *Biotechnol. Histochem.* **66**, 111–116.

Cain, A.J. (1947) The use of Nile Blue in the examination of lipids. *Quart. J. Microsc. Sci.*, **88**, 383–392.

Carlsson, K., Danielsson, P.E., Lenz, R., Liljeaborg, A., Majlof, L. and Åslund, N. (1985) Three-dimensional microscopy using a confocal laser scanning microscope. *Opt. Lett.*, **10**, 53–55.

Chevalley, J. (1975) Rheology of chocolate. *J. Texture Stud.*, **6**, 177–196.

Cogswell, C.J. and Sheppard, C.J.R. (1990a) Confocal brightfield imaging techniques using an on-axis scanning optical microscope, in *Confocal Microscopy* (ed. T. Wilson), Academic Press, London, pp. 213–243.

Cogswell, C.J. and Sheppard, C.J.R. (1990b) Confocal Nomarski differential interference contrast (DIC) microscopy. *Trans. R. Microsc. Soc.*, **1**, 251–255.

Foley, J., Brady, J. and Reynolds, P.J. (1971) The influence of processing on the emulsion stability of cream. *J. Soc. Dairy Techol.*, **24**, 54–58.

Greenspan, P., Mayer, E.P. and Fowler, S.D. (1985) Nile Red: a selective fluorescent stain for intracellular lipid droplets. *J. Cell Biol.*, **100**, 965–973.

Haugland, R.P. (1992) *Handbook of Fluorescent Probes and Research Chemicals*, 5th edn., Molecular Probes, USA.

Heathcock, J.F. (1985) Characterization of milk proteins in confectionery products. *Food Microstruct.*, **4**, 17–27.

Heertje, I. (1993) Microstructural studies in fat research. *Food Struct.*, **12**, 77–94.

Heertje, I., van der Vlist, P., Blonk, J.C.G., Hendrickx, H.A.C.M. and Brackenhoff, G.J. (1987) Confocal scanning laser microscopy in food research: some observations. *Food Microstruct.*, **6**, 115–120.

Heertje, I., Nederlof, J., Hendrickx, H.A.C.M. and Lucassen-Reynders, E.H. (1990) The observation of the displacement of emulsifiers by confocal scanning laser microscopy. *Food Struct.*, **9**, 305–316.

Lewis, D.F. (1986) Features of food microscopy. *Food Microstruct.*, **5**, 1–18.

Liener, I.E., Sharon, N. and Goldstein, I.J. (eds.) (1986) *The Lectins: Properties, Functions and Applications in Biology and Medicine*, Academic Press, London.

Minsky, M. (1957) *Microscopy Apparatus*, United States Patent 3,013,467, Dec. 1961 (filed Nov. 7, 1957).

Pawley, J.B. (ed.) (1990) *Handbook of Biological Confocal Microscopy*, Plenum, New York.

Petráň, M., Hadravský, M., Egger, D. and Galambos, R. (1968) Tandem scanning reflected light microscope. *J. Opt. Soc. Am.*, **58**, 661–664.

Petráň, M., Hadravský, M. and Boyde, A. (1985) The tandem scanning reflected light microscope. *Scanning*, **7**, 97–108.

Tsien, R.Y. and Waggoner, A. (1990) Fluorophores for confocal microscopy: photophysics and photochemistry, in *Handbook of Biological Confocal Microscopy* (ed. J.B. Pawley), Plenum, New York, pp. 169–178.

Vaughan, J.G. (1979) *Food Microscopy* (ed. J.G. Vaughan), Academic Press, London.

White, J.G., Amos, W.B. and Fordham, M. (1987) An evaluation of confocal versus conventional imaging of biological structure by fluorescence light microscopy. *J. Cell Biol.*, **105**, 41–48.

Wilson, T. (1990) *Confocal Microscopy*, (ed. T. Wilson), Academic Press, London.

Yiu, S.H. (1993) Food microscopy and the nutritional quality of cereal foods. *Food Struct.*, **12**, 123–133.

Chapter 3

Scanning Probe Microscopy of Food-Related Systems

M.J. Miles and T.J. McMaster
H.H. Wills Physics Laboratory,
University of Bristol, UK

3.1 General introduction

In carrying out the microscopy of biological systems, there are two major experimental considerations: the inherent sensitivity of the specimen and the source of image contrast. Microscopes which can operate at high resolution in environments appropriate to biological and food systems without damage to the specimen offer major advantages. Scanning probe microscopy (SPM) is the generic name for a family of microscopes which can meet these criteria. The application of such microscopes to food-related substances is the subject of this review. A family tree of the most prominent members of the current SPM family is shown in Fig. 3.1. The

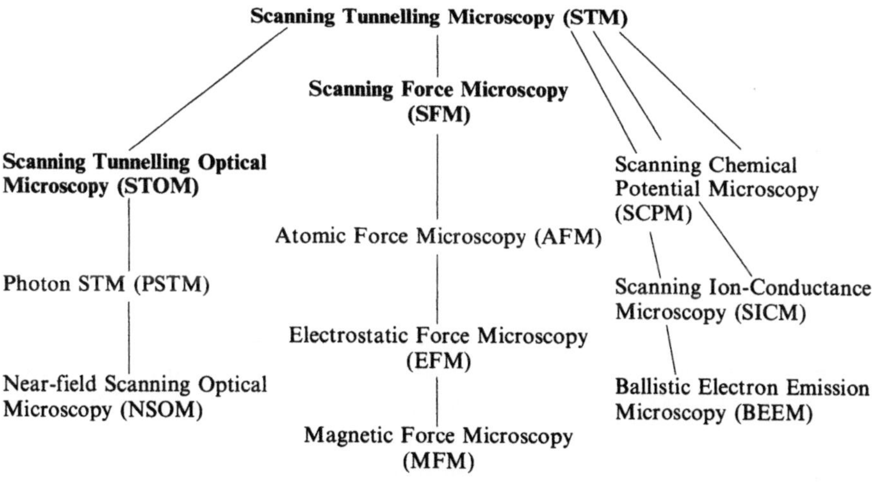

Fig. 3.1 A schematic family tree of the SPM group of techniques. Whilst not exhaustive, it serves to illustrate the development of, and connections between, the principal SPM methods.

unifying characteristic of all SPM techniques is the scanning of a probe close to the specimen surface and the measurement of some interaction between the probe and specimen which has a strong dependence on their separation.

The founding member of this family, scanning tunnelling microscopy (STM), was invented in 1981 by Binnig and Rohrer (1982) who received the 1986 Nobel Prize in physics for this work. In STM, the quantity measured is the current between the probe and the specimen, and so the electrical conductivity of both must be sufficiently high for such a current to be readily detectable. STM is capable of imaging electrical conductors and semiconductors to atomic resolution over a wide range of conditions, including gaseous and liquid environments (Güntherodt and Wiesendanger, 1992). In general, biological molecules are not bulk conductors of electricity, and so there are constraints on the nature and dimensions of the specimens, and on the environmental conditions under which they may be successfully examined by STM. Within these constraints a wide variety of systems can be successfully examined (Marti and Amrein, 1993).

In 1986, another member of the SPM family, atomic force microscopy (AFM), was invented (Binnig *et al.*, 1986). In this technique the interaction measured is the force between the probe and specimen. The restrictions associated with electrical conductivity no longer apply, and so the technique has found widespread applications in the study of biological structures (Yang *et al.*, 1993). A description of the application of the AFM technique and its various modes of operation to biological specimens is presented in Section 3.2.2. Despite the widespread success of AFM, the original technique of STM has not been entirely superseded, and it seems likely that there will be a shift back to STM for particular applications. One of the areas in which STM can provide additional information to AFM is discussed in Section 3.5.2.

Of the current dozen or so other members of the SPM family, near-field scanning optical microscopy (NSOM) is the one being most intensely developed and is the one likely to make the greatest impact in the biological field (Betzig and Trautman, 1992). In NSOM the interaction of the sample material with light conveyed by a sub-wavelength aperture is monitored. Current progress and probable biological applications of NSOM are briefly described in Section 3.2.3.

SPM techniques offer several advantages for the study of biological systems. The most important of these is probably the high-resolution capability of SPM without the need for high vacuum and without the radiation damage associated with electron microscopy. The fact that the SPM techniques are essentially non-destructive means that it is possible to follow processes over a period of time.

A spectrum of applications of AFM and other SPM techniques to food-related biological systems is presented. These range from relatively large structures such as starch granules to sub-molecular organization such as the secondary structure in proteins. Occasionally, it is necessary to step outside food systems to give an example of a particular aspect of SPM which, although of relevance to food applications has not been reported so far for food. Although perhaps more directly related to agriculture than to food, recent progress in the AFM studies of cereal chromosomes, and the possibility of gene mapping, is described in Section 3.4, as well as methods of imaging DNA molecules directly.

3.2 Introduction to scanning probe microscopy techniques

3.2.1 Scanning tunnelling microscopy (STM)

In STM, two electrical conductors, the specimen and the probe, are brought sufficiently close so that the overlap of the electron wavefunctions from each surface results in a detectable tunnel current if a bias voltage (usually < 2 V) is applied between the probe and specimen (see Fig. 3.2). In this quantum-mechanical effect, the electrons tunnel from filled states in the probe to empty states in the specimen, or *vice versa*. The origin of the ultra-high z-sensitivity in STM is the small exponential

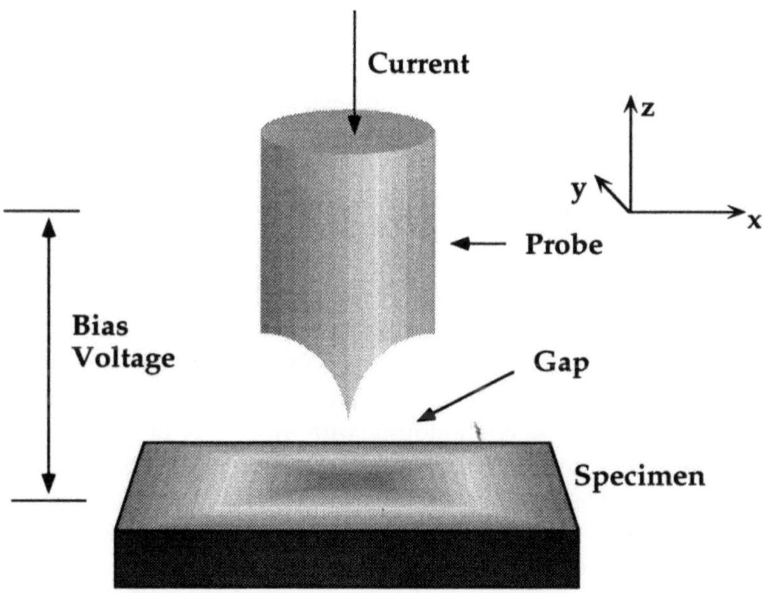

Fig. 3.2 The basic principle of STM. A schematic representation of the scanning metal probe and conducting specimen.

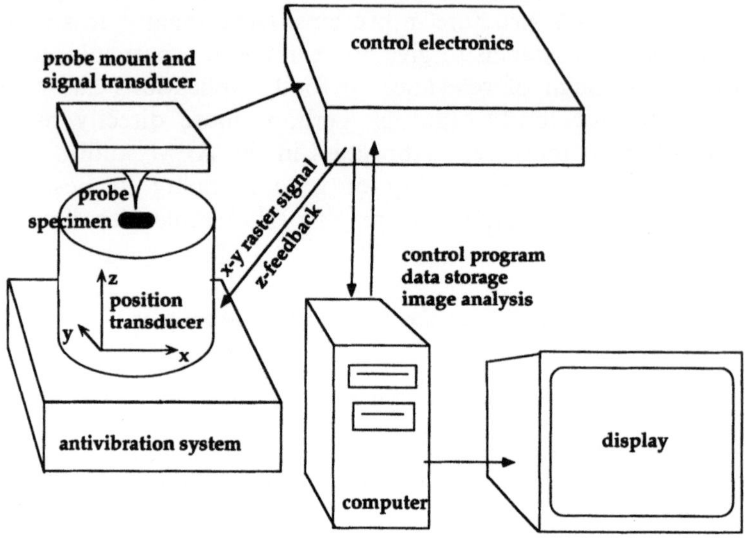

Fig. 3.3 A schematic diagram of a scanning tunnelling microscope showing the principal components required for a high-resolution instrument.

decay length of the wavefunction at the surface. This results in an extremely strong dependence of the tunnel current on the probe–specimen separation. Typically, for a 0.1 nm increase in separation, the current decreases by a factor of ten. This means that the last atom(s) at the extremity of the probe dominates the current, and is the effective size of the tunnelling probe. The principles of STM are well covered in several recent reviews (Güntherodt and Wiesendanger, 1992; Chen, 1993; Bonnell, 1993).

An image of the specimen is obtained by scanning the probe in a raster relative to the specimen surface using piezoelectric transducers and measuring the tunnel current at each point. A schematic diagram of the instrument is shown in Fig. 3.3. The simplest mode of operation is to scan the raster in a fixed plane above the specimen surface. This is known as the constant height mode. The two main disadvantages of this method are (i) that the probe may crash into high features on the surface, and (ii) that, for a specimen surface varying in height by, for example, only 1 nm, the current measurement must cope continuously with 10 orders of magnitude. This mode is most useful for rapid scanning of flat specimens. The most common mode of operation is the constant current mode. A preset value of current is maintained via a feedback loop by moving the probe towards or away from the specimen as it is scanned over the surface. If the specimen surface is homogeneous, then recording the movement of the transducers in three orthogonal directions gives the topographic surface of the specimen.

When the specimen is non-conducting, it must be deposited on a flat conducting substrate. The most widely used substrate is highly oriented pyrolytic graphite (HOPG). Conduction in such specimens depends on the nature and thickness of the specimen. For very thin specimens (< 1 nm), hybrid electronic states of the specimen and substrate are important. For thicker specimens, there is increasing evidence that an adsorbed water layer plays an important role. A detailed discussion of image contrast in STM of biological specimens can be found elsewhere (Lindsay, 1993).

3.2.2 Atomic force microscopy (AFM)

Atomic force microscopy is the principal type of scanning force microscopy (SFM). Two other types of SFM are magnetic force microscopy, MFM, and electrostatic force microscopy, EFM. AFM shares much of its instrumentation, such as the scanning transducers, control electronics, and data storage and display, with STM. However, the quantity measured is the force between the tip of the probe and the specimen, rather than the tunnel current. The normal force is measured by the bending of a micro-fabricated cantilever (Fig. 3.4). The bending of the cantilever is detected by either an optical lever and split photodiode or, less commonly, by an interferometric arrangement. Analogous to the

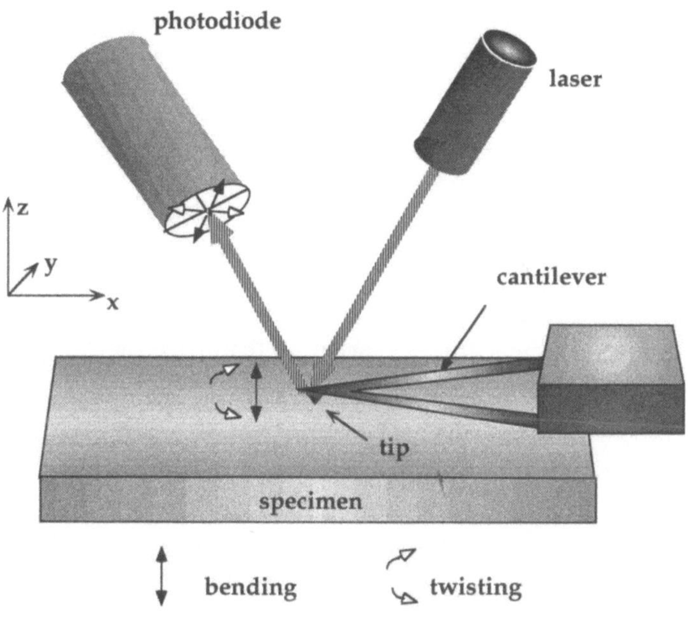

Fig. 3.4 The basic AFM arrangement presented in schematic form. In this illustration, force detection is via a quadrant detector so that both normal and lateral force measurements can be made.

constant current and constant height modes of the STM, the AFM can be used in either constant-force or constant-height (or deflection) modes.

The most common mode of AFM operation has been the constant-force contact mode, in which the probe experiences a repulsive force, and all early work was performed in this mode. In practice, on a typical biological specimen in an ambient atmosphere, the surfaces will usually be contaminated with a layer of water. The surface tension of the surface layer will result in a strong (~100 nN) attractive force, which can often lead to unstable force control necessitating large normal forces to be applied for good image contrast. Imaging under liquid can overcome this problem to a large extent (Drake *et al.*, 1989; Weisenhorn *et al.*, 1989).

Lateral (LFM) or frictional force microscopy (FFM) uses the twist of the cantilever resulting from torsional forces as the sample is scanned in a direction perpendicular to the long axis of the cantilever (Meyer and Amer, 1990). Using a four-quadrant photodiode (Fig. 3.4), the normal and tangential forces can be detected simultaneously. The technique has special applications, in that regions of differing chemical nature, but similar topography, can be distinguished (Overney *et al.*, 1992).

As the probe approaches an ultra-clean surface, it will first experience an attractive (van der Waals) force. This is much weaker than the repulsive forces and special techniques are required for imaging with this force (Martin *et al.*, 1987). This mode has so far had limited application in biology (Wigren *et al.*, 1991). The advantage of such non-contact imaging of biological material is offset by the decreased resolution compared with contact mode studies.

The so-called 'tapping' mode or large amplitude resonance mode (Hansma *et al.*, 1994) has potential to be effective in imaging biological specimens. In this mode, the cantilever is oscillated at its resonant frequency, and as it approaches the specimen it begins to tap the surface. Since this results in a decrease in amplitude of the oscillation, this decrease can be used to map the topography of the surface. Since the probe is in actual contact with the surface for only a small fraction of the oscillation, the shear forces exerted by the probe on the specimen are much reduced from those of normal contact mode.

3.2.3 Near-field scanning optical microscopy (NSOM)

In NSOM, an optical probe is used as a source and/or collector of light interacting with a sample. The probe is a sharpened optical fibre which has been coated to leave only a sub-wavelength aperture at the very end. A schematic diagram of the technique is shown in Fig. 3.5. The probe is brought very close (within the near-field range) to the surface of the specimen. However, the probe is usually controlled not by the optical intensity signal, but by the shear force exerted on the probe by the specimen. At this sub-wavelength distance away from the surface, the

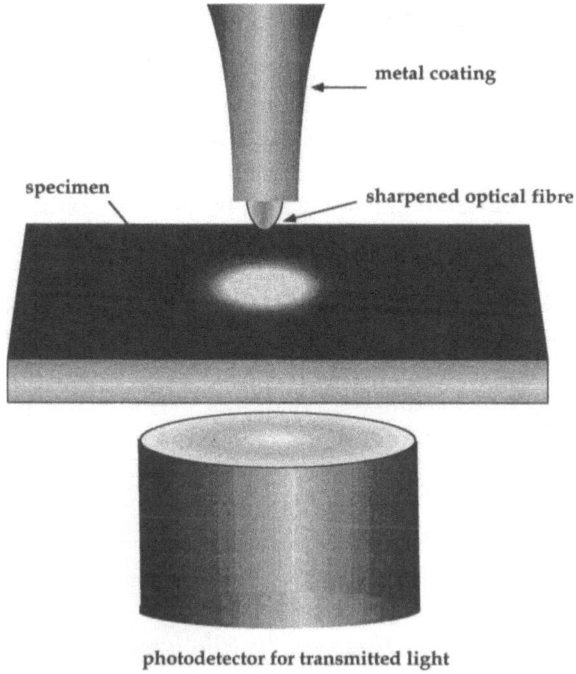

Fig. 3.5 A schematic diagram of the operation of the NSOM technique, with particular emphasis on the probe design.

image resulting from the light emitted from, or collected by, the probe can have a spatial resolution of about 20 nm (Betzig and Trautman, 1992).

The potential applications of NSOM to biology are numerous. The technique could in principle be used wherever far-field (conventional) optical microscopy is currently used with the advantage of higher resolution. Probably the most important immediate application will be in imaging fluorescently labelled specimens (Betzig and Chichester, 1993). However, most optical techniques can be applied to near-field imaging. In particular, the various types of spectroscopy could be used not only to provide local spectroscopic information but also to give high-resolution images in which particular chemical species are identifiable.

3.3 Carbohydrates

SPM imaging of carbohydrates ranges in scale from AFM of whole starch granules (Thomson *et al.*, 1994) to STM of individual molecules (Miles *et al.*, 1990, 1991b). Starch granules are, of course, easily visible by conventional optical techniques. Imaging of structures which have dimensions of a few micrometres in each of the principal directions causes some minor problems because of the limited range of the transducers which translate the probe. This is the case for starch granules (Fig. 3.6; *see Colour section*); the thickness of this particular type of granule was greater than the range of the *z*-transducer. It was therefore necessary to embed the specimen so that only part of the granule was exposed (Thomson *et al.*, 1994). From the point of view of stability during scanning, the rigidity of starch in its dry state makes it a good candidate for AFM. Similarly, the stiffness of cellulose fibrils from cotton has the consequence of stable AFM imaging (Fig. 3.7). AFM images at the molecular scale have been reported for cellulose microfibrils from *Valonia ventricosa*, a green alga (Hanley *et al.*, 1992).

Several cellulose derivatives have been studied by both STM and AFM. These include the viscosity-modifying polysaccharide hydroxypropylcellulose (HPC). Under suitable conditions, HPC exists in a cholesteric liquid-crystal phase. Removal of the solvent, water, results in a banded structure, which is proposed to arise as a stress-relief mechanism. AFM imaging of the banded structure in HPC (Fig. 3.8) has recently been reported (Fischer *et al.*, 1994). The image suggests a three-dimensional nature to the banding structure, at least at the surface of the specimen. Tang *et al.* (1993) have found microfibrillar structures in AFM images from dried HPC films *in vacuo*. Slow-drying of dilute solutions of liquid-crystal forming molecules, such as HPC, can result in monolayer rafts of aligned molecules. Such rafts represent suitable specimens for high-resolution STM, as molecular stability is provided by neighbouring molecules, and the layer is sufficiently thin to allow adequate electron transport.

The polysaccharide xanthan, widely used in the food and oil-recovery industries for its remarkable solution properties, has been imaged using STM both at the molecular level in the uncoated state (Miles *et al.*, 1991b; Gunning *et al.*, 1993) and with metal coating (Wilkins *et al.*, 1993). The latter approach has been demonstrated to be a more reliable method of imaging. Its disadvantage is that sub-molecular structure cannot be revealed owing to the limit on resolution set by the grain size of the metal coating. However, there is significant improvement in resolution over that achieved on the same specimens shadowed for transmission electron microscopy (TEM), because the glancing-angle shadowing used to increase contrast in TEM specimens produces larger grains than the method of evaporating the metal perpendicularly to the specimen surface that can be used for STM. The coating technique has also been successfully applied to the galactomannan polysaccharide locust bean gum (LBG). Again individual molecules could be reproducibly imaged. The seaweed polysaccharides are also of importance to the food industry. STM images of air-dried specimens of ι- and κ-carrageenan (Lee *et al.*, 1992) have shown structures of dimensions approximately consistent with those determined from X-ray diffraction.

The application of AFM to the imaging of individual and interacting polysaccharide molecules is only just beginning. Experience from AFM of

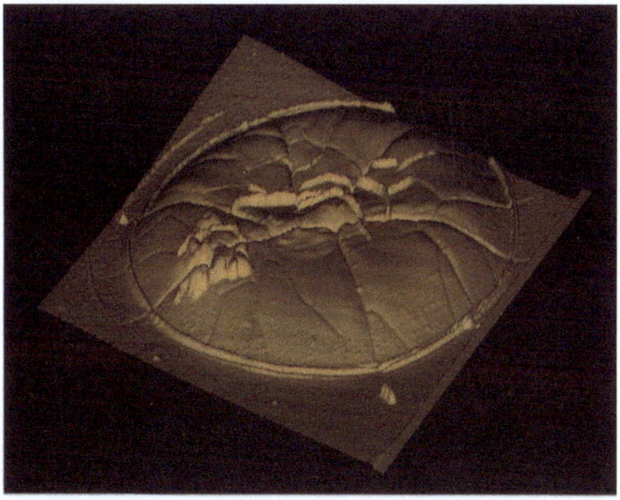

Fig. 3.6 A 30 × 30 µm AFM scan of a native embedded starch granule. Only the top surface is visible and it shows various cracks and bumps. The height difference between the metal embedding medium and the top of the granule is 250 nm. Data were obtained in this laboratory (Thomson *et al.*, 1992).

Fig. 3.7 Two three-dimensional AFM images of cellulose fibres from cotton.

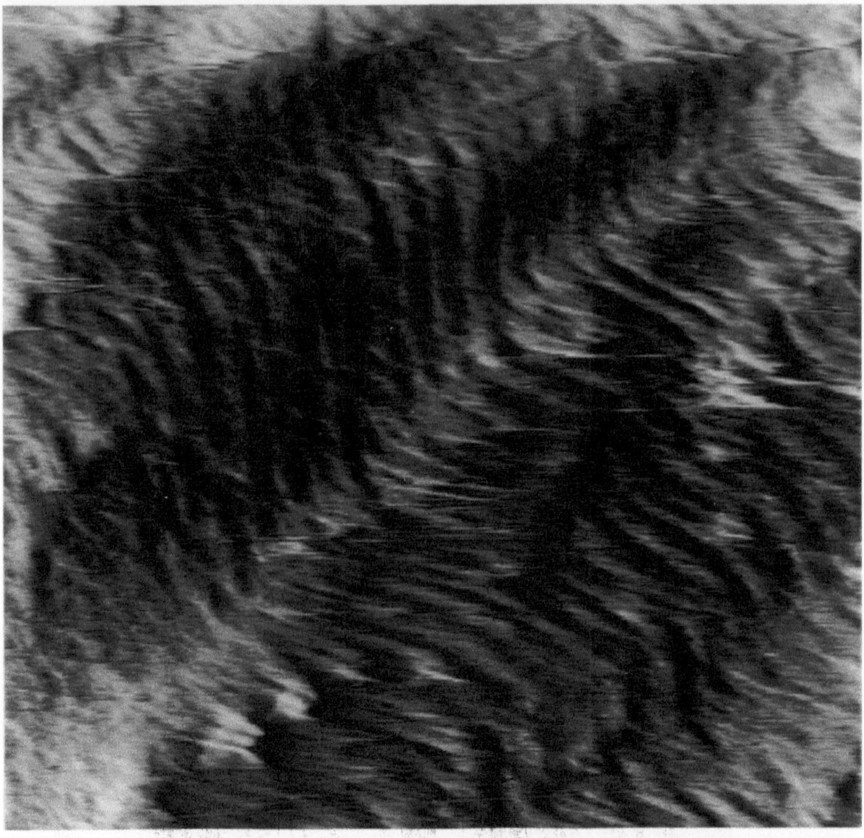

Fig. 3.8 AFM scan of the surface of an HPC film. The scan area is 9.5 μm, and the height difference represented by the black-to-white greyscale is 300 nm. Data from Fischer *et al.* (1994).

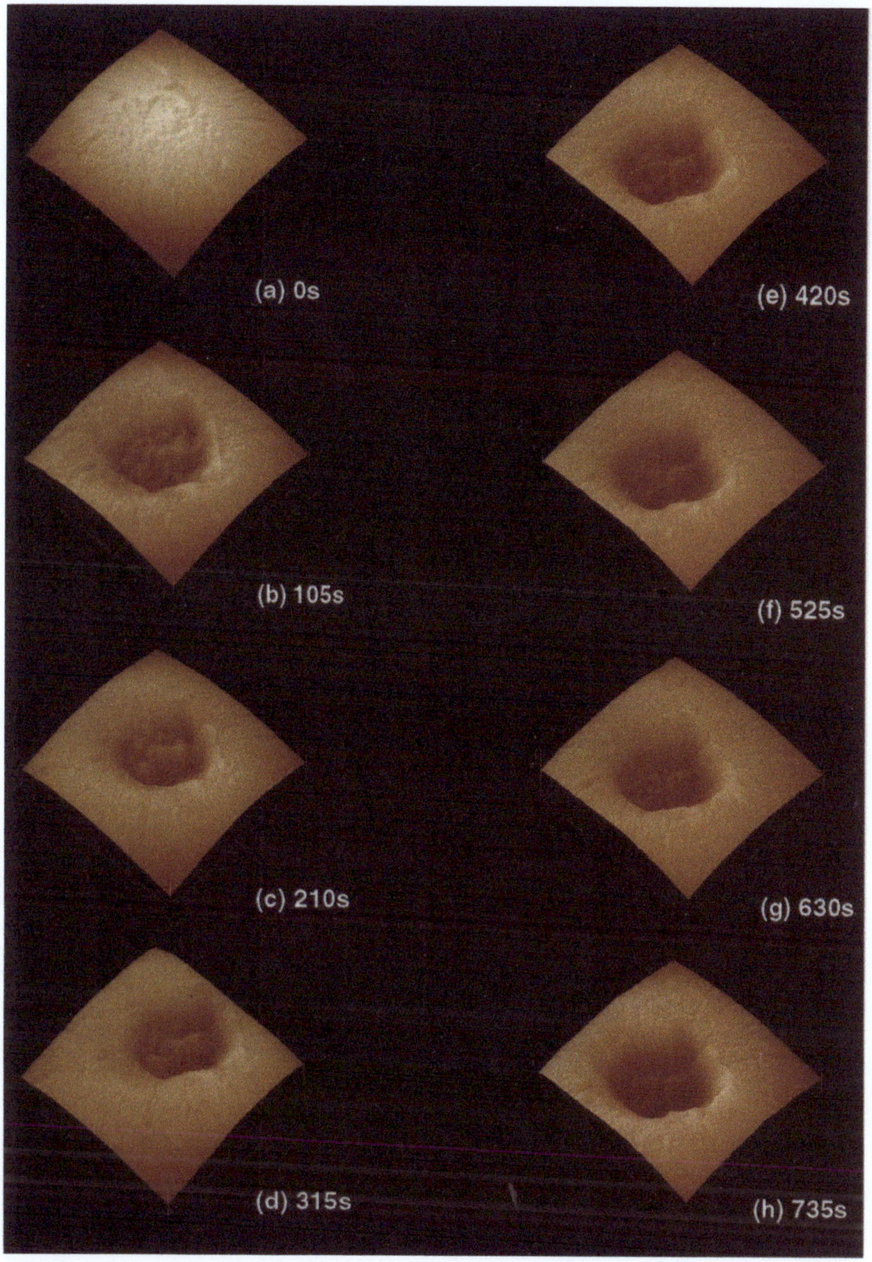

Fig. 3.9 A series of 4 × 4 μm AFM images in three-dimensional format showing (a) the top of a starch granule imaged in air, (b) the same region of the granule imaged under water, and (c)–(h) the progressive attack of α-amylase on the top of this starch granule. The hole can clearly be seen to deepen with the passage of time. Data from Thomson *et al*. (1994).

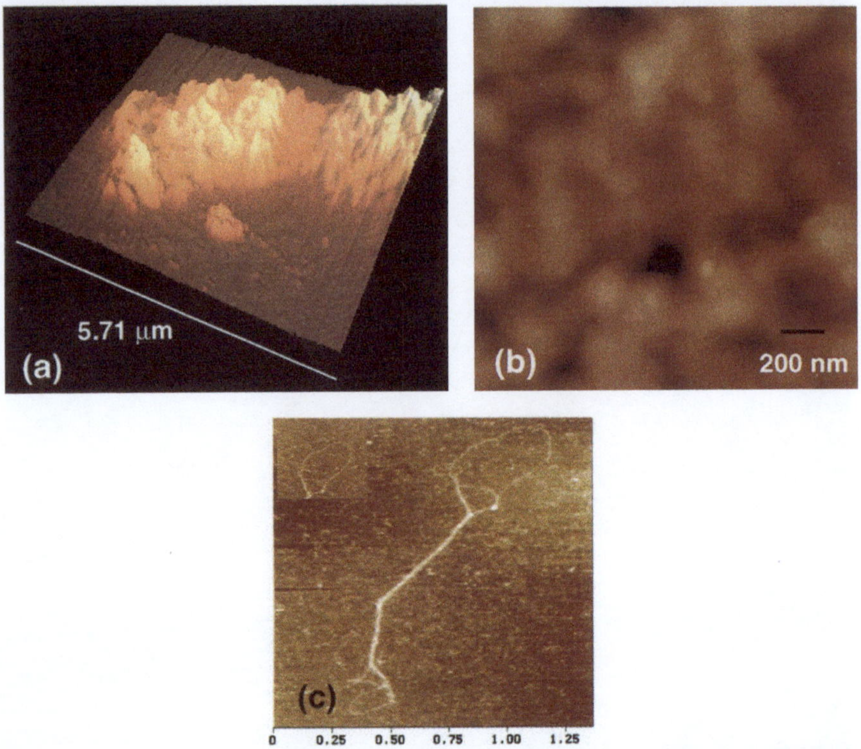

Fig. 3.10 (a) Three-dimensional AFM image of part of a chromosome of *Triticum tauschii*. The centromere is visible as a depressed region, and the satellite is to the extreme left of the chromosome. At this stage of condensation, the individual arms of the chromosome are not yet visible. (b) High-resolution detail of the centromeric region demonstrating that features as small as 10–15 nm can be resolved. (c) AFM image of an individual plasmid DNA molecule from a bacterium. The circular molecule has coiled up in some places to form thick 'ropes', but the naked duplex can be seen at some points, particularly at the top and bottom loops. The inset (top left) shows the effect of increased force during imaging on one of these loops. Chromosome data from McMaster and Winfield (unpublished), and DNA data from McMaster and Miles (unpublished).

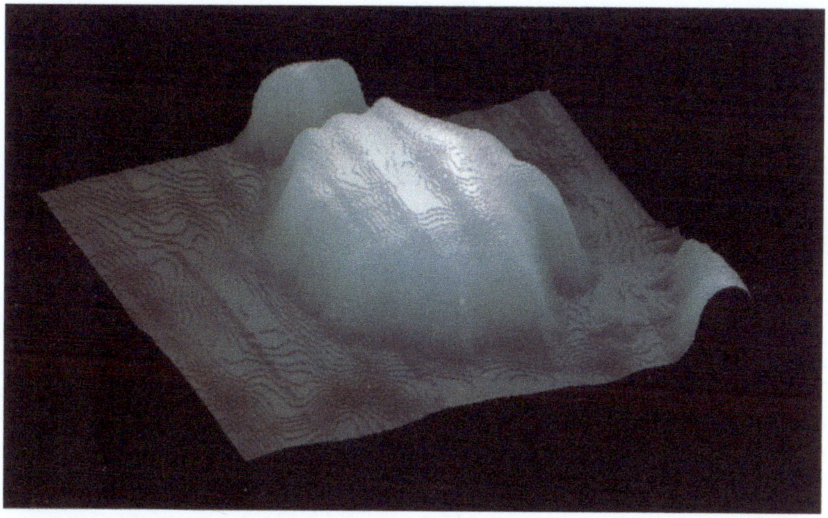

Fig. 3.11 Individual vicilin molecule as imaged by STM. The molecule is ~10 nm across and 3 nm high. The flat-topped polygonal structure agrees with predictions based on X-ray scattering results (I'Anson *et al.*, 1988). In the STM experiment, the molecule was observed to change shape as a function of hydration. Data from Welland *et al.* (1989).

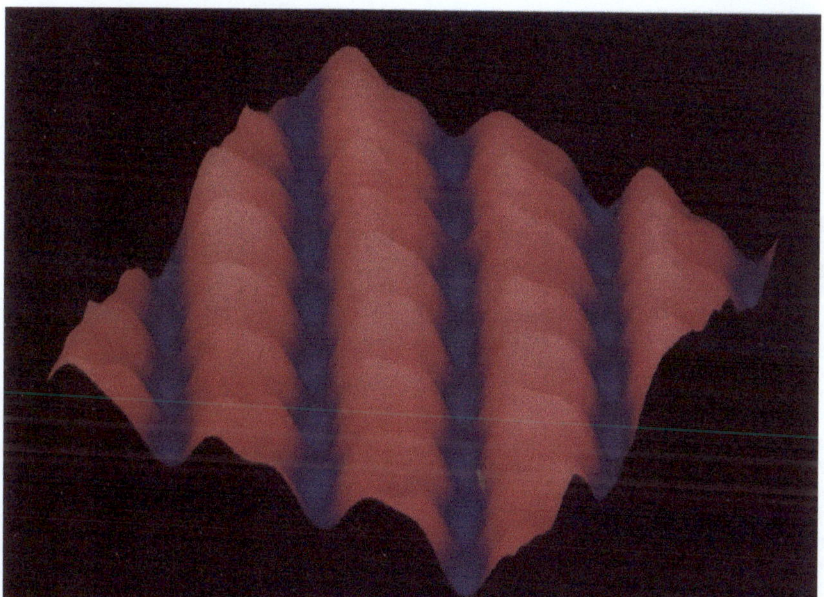

Fig. 3.12 A portion of an array of rod-like HMW molecules on graphite imaged by STM with typical values of tunnel current and bias voltage of 0.8 V and 70 pA, respectively. There is a periodicity along the molecules of 1.49 nm, and a lateral spacing of the molecules of 3.1 nm. Data from Miles *et al.* (1991a).

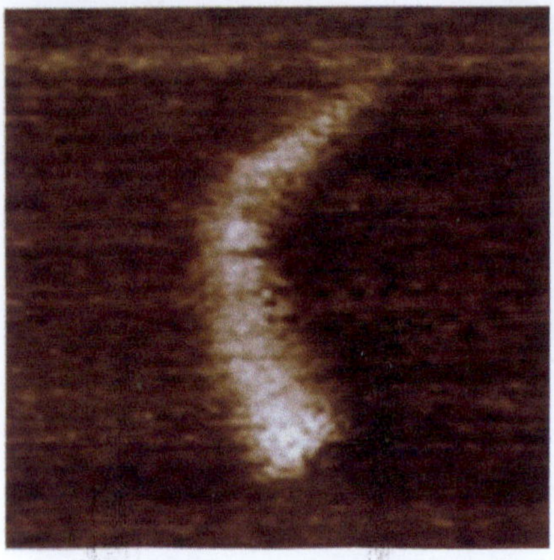

Fig. 3.13 An individual γ-gliadin molecule from wheat imaged by STM. The tunnelling conditions were bias voltage of 1 V and tunnel current of 10 pA. The scan area was 27.8 nm and the vertical height range was 0.8 nm. Data from Thomson *et al.* (1992).

Fig. 3.14 (a) STM image of uncoated, hydrated collagen I fibrils. The fibrils can be seen to be composed of 8 nm diameter microfibrils, frequently proposed as the fundamental structural unit of collagen. Maximum height between dark and bright in the image is 12.5 nm. (Tunnel current was 10 pA; bias voltage was 1500 mV.) (b) A thinner sheet of collagen I filaments, with some kinked regions. This microkinking had previously been predicted by a model based on the liquid-crystalline nature of collagen. The asymmetric banding pattern is correlated with TEM staining specific to regions of charged amino acids. (Tunnel current was 50 pA; bias voltage was 500 mV.) (c) AFM image of pN-collagen tapes on an EM grid. Data from Gathercole *et al.* (1993).

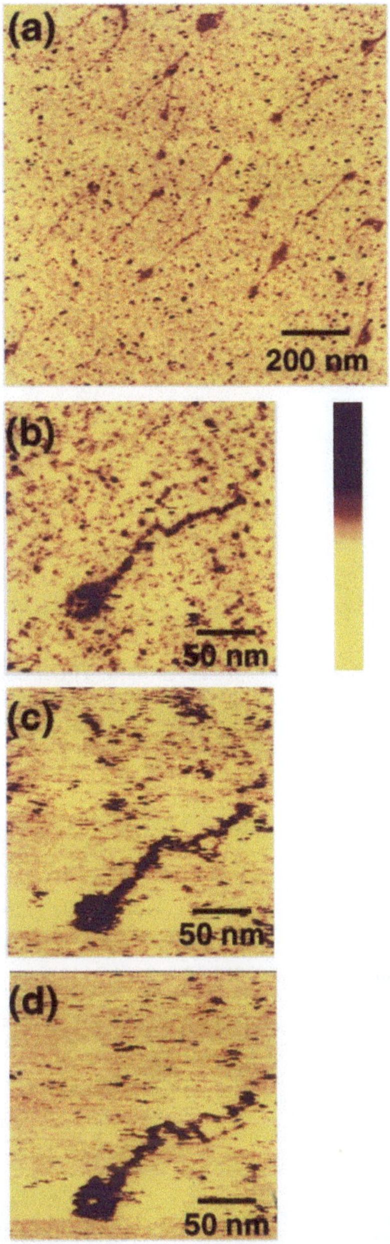

Fig. 3.15 Sequence of AFM images of myosin molecules. (a) Spread of individual molecules. (b)–(d) Series showing the effect of increasing the normal force between tip and sample on the coiled-coil structure of the myosin tail. The colour bar shows the height information and corresponds to a z-range of 4.1 nm in (a) and 2.3 nm in (b)–(d). Data from Hallett *et al.* (1995) (this laboratory).

Fig. 3.16 A 13.6 nm AFM image of gas vesicle protein. The image has been Fourier-filtered to remove random noise. The ribs of this protein structure run near-vertical in the image, and, within them, the β-sheet structure can be observed as U-shaped loops. Data from McMaster *et al.* (1995).

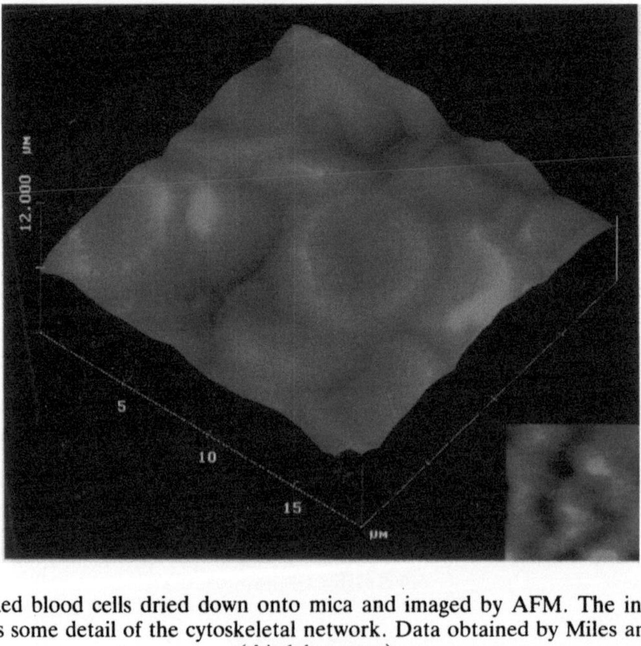

Fig. 3.17 Red blood cells dried down onto mica and imaged by AFM. The inset (bottom right) shows some detail of the cytoskeletal network. Data obtained by Miles and Ashmore (this laboratory).

molecules of similar size and charge, for example DNA, indicates that this could be a fruitful approach for high-resolution studies of uncoated molecules. The cyclodextrins form a group of important small molecule carbohydrates. Their importance lies in their ability to complex other molecules within their toroidal structure and to encapsulate their guest molecules, which are usually hydrophobic in character, in aqueous environments. The hope of imaging the cyclodextrin molecule with its complexed 'guest' has not been realized. Early STM attempts (Miles *et al.*, 1990) used a tosylated β-cyclodextrin in order to enhance its adsorption to the graphite substrate via the hydrophobic effect. The dominant structure in these images was interpreted as the tosyl groups, which, from STM results on aromatic-containing molecules, might be expected to appear bright (or high) owing to their aromatic nature (Smith, 1991). Subsequent imaging of α-cyclodextrin (Shigekawa *et al.*, 1991) has also not provided submolecular structural information of the cyclodextrin molecule itself.

3.3.1 Enzymic degradation of starch

The capability of SPM to follow biological processes in real time at high resolution has been realized in the case of the enzymic degradation of an ungelatinized starch granule (Thomson *et al.*, 1994). The structure of starch granules following enzymic attack has been examined previously in a scanning electron microscope (SEM) using metal coated specimens (Bhat *et al.*, 1983). These images show small bore holes in the surface of the granules. Even in an environmental SEM, it had not been possible to follow the process of enzymic attack in real time because of radiation damage to the enzyme from the electron beam.

Figure 3.9(a) shows an AFM image of a region of a starch granule in air. The uppermost region of the granule exhibits some damage, which was probably incurred during sample preparation. On scanning the same area under water, some swelling of the granule is evident (Fig. 3.9(b)). With the addition of the enzyme α-amylase to the water surrounding the granule, higher magnification images showed clusters of the enzyme molecules adsorbed to the granule surface. A hole appeared in the damaged region of the granule (Fig. 3.9(c)). In the subsequent images taken over 15 min (Fig. 3.9(d–h)), the hole became deeper and remained about the same diameter. The three-dimensional nature of the AFM image has allowed the rate of local enzymic degradation of the starch to be calculated.

3.4 Chromosomes, DNA and bases

Metaphase chromosomes provide a relatively rigid structure for AFM imaging. The first chromosomes to be imaged with this technique were human chromosomes (Rasch *et al.*, 1993; McMaster *et al.*, 1994;

Fritzsche *et al.*, 1994) and Chinese hamster chromosomes (De Grooth and Putman, 1992). Recently, high-resolution AFM images have been obtained of plant chromosomes (Winfield *et al.*, 1995). Figure 3.10 shows images of wheat chromosomes. The resolution of distinct features in these images is about 15 nm. Banded and hybridized chromosomes have also been imaged, and the possibility of gene mapping by AFM exists.

The use of SPM to image DNA molecules and to read their base sequence has been a goal sought after by many groups. In retrospect, many of the early STM images reported to be of DNA molecules, some of which appeared to approach atomic resolution, can now be recognized as structures associated with the graphite substrate (Clemmer and Beebe, 1991). It is perhaps unsurprising that some artefactual images have arisen in the early days of these new microscopy techniques. Heckl and Binnig (1992) have categorized many of these artefacts associated with grain boundaries and surface steps of the graphite substrate. AFM imaging using mica as a substrate appears to be less prone to artefacts than imaging using graphite. AFM imaging of individual double-stranded DNA molecules has now become routine. The plasmid DNA illustrated in Fig. 3.10 shows regions of supercoiling as well as individual double-stranded regions. Guckenberger *et al.* (1994) have demonstrated, by using < 1 pA of current in a 70% relative humidity environment, that STM can give very similar images of plasmid DNA to those obtained by AFM. This is a most exciting development which revitalizes the application of STM to biological systems.

The potential of SPM to read the DNA base sequence at a rate competitive with existing biochemical methods has not been realized. Probably the single most important problem to be overcome in such a method is a reliable means of stretching out the DNA molecule so that the individual bases or base pairs are accessible to the scanning probe in a reproducible orientation and manner. Heckl *et al.* (1991) and Allen *et al.* (1992) have demonstrated that when monolayers of a particular base are deposited on HOPG, each base type has a characteristic shape and so can be identified. However, the task of arranging the bases in a similar manner while still attached to the sugar-phosphate backbone remains.

3.5 Proteins

3.5.1 Seed storage proteins

Proteins are the class of biological macromolecule most studied by SPM (Amrein *et al.*, 1993; Miles and McMaster, 1993; Morris, 1994). Many of these studies have been on food-related proteins. One of the first SPM studies of a protein was the STM imaging of the seed-storage protein vicilin (Welland *et al.*, 1989). Figure 3.11 shows one of the images obtained in air using the amorphous carbon surface of an X-ray mirror as the substrate.

The seed storage proteins from wheat have also been studied by STM (Miles *et al.*, 1991a; Thomson *et al.*, 1992). STM imaging of the high molecular weight subunit protein, which plays an important role in the elastic properties of dough, has revealed a helical structure to the molecule which is consistent with the β-spiral structure proposed from the amino acid sequence of the protein (Miles *et al.*, 1991a). The long repetitive sequence of the molecule associated with the β-spiral confers a rod-like nature to the molecule. This allows the specimen to be prepared as a raft-like monolayer of molecules, which, having a hydrophobic character, is adsorbed to the HOPG substrate (Fig. 3.12).

The hydrophobic nature of regions of the γ- and ω-gliadin proteins from wheat has allowed adsorption of individual protein molecules to HOPG, as well as the formation of monolayers of γ-gliadin (Thomson *et al.*, 1992). The STM images of individual molecules of γ-gliadin show them to be tadpole-like in shape (Fig. 3.13). This is consistent with predictions from the amino acid sequence, which suggest a hydrophobic repetitive tail and a non-repetitive hydrophilic charged head. The ω-gliadin appears as a rod-like molecule. This is also consistent with the repetitive amino acid sequence. α-Gliadin appears in the STM images as a globular structure, the dimensions of which are larger than might be expected from its molecular weight. Two effects account for this. Firstly, the size is increased by the adsorption of a stable water layer to this hydrophilic protein. Secondly, and more generally, the finite probe size leads to a broadening of the actual structure. This latter effect has been well documented (Bustamante *et al.*, 1993). Technologies to grow sharper tips to minimize this effect have been developed (Keller and Chih-Chung, 1992).

In the seed, these kinds of proteins are stored in protein bodies measuring about 1 μm in diameter. AFM has been used to image these bodies (Thomson, 1994), but it has so far revealed no new information on either the packing of the protein molecules or the nature of the surface of these structures.

3.5.2 Collagen

The techniques of STM and AFM have been used to image hydrated fibrous assemblies of collagen I (Gathercole *et al.*, 1993). For STM, the collagen was allowed to assemble on HOPG and imaged at a relative humidity of about 55%. This allowed relatively thick (~ 20–30 nm) layers of collagen to be successfully imaged. At humidities less than ~ 35%, the STM tip crashed into the specimen, and at higher humidities the images became too noisy. The image in Fig. 3.14(a) shows collagen fibrils consisting of microfibrils of about 8 nm in cross-section. Arrays of these microfibrils show a 67 nm repeat distance along their length. In thinner regions of the specimen, a detailed contrast variation, which depends on the humidity, can be observed along the microfibrils (Fig. 3.14(b)). This repeating pattern of contrast was found to correspond in both dimensions and sequence to the established banding pattern observed in TEM on stained collagen specimens (Meek *et al.*, 1979). The regions of high-current contrast corresponded to heavy-metal stained regions of the TEM images. These regions are associated with clusters of hydrophilic, charged amino acid side chains along the staggered molecule microfibril (Doyle *et al.*, 1974). This provides further evidence that adsorbed water is an important factor in the STM imaging of insulators. It also demonstrates that the technique can be used to distinguish between hydrophobic and hydrophilic regions of a specimen without the need for sample preparation techniques such as staining.

AFM imaging has demonstrated (Gathercole *et al.*, 1993) the 67 nm D-period repeat along collagen fibrils, but it has not detected the detailed charge pattern revealed by STM (Fig. 3.14(c)).

3.5.3 Myosin

The muscle protein myosin is ubiquitous within eukaryotic cells, and, along with the protein actin, it is involved in a wide range of motile processes, from microscopic movement of cells and organelles within them to the large-scale contraction of muscle tissue. AFM has been used to study the conformation and structure of myosin molecules (Hallett *et al.*, 1995). The highest quality images were obtained with the myosin adsorbed to a mica substrate and imaged in a propanol environment. The preparation method used was similar to that used to prepare TEM specimens. The appearance of the molecules in the AFM images (Fig. 3.15(a)) is similar to that seen in TEM (Slayter and Lowey, 1967; Walker *et al.*, 1985). The characteristic shape of the myosin molecule is comprised of two head regions and a tail (~156 nm long, 2 nm wide) comprised of two α-helices twisted around each other to form a super-helical structure. Although there has been no metal coating of the specimen, there is an underlying granular nature to these myosin images (Fig. 3.15) which limits the achievable resolution of the molecules themselves. This granular material could not be easily removed by washing the specimen, and it seems to be inherent in the process of preparing the myosin. It illustrates the fact that small amounts of impurity present during specimen preparation can be a serious problem in AFM imaging, as such material will be deposited on the surface of the specimen and substrate. Unlike images obtained by transmission microscopy, it is only the surface that is imaged in AFM.

3.5.4 Controlled modification of a molecule

In imaging the myosin molecules in Figs 3.15(a) and 3.15(b), the force exerted normal to the surface by the probe on the specimen was *ca.* 1 nN.

When this force was increased to 4 nN in order to investigate the effect on the protein conformation, the images shown in Figs 3.15(c) and 3.15(d) were obtained. The effect is remarkable, as first a molecular loop has been teased out from the coiled-coil α-helical tail (Fig. 3.15(c)). On continued scanning, the loop becomes extended so that the helices which comprise the tail become separated (Fig. 3.15(d)). Further scanning has resulted in the dissection of the myosin molecule at several points. This is a graphic demonstration of the ability of an SPM probe to manipulate molecular structures at the nanometre scale.

3.5.5 Imaging of processes in proteins

As discussed above, a major advantage of SPM is its ability to follow biological processes in real time, such as the enzymic degradation of starch. In this section, two examples of real-time imaging of biological processes concerning structural changes to proteins are described.

The first is concerned with AFM imaging of the network-forming process involved in blood clotting (Drake *et al.*, 1989). The initial images show a mica surface under a solution of the protein fibrinogen. Some fibrinogen molecules are seen adsorbed to the surface. On the addition of thrombin, the trigger for the clotting process, a region of the fibrinogen molecules is cleaved, so that polymerization and subsequent fibrin network formation occurs, and is followed in a series of AFM images over a 15-min period.

The second example is the direct observation of a change in dimensions of individual molecules of the enzyme lysozyme during binding to the oligoglycoside substrate molecules (Radmacher *et al.*, 1994). In this experiment, the molecules were first bound to a mica surface and intially imaged under water. Using the AFM in tapping mode in the liquid, changes in the height of the enzyme molecules were monitored with time, first without the presence of the lysozyme substrate, and then with the substrate present in the water. In the presence of the substrate, a change in height of 1 nm for a duration of 50 ms was found with the probe over the lysozyme molecules. In the presence of the enzyme inhibitor chitobiose, this height change was not observed. The simplest interpretation is that the probe is detecting conformational changes in lysozyme during hydrolysis.

3.5.6 Ultra-high resolution of protein structure

In general, on extremely flat surfaces such as the cleaved faces of a crystal, SPM can give ultra-high resolution images. This is also the case for biological structures. Sub-molecular resolution has been achieved on the surface of an amino acid crystal (Manne *et al.*, 1993). However, on a naturally occurring biological structure, the highest resolution so far achieved has been on the surface of gas vesicles from the cyanobacterium *Anabaena flos-aquae* (McMaster *et al.*, 1995). These vesicles consist of a hollow cylinder closed at each end by a conical cap. The cylinder and caps are formed from ribs of protein in the β-sheet conformation. AFM imaging, under propanol, of the collapsed cylinders has revealed not only the ribbed structure but also the β-sheet structure to a resolution better than 0.5 nm, so that the adjacent molecular arms of the β-sheet structure have been clearly delineated (Fig. 3.16).

3.6 Cells

The first AFM images of cells were obtained for red and white blood cells dried down onto a substrate (Gould *et al.*, 1990). Figure 3.17 shows AFM images of some human red blood cells deposited on mica without further preparation. The inset shows a higher magnification image of the cell surface on which evidence of the cytoskeletal structure can be seen. Häberle *et al.* (1991) have developed an AFM technique for imaging living cells. Their published image of red blood cells, kept alive under appropriate physiological conditions, has a resolution of up to *ca.* 10 nm. Häberle *et al.* (1992) subsequently imaged living cultured monkey kidney cells under normal growth conditions with a reproducible resolution of about 10 nm. They added a suspension of pox virus, and observed the effect of internal cellular processes on the changing morphology of the cell surface. Firstly, almost immediately, the cell surface softened. (This demonstrates the advantage of the capability of AFM to make local mechanical property measurements.) Then, after about 2 h, protrusions, which lasted only a few minutes, appeared on the cell membrane. After about 20 h, the AFM images apparently showed the exocytosis of progeny of the original viruses. Thus, a remarkable biological process on the scale of tens of nanometres was observed in real time in a living system.

Molecular growth of actin filaments has been observed on the surface of a living glial cell (Henderson *et al.*, 1992). In an elegant experiment, the AFM tip was first used to scratch away part of the cell structure, and then in more normal low-force imaging, the cellular repair process was recorded.

3.7 Conclusions and future prospects

The applications of SPM to biological systems discussed briefly here represent, of course, only a sample of the progress made using these new microscopy techniques. SPM has a major advantage in its ability to work in liquid environments and still produce high-resolution images. A physiological liquid environment allows specimens to be maintained in their native state, or simply kept 'alive'. This allows biochemical changes to be followed in real time, from changes in enzyme conformation, through the enzymic degradation of substrate, to the reproduction and exocytosis of viruses.

A range of techniques is being developed for imaging biological structures. For soft specimens, the non-contact and 'tapping' modes are important. The force modulation mode will find many applications in determining local modulus of specimens. Lateral force measurements have already demonstrated their ability to distinguish between regions of different chemical nature. Continued increases in force sensitivity and in scan rates will bring many benefits for biological investigations, particularly for studies involving 'large' soft structures such as living cells.

Scanning near-field optical microscopy is a type of SPM which has great potential for biological studies. Current developments favour a combination of NSOM and AFM. Essentially, NSOM can be configured to give the same facilities as a conventional far-field optical microscope, but with sub-wavelength resolution. It offers many exciting opportunities for fluorescence or infrared spectroscopic imaging of biological systems. With new types of SPM continuing to be invented, and new techniques being developed for existing SPM methods, it is clear that there are exciting future possibilities for the study of an increasing wide range of food-related biological systems, problems and processes.

References

Allen, M.J., Balooch, M., Subbiah, S., Tench, R.J., Balhorn, R. and Siekhaus, W. (1992) Analysis of adenine and thymine adsorbed on graphite by scanning tunnelling and atomic force microscopy. *Ultramicroscopy*, **42–44**, 1049–1053.

Amrein, M., Gross, H. and Guckenberger, R. (1993) STM of proteins and membranes, in *STM and SFM in Biology* (eds. O. Marti and M. Amrein), Academic Press, San Diego, CA, pp. 127–176.

Betzig, E. and Chichester, R.J. (1993) Single molecules observed by near-field scanning optical microscopy. *Science*, **262**, 1422–1425.

Betzig, E. and Trautmann, J.K. (1992) Near-field optics – microscopy, spectroscopy, and surface modification beyond the diffraction limit. *Science*, **257**, 189–195.

Bhat, U.R., Paramahans, S.V. and Tharanathan, R.N. (1983) Scanning electron-microscopy of enzyme-digested starch granules. *Stärke*, **35**, 261–265.

Binnig, G. and Rohrer, H. (1982) Scanning tunnelling microscopy. *Helv. Phys. Acta*, **55**, 726–735.

Binnig, G., Quate, C.F. and Gerber, C. (1986) Atomic force microscope. *Phys. Rev. Lett.*, **56**, 930–933.

Bonnell, D.A. (ed.) (1993) *Scanning Tunnelling Microscopy and Spectroscopy*, VCH Publishers, Cambridge.

Bustamante, C., Keller, D. and Yang, G. (1993) Scanning force microscopy of nucleic-acids and nucleoprotein assemblies. *Current Opinion Struct. Biol.*, **3**, 363–372.

Chen, C.J. (ed.) (1993) *Introduction to Scanning Tunnelling Microscopy*, Oxford University Press, Oxford.

Clemmer, C.R. and Beebe, T.P. (1991) Graphite – a mimic for DNA and other biomolecules in scanning tunneling microscope studies. *Science*, **251**, 640–642.

De Grooth, B.G. and Putman, C.J.A. (1992) High-resolution imaging of chromosome-related structures by atomic force microscopy. *J. Microsc.*, **168**, 239–247.

Doyle, B.B., Hulmes, D.J.S., Miller, A., Parry, D.A.D., Piez, K.A. and Woodhead-Galloway, J. (1974) *Proc. R. Soc. London, Ser.* B, **187**, 37–46.

Drake, B., Prater, C.F., Weisenhorn, A.L., Gould, S.A.C., Albrecht, T.R., Quate, C.F., Cannell, D.S., Hansma, H.G. and Hansma, P.K. (1989) Imaging crystals, polymers, and processes in water with the atomic force microscope. *Science*, **243**, 1586–1589.

Fischer, H., Miles, M.J. and Odell, J.A. (1994) Atomic force microscopy of the banded structure of lyotropic polymers. *Macromol. Chem. Phys., Rapid Commun.* **15**, 815–821.
Fritzsche, W., Schaper, A. and Jovin, T.M. (1994) Probing chromatin with the scanning force microscope. *Chromosoma*, **103**, 231–236.
Gathercole, L.G., Miles, M.J., McMaster, T.J. and Holmes, D.F. (1993) Scanning probe microscopy of collagen I and pN-collagen assemblies and the relevance to scanning tunnelling microscopy contrast generation in proteins. *J. Chem. Soc. Faraday Trans.*, **89**, 2589–2594.
Gould, S.A.C., Drake, B., Prater, C.B., Weisenhorn, A.L., Manne, S., Hansma, H.G., Hansma, P.K., Massie, J., Longmire, M., Elings, V., Dixon Northern, B., Mukergee, B., Peterson, C.M., Stoeckenius, W., Albrecht, T.R. and Quate, C.F., (1990) From atoms to integrated circuit chips, blood cells, and bacteria with the atomic force microscope. *J. Vacuum Sci. Technol.*, **A8**, 369–373.
Guckenberger, R., Heim, M., Cevc, G., Knapp, H., Wiegräbe, W. and Hillebrand (1994) STM of insulators and biological specimens, based on lateral conductivity of ultrathin water films. *Science*, **266**, 1538–1540.
Gunning, A.P., McMaster, T.J. and Morris, V.J. (1993) Scanning tunnelling microscopy of xanthan gum. *Carbohydr. Polym.*, **21**, 47–51.
Güntherodt, H.-J. and Wiesendanger, R. (eds.) (1992) *Scanning Tunnelling Microscopy I*, Springer-Verlag, Berlin.
Häberle, W., Hörber, J.K.H. and Binnig, G. (1991) Force microscopy on living cells. *J. Vacuum Sci. Technol.*, **B9**, 1210–1213.
Häberle, W., Hörber, J.K.H., Ohnesorge, F., Smith, D.P.E. and Binnig, G. (1992) *In situ* investigations of single living cells infected by viruses. *Ultramicroscopy*, **42–44**, 1161–1167.
Hallett, P., Offer, G. and Miles, M.J. (1995) Atomic force microscopy of the myosin molecule. *Biophys. J.*, in press.
Hanley, S.J., Giasson, J., Revol, J.-F. and Gray, D.G. (1992) Atomic force microscopy of cellulose microfibrils – comparison with transmission electron-microscopy. *Polymer*, **33**, 4639–4642.
Hansma, P.K., Cleveland, J.P., Radmacher, M., Walters, D.A., Hillner, P.E., Bezanilla, M., Fritz, M., Vie, D., Hansma, H.G., Prater, C.B., Massie, J. Fukunaga, L., Gurley, J. and Elings, V. (1994). Tapping mode atomic force microscopy in liquids. *Appl. Phys. Lett.*, **64**, 1738–1740.
Heckl, W.M. and Binnig, G. (1992) Domain-walls on graphite mimic DNA. *Ultramicroscopy*, **42–44**, 1073–1078.
Heckl, W.M., Smith, D.P.E., Binnig, G., Klagges, H., Hänsch, T.W. and Maddocks, J. (1991) Two-dimensional ordering of the DNA-base guanine observed by scanning tunneling microscopy. *Proc. Natl. Acad. Sci. USA*, **88**, 8003–8005.
Henderson, E., Haydon, P.G. and Sakaguchi, D.S. (1992) Actin filament dynamics in living glial-cells imaged by atomic force microscopy. *Science*, **257**, 1944–1946.
I'Anson, K.J., Miles, M.J., Bacon, J.R., Carr, H.J., Lambert, N., Morris, V.J. and Wright, D.J. (1988) Structure of the 7S globulin (vicilin) from pea (*Pisum sativum*). *Int. J. Biol. Macromol.*, **10**, 311–318.
Keller, D.J. and Chih-Chung, C. (1992) Imaging steep, high structures by scanning force microscopy with electron beam deposited tips. *Surf. Sci.*, **268**, 333–339.
Lee, I., Atkins, E.D.T. and Miles, M.J. (1992) Visualization of the algal polysaccharide carrageenan by scanning tunneling microscopy. *Ultramicroscopy*, **42–44**, 1107–1112.
Lindsay, S.M. (1993) Biological applications of the scanning probe microscope, in *Scanning Tunnelling Microscopy and Spectroscopy* (ed. D.A. Bonnell), VCH Publishers, Cambridge, pp. 335–408.
Manne, S., Cleveland, J.P., Stucky, G.D. and Hansma, P.K. (1993) Lattice resolution and solution kinetics on surfaces of amino-acid crystals – an atomic force microscope study. *J. Crystal Growth*, **130**, 333–340.
Marti, O. and Amrein, M. (eds.) (1993) *STM and SFM in Biology*, Academic Press, San Diego, CA.
Martin, Y., Williams, C.C. and Wickramasinghe (1987) Atomic force microscope force mapping and profiling on a sub-100 Å scale. *J. Appl. Phys.*, **61**, 4723–4729.
McMaster, T.J., Hickish, T., Min, T., Cunningham, D. and Miles, M.J. (1994) Application of scanning force microscopy to chromosome analysis. *Cancer, Genet. Cytogen.*, **76**, 93–95.
McMaster, T.J., Miles, M.J. and Walsby, A. (1995) Direct observation of a protein secondary structure by AFM, *Biophys. J.*, in press.
Meek, K.M., Chapman, J.A. and Hardcastle, R.A. (1979) *J. Biol. Chem.*, **254**, 10710–10714.

Meyer, G. and Amer, N.M. (1990) Simultaneous measurement of lateral and normal forces with an optical-beam-deflection atomic force microscope. *Appl. Phys. Lett.*, **57**, 2089–2091.

Miles, M.J. and McMaster, T.J. (1993) Protein assemblies and single molecules imaged by STM, in *STM and SFM in Biology* (eds. O. Marti and M. Amrein), Academic Press, San Diego, CA, pp. 177–204.

Miles, M.J., McMaster, T.J., Carr, H.J., Tatham, A.S., Shewry, P.R., Field, J.M., Belton, P.S., Jeenes, D., Hanley, B., Whittam, M., Cairns, P., Morris, V.J. and Lambert, N. (1990) Scanning tunnelling microscopy of biomolecules. *J. Vacuum Sci. Technol.*, **A8**, 698–702.

Miles, M.J., Carr, H.J., McMaster, T.J., I'Anson, K.J., Belton, P.S., Morris, V.J., Field, J.M., Shewry, P.R. and Tatham, A.S. (1991a) Scanning tunnelling microscopy of wheat seed storage protein reveals details of an unusual supersecondary structure. *Proc. Natl. Acad. Sci. USA*, **88**, 68–71.

Miles, M.J., Lee, I. and Atkins, E.D.T. (1991b) Molecular resolution of polysaccharides by scanning tunnelling microscopy. *J. Vacuum Sci. Technol.*, **B9**, 1206–1209.

Morris, V.J. (1994) Biological applications of scanning probe microscopies. *Prog. Biophys. Mol. Biol.*, **61**, 131–185.

Overney, R.M., Meyer, E., Frommer, J., Brodbeck, D., Lüthi, R., Howald, L., Güntherodt, H.-J., Fujihira, M., Takano, H. and Gotoh, Y. (1992) Friction measurements on phase-separated thin-films with a modified atomic force microscope. *Nature* (*London*), **359**, 133–135.

Radmacher, M., Fritz, M., Hansma, H.G. and Hansma, P.K. (1994) Direct observation of enzyme activity with the atomic force microscope. *Science*, **265**, 1577–1579.

Rasch, P., Wiedemann, U., Wienberg, J. and Heckl, W.M. (1993) Analysis of banded human-chromosomes and *in situ* hybridization patterns by scanning force microscopy. *Proc. Natl. Acad. Sci. USA*, **90**, 2509–2511.

Shigekawa, H., Morozumi, T., Komiyama, M., Yoshimura, M., Kawazu, A. and Saito, Y. (1991) Scanning tunnelling microscopy on cyclodextrin inclusion complexes. *J. Vacuum Sci. Technol.*, **B9**, 1189–1192.

Slayter, H.S. and Lowey, S. (1967) Structure of the myosin molecule visualized by electron microscopy. *Proc. Natl. Acad. Sci. USA*, **58**, 1611–1618.

Smith, D.P.E. (1991) Defects in alkylcyanobiphenyl molecular-crystals studied by scanning tunnelling microscopy. *J. Vacuum Sci. Technol.*, **B9**, 1119–1125.

Tang, S.L., McGhie, A.J. and Suna, A. (1993) Molecular-resolution imaging of insulating macromolecules with the scanning tunnelling microscope via a non-tunnelling, electric-field-induced mechanism. *Phys. Rev.*, **B47**, 3850–3856.

Thomson, N.H., (1994) *Scanning Probe Microscopy of Seed-Storage Components*. Ph.D. thesis, University of Bristol.

Thomson, N.H., Miles, M.J., Tatham, A.S. and Shewry, P.R. (1992) Molecular images of cereal proteins by STM. *Ultramicroscopy*, **42–44**, 1204–1213.

Thomson, N.H., Miles, M.J., Ring, S.G., Shewry, P.R. and Tatham, A.S. (1994) *J. Vacuum Sci. Technol.*, **B12**, 1565–1568.

Walker, M., Knight, P. and Trinick, J. (1985) Negative staining of myosin molecules. *J. Mol. Biol.*, **184**, 535–542.

Weisenhorn, A.L., Hansma, P.K., Albrecht, T.R. and Quate, C.F. (1989) Forces in atomic force microscopy in air and water. *Appl. Phys. Lett.*, **54**, 2651–2653.

Welland, M.E., Miles, M.J., Lambert, N., Morris, V.J., Coombs, J.H. and Pethica, J.B. (1989) Structure of the globular protein vicilin revealed by scanning tunnelling microscopy. *Int. J. Biol. Macromol.*, **11**, 29–32.

Wigren, R., Elwing, H., Erlandsson, R., Welin, S. and Lundström, I. (1991) Structure of absorbed fibrinogen obtained by scanning force microscopy. *FEBS Lett.*, **280**, 225–228.

Wilkins, M.J., Davies, M.C., Jackson, D.E., Mitchell, J.R., Roberts, C.J., Stokke, B.T. and Tendler, S.J.B. (1993) Comparison of scanning tunnelling microscopy and transmission electron-microscopy image data of a microbial polysaccharide. *Ultramicroscopy*, **48**, 197–201.

Winfield, M., McMaster, T.J., Karp, A. and Miles, M.J. (1995) Atomic force microscopy of plant chromosomes. *Chromosome Res.*, **3**, 128–131

Yang, J., Tamm, L.K., Somlyo, A.P. and Shao, Z. (1993) Promises and problems of biological atomic-force microscopy. *J. Microsc.*, **171**, 183–198.

Chapter 4

Electron Spin Resonance Spectroscopy for Detection of Irradiated Food

R. Gray
Department of Agriculture for Northern Ireland, Belfast, Northern Ireland, UK

4.1 Introduction

With the increasing use of the process of food irradiation in national and international food chains, it has become clear that internationally accepted methods for the detection of foods treated by irradiation are required. A conference in 1988 under the title of 'The Acceptance, Control of and Trade in Irradiated Food' recommended that

> governments should encourage research into methods of detection of irradiated foods so that administrative control of irradiated food once it leaves the facility can be supplemented by an additional means of enforcement (Anon., 1989).

The development of accepted detection procedures not only gives enforcement authorities the ability to check that products are correctly labelled, but also gives the consumer confidence that adequate independent controls are available. Considerable international effort has since been directed towards these goals and substantial progress has been made in a number of directions. The detection method that is, to date, the most internationally accepted is electron spin resonance (ESR) spectroscopy.

4.2 Uses of irradiation in the food industry

The use of ionizing radiation for the preservation of food is not a new technology. In fact, about 90 years ago, a patent was issued in the United Kingdom which detailed the use of the process for the preservation of foods, especially cereals. Despite this initial interest in the technology, progress was hindered because of the limited availability of suitable sources of ionizing radiation.

About 1950, the isotope ^{60}Co, which generates gamma photons, and machines producing high energy electrons, began to become available, and as a consequence extensive research programmes commenced in the United States. Since then, many countries throughout the world have been involved in evaluation of the technology for the preservation of a wide range of foods and the process is used commercially in a number of countries.

In combination with good hygienic practices, the irradiation process is effective in enhancing food safety by reducing the numbers of pathogenic micro-organisms including *Listeria*, *Salmonella* and *Campylobacter* which are often implicated in food poisoning outbreaks involving poultry, meat, fish and shellfish. The process also kills spoilage micro-organisms and so the shelf-life of these foods can be extended. Ripening of fruits can be delayed by irradiation and the technology provides an alternative to the chemicals which have been or are still being used to decontaminate spices and herbs, disinfest cereals and tropical fruit, and inhibit sprouting in tubers and bulbs such as potatoes and onions.

As well as the effects of the technology on the chemical, microbiological, sensory and nutritional quality of foods, the safety of irradiated foods has been repeatedly evaluated by a number of independent expert groups. In 1981, the Food and Agriculture Organization (FAO)/International Atomic Energy Agency (IAEA)/World Health Organization (WHO) Joint Expert Committee concluded that

> the irradiation of any food commodity, up to an overall average dose of 10 kGy, presents no toxicological hazard and introduces no special nutritional or microbiological problems (Anon., 1981).

Similar conclusions were reached by the UK Advisory Committee on Irradiated and Novel Foods (ACINF) in their report on the safety and wholesomeness of irradiated food (Anon., 1986).

Following the favourable response of this committee, steps were taken to amend the existing UK legislation which permitted only the radiation sterilization of foods for people whose immune response was compromised. New regulations came into force in early 1991 which permitted the irradiation of seven groups of foods under strictly controlled conditions (Anon., 1990a) (Table 4.1).

Table 4.1 Foods for which UK legislation permits treatment with ionizing radiation

Food group	Maximum dose (kGy)
Bulbs and tubers	0.2
Cereals	1.0
Vegetables	1.0
Fruit and mushrooms	2.0
Fish and shellfish	3.0
Poultry and meat	7.0
Spices and condiments	10.0

(From Anon., 1990a).

Scientific evidence has shown that on health grounds irradiated foods need not be labelled, but it is generally felt that consumers should be able to either choose or avoid irradiated foods. In the UK, surveys carried out using questionnaires indicated that consumers harbour considerable reservations about the consumption of food treated with ionizing radiation (Anon., 1990b). On the other hand, in countries where test marketing of irradiated foods had been undertaken, and where consumers had been given unbiased information about the process, the response was very favourable (Bruhn and Noell, 1987; Anon., 1988; Marcotte, 1992). Nevertheless, it was accepted that in order to enforce labelling regulations, a method (or methods) to detect irradiated food was required and the availability of such a method would help to promote international trade in irradiated foods.

4.3 Development of detection methods

When food is treated with ionizing radiation, the changes which occur are minimal and similar to those induced by other processes, such as cooking. As a consequence, considerable scientific effort has been necessary to develop methods which fulfil the technical and practical criteria considered necessary for an effective identification method, i.e. the test must be able to discriminate between treated and untreated samples, it must be reliable and practical, and preferably it should be able to give an estimate of the absorbed dose. It is generally accepted that several techniques will be needed to cover the range of foods which are likely to be treated with irradiation. One technique which has the ability to detect irradiation treatment qualitatively in a number of foods and which also has the potential to give an estimate of absorbed dose is ESR spectroscopy.

4.4 Principles of electron spin resonance

When foodstuffs are subjected to ionizing radiation, the absorption of some of the incident energy can lead to the ejection of electrons from chemical bonds. These free electrons may immediately recombine or may form a number of highly reactive entities, each of which can interact with the foodstuff to produce a stable compound. In the majority of foods, the moisture content is sufficiently high to ensure that these radicals are rapidly neutralized. However, in food components and packaging materials of high dry matter, e.g. bone, seeds, shells and cellulose, the radicals are trapped in an environment which ensures a relatively long lifespan. These are the radicals that can be detected by ESR spectroscopy.

Spectroscopy is the measurement and interpretation of the energy differences between atomic or molecular states, each of which have discrete energy values; in other words it monitors the net absorption of energy from a radiation field when molecules change their energy state. According to Planck's Law, energy will be absorbed if the difference in energy is $\Delta E = h\nu$, where h is Planck's constant and ν is the frequency of the radiation. The energy differences studied by ESR spectroscopy are predominantly due to the interaction of unpaired electrons with the combined effects of a strong magnetic field and a source of microwave energy.

ESR spectroscopy detects species with unpaired electrons, and, as electrons can only have one of two spin states and are normally paired, any species, except a transition metal or rare earth ion, with an odd number of electrons is referred to as a free radical. These species are highly reactive and normally very short-lived surviving for only milliseconds in the liquid phase. Electrons may be visualized as spinning negative charges and, since a moving charge generates a magnetic field, each electron in effect acts as a minute bar magnet. In species with an even number of electrons, the effects of the individuals in each pair are cancelled out. However, in the case of free radicals, having an odd number of electrons, the magnetic effects do not cancel and the species is said to be paramagnetic. ESR spectroscopy only detects the presence of these unpaired electrons, and this means that even very small populations of free radicals can be detected in biological samples. If an external magnetic field is applied to such paramagnetic species, the unpaired electrons, which are normally randomly orientated, become aligned to the external field and can only occupy one of two states: (a) parallel to the external field (lower energy state), or (b) anti-parallel to the external field (higher energy state). The most basic equation for ESR, derived from quantum mechanics, is

$$E = g\mu_B H M_s,$$

where E is the energy level, g is the g-factor or proportionality constant, μ_B is the Bohr magneton (unit of electromagnetic moment), H is the magnetic field strength, and M_s is the moment of electron spin (which can be $+\frac{1}{2}$ or $-\frac{1}{2}$). Therefore the difference between two energy states, ΔE, is

$$\Delta E = g\mu_B H.$$

It follows that (a) if there is no magnetic field present then there will be no difference between the energy levels of the two spin states, and (b) as the magnetic field strength is increased the difference in energy level of the spin states increases linearly.

The normal procedure for most forms of spectroscopy would be to apply a constant magnetic field and scan the frequency of the radiation looking for resonances. But, because of the limitations of microwave electronics, it is easier to maintain a constant frequency and scan the magnetic field.

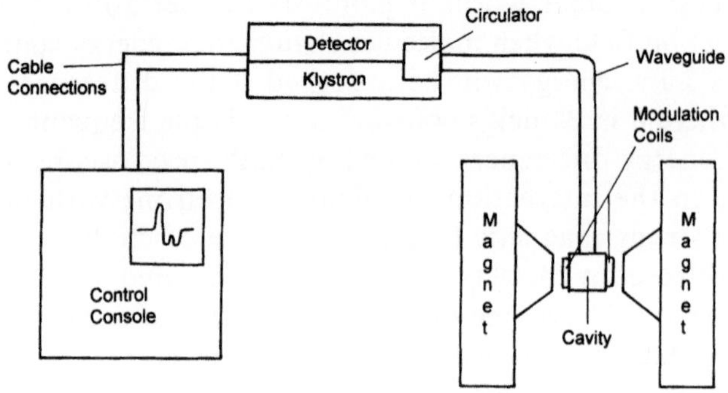

Fig. 4.1 Diagrammatic representation of a modern ESR spectrometer.

Thus, electrons can be made to resonate between these two states by the application of microwave energy, i.e. they can absorb energy and jump from the lower energy state to the higher energy state. Because this is essentially a resonance situation, under normal circumstances a similar number of excited electrons will release energy and return to the lower energy state. However, there is always a larger population in the lower energy state, and as a consequence resonance will give rise to a net absorption of energy. It is this net absorption of energy that is detected and amplified to give the ESR signal.

Modern ESR spectrometers consist of four basic elements or components: a radiation source, some form of sample containment, a magnet and the detection equipment (see Fig. 4.1).

4.4.1 The radiation source

ESR spectrometers operate in the microwave region and principally in the X-band (9.5 GHz) although instruments operating in the K-band (24 GHz) and Q-band (35 GHz) are also available. One of the useful properties of microwaves is that a polarized beam may be transmitted along a rectangular metal pipe, referred to as a 'wave guide'. Even bends can be negotiated provided they are not too sharp, and as a consequence the different components of a microwave system are connected by metal wave guides.

The polarized beam of microwaves is normally produced by a unit called a Klystron, but recently another type of source called a Gunn diode has been developed for the smaller 'bench top' spectrometers. The output from either of these microwave sources must have a controllable intensity. This is regulated using an attenuator, and the attenuated monochromatic beam of microwaves is then directed down the wave guide to the cavity.

4.4.2 The cavity

The cavity is a rectangular or cylindrical box on the end of the wave guide which acts as a resonator in a similar manner to an organ pipe resonating sound waves. Thus, it has the capacity to magnify the intensity of the microwaves provided that the incident microwaves are of the correct frequency, i.e. the *resonant frequency* of the cavity.

One of the characteristics of electromagnetic waves, including microwaves, is that they consist of two fields, the electric field and the magnetic field, and these two fields are always perpendicular to each other. As a consequence, cavities are designed to give maximum absorption of microwaves by ensuring that the sample is placed at a point of high magnetic field and low electric field. A common basic design would take the form of that in Fig. 4.2. Since the operation of the cavity is optimized at the resonant frequency, the output of the Klystron is locked at this frequency by an automatic frequency control unit. (This is also another reason why the microwave frequency is 'fixed' and the magnetic field strength scanned.)

All modern ESR cavities are reflection cavities, which means that there are both incident and reflected microwaves travelling along the wave guide. The microwave bridge must be able to separate these two and to direct the reflected microwaves to the detector unit. This is accomplished using a unit called the circulator which is situated along the wave guide and is able to direct the incident wave from the source to the cavity while

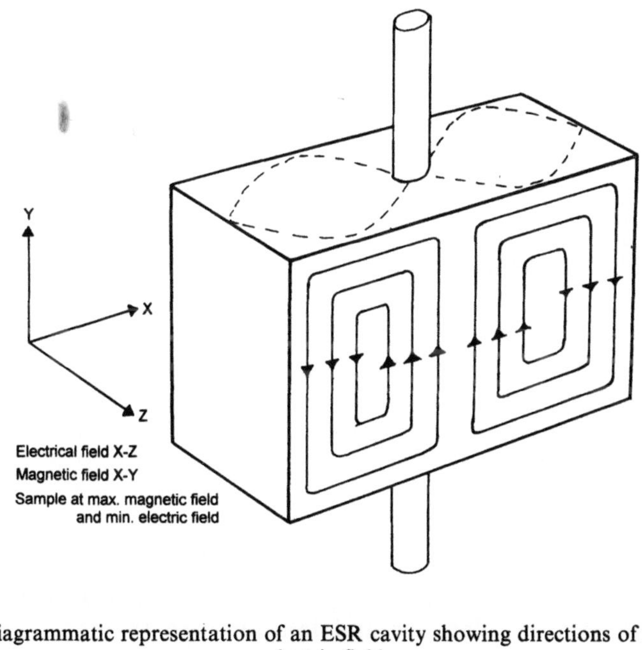

Fig. 4.2 Diagrammatic representation of an ESR cavity showing directions of magnetic and electric fields.

at the same time ensuring that the reflected waves are all directed towards the detector.

4.4.3 The magnet

Investigations using the X-band frequency require a magnetic field strength of the order of 350 mT which then allows the size of the magnet in the spectrometer to be calculated. Modern electromagnets combined with highly regulated power supplies can easily cope with field strengths of this order.

4.4.4 The detector assembly

The intensity of the reflected microwaves directed to the detector is measured by a special diode which acts as a microwave rectifier converting the microwave energy into a direct current. In order to remove the considerable noise detected along with the microwave signal, a technique known as *phase sensitive detection* is employed. Basically this means that the signal of interest is made to appear at only one frequency, thus allowing all other frequencies (noise) to be filtered out. This technique requires the addition of two oscillation coils, one on each side of the resonant cavity, which modulate the magnetic field strength at a given frequency, usually 100 kHz. When the free radicals in the sample under investigation are at or near resonance, the detected signal will be modulated at the given modulation frequency. This modulated signal is amplified and used as one input of a phase sensitive detector whose other input is a sub-signal of that driving the modulation coils. The end result is that the filtered output of the phase sensitive detector is proportional to the amplitude of the input to the diode.

4.5 Factors affecting spectral derivation

Derivation of spectra using a fully operational modern spectrometer is almost entirely automated, but, depending on the use to which the spectra are to be put, a number of factors must be taken into consideration as each may have a more or less significant effect on the end result.

4.5.1 ESR tubes

Samples of irradiated product, being generally in the powder or crystalline form, are placed in ESR tubes, made either of pyrex or quartz glass, of 4–6 mm internal diameter and long enough to more than fill the cavity. Care should be taken, especially with the cheaper pyrex tubes, that there are no inherent ESR signals in the glass itself; any such tubes should be rejected.

4.5.2 *Sample integrity*

Every effort should be made to ensure that only one physical component is introduced into the ESR cavity. For example, in the case of chicken bone, all traces of cartilage, marrow and connective tissue should be removed as each will have been exposed to the irradiation process and indeed some (e.g. marrow) may have their own intrinsic ESR signal, whether or not they have been irradiated.

4.5.3 *Moisture content*

Since the free radicals induced by irradiation of food are very short-lived, except in relatively dry crystalline or fibrous substances, most substrates used for this type of ESR analysis are naturally of a relatively high dry matter content. Nevertheless, since the exposure to microwave radiation of samples containing moisture will lead to some degree of heating, the dry matter content of the sample should be as high as possible. There is ample evidence that simple high temperature drying produces additional free radicals which in turn will complicate the isolation of the spectrum due to irradiation. Therefore, as a general rule, samples should be freeze-dried or dried in a vacuum oven and not oven-dried, especially if the signal is to be quantified. Also, in the case of quantification, it is desirable to determine the sample dry matter once ESR analysis is completed and to correct for any differences between samples. On the other hand, rapid qualitative analysis may be aided by microwave drying provided reduced levels of power are used in the microwave oven.

4.5.4 *Physical form of the sample*

Some samples (e.g. strawberry achenes) may, in their natural form, be small enough for introduction into the ESR tube. In the case of bone or the exoskeleton from *Crustacea*, the samples may need to be reduced in size. This can be achieved by either using fragments or by grinding. Since the latter process involves frictional forces, which in turn generate heat, care should be exercised to minimize any rise in temperature of the sample during sample preparation. This is most effectively done by using a mill with a high cooling air throughput.

4.5.5 *Quantity of sample in the cavity*

For qualitative identification of irradiation treatment, the amount of sample placed in the cavity need not actually be known, but the greater the quantity of sample present, the stronger will be the signal. On the other hand, if the effect of irradiation on the population of free radicals is to be quantified, the amount of each sample exposed to the microwaves in the cavity must be known. Only part of the ESR tube in the cavity is actually exposed to the microwaves, and the relevant part or active length can be estimated for each cavity by the sequential introduction into the ESR tube of small quantities of a weak solution of diphenyl picryl hydrazyl (DPPH), an intense source of unpaired electrons. Stevenson and Gray (1989a) have measured the active length as follows. A small amount of sample was introduced into the ESR tube, and the tube plus sample was weighed. The sample was compacted by tapping the tube on a hard surface, and the distance from the base of the tube to the top of the sample was measured using digital callipers. Sufficient sample was then added to the tube to exceed the amount required to fill the active length, and the sample was weighed, compacted and measured again. From these values the fresh weight of sample in the active length was calculated.

4.5.6 Spectrometer settings

(*a*) *Microwave power.* Basically, ESR spectra show that fraction of the energy of the microwave beam absorbed at resonance, and the signal intensity is proportional to the differences in population of the two spin states. As the microwave power is increased, so the magnitude of the ESR signal will increase. If the microwave power is great enough, it will increase the population of electrons in the higher energy state, thus reducing the difference between the two populations and hence the power absorbed. This is known as power saturation. To ensure that a sample is not being saturated, the power should be temporarily decreased and the intensity should then decrease by the square root of the microwave power, i.e. when the signal is not saturated the signal intensity is directly related to the square root of power.

(*b*) *Modulation amplitude.* Modulation amplitude directly affects line width, and if the modulation amplitude is greater than the line width of the ESR signal to be detected, the ESR signal will broaden and become distorted. Therefore, as a simple rule, the modulation amplitude should not exceed the width of the narrowest line.

(*c*) *Filtering.* Virtually all commercial spectrometers have a variable factor termed the *time constant.* This is a system of noise filtering which works by slowing down the response time of the spectrometer, and so as the time constant is increased the noise level will drop. However, if the time constant is too high, the signal may be distorted or even completely filtered out. If a long time constant is needed in order to see a weak signal, then the scan time must be increased. A useful guide is to ensure that the time taken to scan through a single ESR peak should be at least ten times larger than the time constant.

4.6 Application of ESR spectroscopy to detection of irradiated foods

ESR methods have been developed for a number of products, including food containing bone or shell, fresh and dried fruits and vegetables, and spices and nuts. Although the vast majority of the work has dealt with primary food products, research has shown that the procedure can be used to detect the presence of irradiated food components such as mechanically recovered meat (MRM) in secondary food products. Also a number of materials used to package food can give a radiation-induced signal which would indicate that the packaging, and possibly the food contained within it, had been irradiated.

4.6.1 Foods with bone present

As early as 1955, Gordy and coworkers showed that, when bone was treated with ionizing radiation, free radicals were trapped in the crystal structure of the bone which could be detected by ESR spectroscopy. Trapped free radicals in bone have also been used to date archaeological samples and are frequently used to determine the level of radiation exposure in accidents involving sources of radiation. These techniques and other related applications are covered in a recent book by Ikeya (1993).

Detection methods which use bone samples examine only the mineralized tissue, thereby avoiding the additional free radical species present in the marrow. Two predominant paramagnetic species are generated in mineralized tissues following irradiation at room temperature (Ostrowski *et al.*, 1980). The species derived from the organic material, probably from collagen, is characterized by a symmetric doublet, but it is unstable and hence cannot be used to identify irradiated bone containing food. On the other hand, a much more stable paramagnetic centre localized in the crystalline hydroxyapatite gives an asymmetric singlet.

The ESR signal induced in irradiated fragments of bone or ground bone is different in shape to the signal present in unirradiated bone (Fig. 4.3). It is also independent of the origin of the bone, as the same signal can be found in irradiated samples of bone from chicken, duck, turkey, goose, whiting, pork, beef and frog legs. Research has shown that the characteristic ESR signal is not generated by such common practices as grinding (Stevenson and Gray, 1989a) and cooking (Gray and Stevenson,

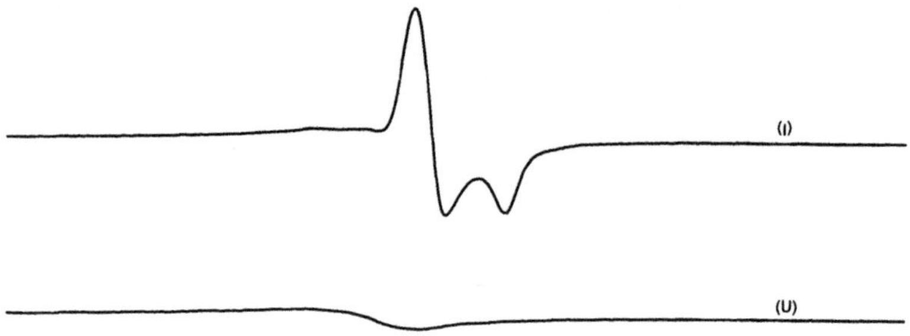

Fig. 4.3 ESR spectra of irradiated (I) and unirradiated (U) chicken bone.

1989a). The signal is also sufficiently stable during storage (Stevenson and Gray, 1989b) and following cooking (Gray and Stevenson, 1989a) to be used, in both raw and cooked product, as an indication of irradiation treatment.

The ideal detection method should not only give an indication of irradiation treatment but should also be able to provide an estimate of the absorbed dose. Thus, considerable effort has been directed towards studying the variables which might influence the intensity of the radiation-induced signal and the conditions under which it might no longer be detectable.

(a) Factors affecting signal intensity. Variables associated with several aspects of the processing chain, from primary production through irradiation processing and storage conditions to consumption, have been examined using mainly chicken bone. The intensity of the ESR signal can be quantified by integration of the area under the absorption spectra (i.e. double integration of the first derivative or normally displayed spectrum) (Stevenson and Gray, 1989a, b) or it can be estimated by the measurement of peak height (the distance between the maximum and minimum of a particular spectral peak) (Dodd *et al.*, 1988). Research has shown that the method used to prepare the bone for analysis can greatly influence the signal intensity. In comparison to fragmented bone, which had been air-dried, combinations of freeze-drying, oven-drying, microwave-drying and grinding were found to produce considerable variation in signal intensity (Table 4.2). Although the most intense signals have been obtained from samples which were fragmented and freeze-dried, most research has been performed using freeze-dried and ground samples because of the greater homogeneity of these samples. The possibility of using ESR spectroscopy to provide an estimate of the absorbed dose was demonstrated when it was shown that the intensity of the signal induced in chicken bone increased linearly with irradiation dose over the range

Table 4.2 Effect of sample preparation on the relative ESR signal strengths from irradiated chicken bones

Sample preparation method	Relative signal intensity			
	Basic	Control	DM conc.	1 g DM
Fragmented and fresh	1.138	0.955	0.949	6.889
Fragmented and freeze-dried	1.813	1.467	1.466	9.965
Fragmented and oven-dried	1.888	0.830	0.794	5.188
Microwaved and ground	2.317	1.160	1.174	6.643
Freeze-dried and ground	2.846	1.560	1.591	9.390
Oven-dried and ground	2.945	0.983	0.903	5.005
SEM	0.0757	0.0811	0.0593	0.3883
Significance	***	***	***	***

*** $P < 0.001$; DM, dry matter; SEM, standard error of mean; basic, uncorrected integral; control, unirradiated spectrum subtracted; DM conc., corrected for differences in sample DM; 1 g DM, corrected to a standard 1 g DM in cavity (from Stevenson and Gray, 1989a).

2.5–10 kGy (Stevenson and Gray, 1989b). Also the intensity of the signal induced by a given irradiation dose appeared to be species dependent, being much lower for fish (Stewart *et al.*, 1991) than chicken.

Since the trapped radicals are associated with the hydroxyapatite fraction of bone, the concentrations of calcium and phosphorus might be expected to affect signal intensity. However, using birds from 4 to 8 weeks of age, it was shown that the increase in ESR signal strength with increasing age was not due to a change in mineral composition since the latter was consistent over the age range investigated (Gray *et al.*, 1990a). Work published by Ostrowski *et al.* (1981) has shown that the degree of crystallinity in bone structure can influence the ESR signal intensity. The crystallinity of a bone can be defined as the ratio of the crystalline fraction of tissue mineral to the total amount of mineral, and it can be estimated by calculating the ratio of the concentration of stable free radicals induced by a saturating dose of ionizing radiation to the ash content of the tissue. To calculate this ratio the bones in question are given a saturating dose of gamma photons (120 kGy) so ensuring that the maximum population of free radicals is present. The magnitude of this population, per unit mass of bone, is then measured using ESR spectroscopy and the ash content of the sample is determined. The ratio of the signal intensity to the ash content is taken as the crystallinity coefficient, and this allows comparison of bone crystallinity between samples. In work with chicken bone, it has been demonstrated that as birds age there is a linear increase in crystallinity coefficient (Table 4.3) (Gray *et al.*, 1990).

When bones from different sites within the carcass of an irradiated chicken are examined, the intensity of the ESR signal varies depending on the part of the skeleton from which the bones are removed (Fig. 4.4). Correction of the various signal intensities for any variation in calcium or phosphorus content of the bones did not remove the effect, but there

Table 4.3 Effect of age on the ESR signal strength in irradiated chicken bone

Age (weeks)	ESR signal intensity				
	Dose corrected	Ash corrected	Ca corrected	P corrected	Crystallinity coefficient
4	1.913	1.946	1.787	1.761	2.32
5	2.021	2.036	1.997	1.840	2.42
6	2.099	2.107	1.912	1.854	2.48
7	2.323	2.304	2.227	2.008	2.55
8	2.489	2.464	2.367	2.215	2.60
SEM	0.0638	0.0630	0.0752	0.0604	0.080
CV%	11.8	11.6	14.6	12.5	13.0
Significance	***	***	***	***	***

*** $P < 0.001$; SEM, standard error of mean; CV%, percentage coefficient of variation (from Gray *et al.*, 1990).

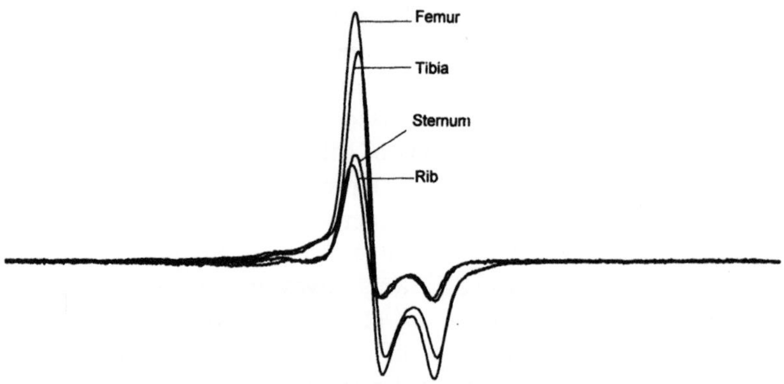

Fig. 4.4 ESR spectra from different bones (femur, tibia, sternum and rib) within the chicken carcass each having received the same dose of gamma photons.

was a strong linear relationship between the ESR signal intensity and the crystallinity coefficient. More crystalline bones, such as the tibia and femur of the leg and the humerus, gave a greater response to a given irradiation dose than the more cartilaginous and less crystalline rib and sternum (Gray and Stevenson, 1990).

Chicken can be retailed either frozen or chilled and irradiation can be applied under both of these temperature regimes. In samples irradiated in the frozen state, the movement of free radicals is hindered and, as expected, fewer radicals are trapped in the bone. Consequently, the intensity of the signal induced in bone from irradiated frozen chicken was lower than in samples from chilled poultry carcasses given the same irradiation dose (Stevenson and Gray, 1990) (Fig. 4.5).

In commercial practice, chicken might be cooked either before or after irradiation, and it is important to establish that the cooking procedure neither completely destroys the radiation-induced signal nor generates

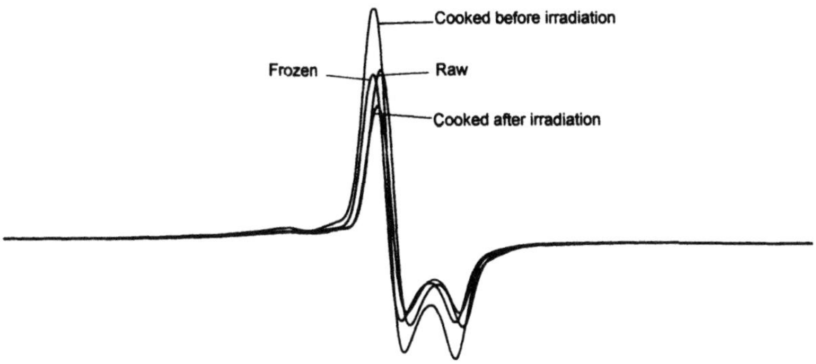

Fig. 4.5 ESR spectra from irradiated chicken bone showing the effect of different processing procedures.

radicals which can interfere with the signal induced by irradiation. It has been shown that the cooking process does not generate signals similar in shape to those induced by irradiation, but that it does affect signal intensity, as a 23% reduction in signal intensity has been observed when chicken is cooked after irradiation (Gray and Stevenson, 1989a). On the other hand, if chicken is cooked before irradiation, the concentration of free radicals is significantly increased (Fig. 4.5) (Stevenson and Gray, 1990). It has been suggested that, because the water content of cooked bones is reduced, there is less opportunity for free electrons to react with water. As a consequence, more free radicals are trapped in bone thereby giving an enhanced ESR signal (Davidson, 1988). Alternatively, the observable change in the structure of the bone and, in particular, the loss of collagen and consequent relative increase in calcium content, may contribute to the greater signal intensity (Dodd *et al.*, 1992).

The rate at which an irradiation dose is applied does not appear to affect the intensity of the induced signal significantly (Gray and Stevenson, 1991). Yields of radicals induced using a linear accelerator producing 10 MeV electrons (dose rate = 600 kGy h^{-1}) were similar to those produced by ^{60}Co gamma photons (dose rate range 0.2 – 20 kGy h^{-1}). Consequently, the ESR technique can be used to identify products treated using either of the major irradiation sources which are permitted for the treatment of food (Anon., 1984).

The stability of the trapped radicals appears to be similar in all types of chicken bone (Gray and Stevenson 1990). Storage at chill temperatures (+5 °C) was found to cause a small decay in signal intensity (Stevenson and Gray, 1989b) while storage at −20 °C for 10 months caused no apparent change in free radical concentration (Dodd *et al.*, 1992).

(b) Quantification of absorbed dose. The demonstration that the ESR signal strength is affected to some extent by a number of processing variables suggested that there may be difficulties in accurately estimating the dose received by samples if their processing history was unknown. Re-irradiation has been suggested as a way of estimating dose which does not require a detailed knowledge of the sample history. However, there are still likely to be difficulties encountered with cooked samples or those irradiated in the frozen state (Dodd *et al.*, 1992). The procedure of re-irradiation was first used by Dodd and his co-workers to estimate the dose received by chicken, pork and fish bones (Dodd *et al.*, 1988) and has subsequently been used in a number of blind trials (Desrosiers *et al.*, 1990; Raffi *et al.*, 1992) with varying degrees of success. The procedure is essentially a standard addition technique whereby the sample is irradiated several times with the signal strength being recorded after each re-irradiation. An estimate of the original absorbed dose is then extrapolated from a graph of signal height versus applied dose. However, the form of the mathematical function used to fit the data is critical to the accuracy of the dose estimate (Desrosiers *et al.*, 1991). A practical drawback to the re-irradiation technique is the need to have access to an irradiator and it has been proposed that the derivation of a dose response calibration curve could provide an alternative procedure for the quantitative evaluation of samples of unknown background (Bordi *et al.*, 1993).

(c) Detection in secondary products. As well as using ESR spectroscopy to identify primary food products containing bone, the technique has also been successfully used to identify irradiated MRM. This product is manufactured by hydraulically removing meat from carcass components and normally contains less than 1% bone by weight. As a consequence of the process used to obtain the MRM, the end product may have an unacceptably high bacterial count and is therefore a prime candidate for treatment with ionizing radiation. The small quantity of bone in the MRM may be extracted by digestion with alcoholic potassium hydroxide at a temperature of 85°C for 90 min and subsequent vacuum filtration (Gray and Stevenson, 1989b). The fact that an ESR signal can be detected in the bone fragments extracted from irradiated MRM after such a severe digestion process is a further indication of the stability of the free radicals trapped in the bone. The ability to detect a radiation-induced ESR signal in extracted bone fragments has enabled the identification of an irradiated frozen block of MRM. Samples from an unirradiated block were given doses of 3, 5 and 7 kGy and the magnitude of the ESR signals plotted against absorbed dose. By substituting the signal intensity of a sample from the irradiated block of MRM into this relationship, an accurate estimate of the dose received by the unknown block of MRM was determined (Gray and Stevenson, 1989b).

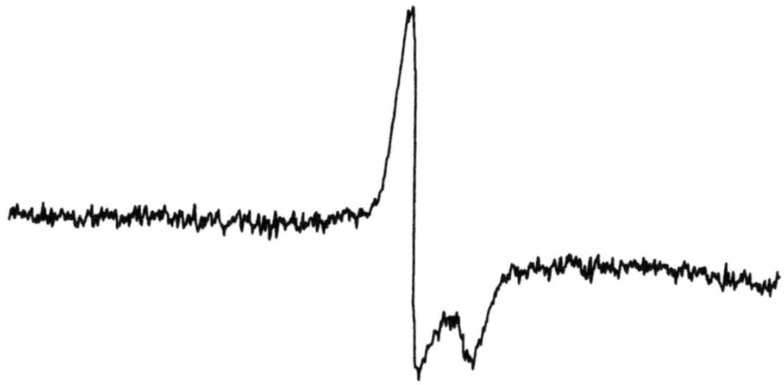

Fig. 4.6 ESR spectrum from bone fragments extracted from a burger containing MRM.

Recently this work has been extended to detect the presence of irradiated MRM in tertiary food products (i.e. burgers), where as little as 3% of the MRM had been included in the product (Fig. 4.6). However, this work will almost certainly only allow qualitative detection of irradiation, as the small sample size and consequent high signal noise will make quantification extremely difficult.

4.6.2 Foods with shells present

A radiation-induced signal can be detected in the exoskeleton of Norway lobster (*Nephrops norvegicus*) and other species of prawn and shrimp.

(a) Norway lobster. In the case of Norway lobster, the signal in both irradiated and unirradiated cuticle is complex because of the presence of the six resonance peaks due to the natural ESR spectrum of Mn^{2+}. In the irradiated samples there is an additional free radical peak in the centre of the Mn^{2+} signal at 349.5 mT (Fig. 4.7). This signal is more easily seen when the effect of the Mn^{2+} signal is removed using the software associated with the spectrometer. The resulting peak appears to be specific for irradiation, as it was neither generated nor completely destroyed by everyday processes such as chilling, freezing or cooking (Fig. 4.8). As for bone, there was found to be a linear relationship between the intensity of the radiation-induced signal and irradiation dose over the range 1 to 5 kGy; so the ESR method had potential for the quantification of dose (Stewart *et al.*, 1992). However, the fact that the signal decayed during storage (Stewart *et al.*, 1992) and was influenced by cooking (Fig. 4.8) could affect the accuracy with which the absorbed dose can be estimated. It has also been found that the shape of the radiation-induced signal derived from different components of the

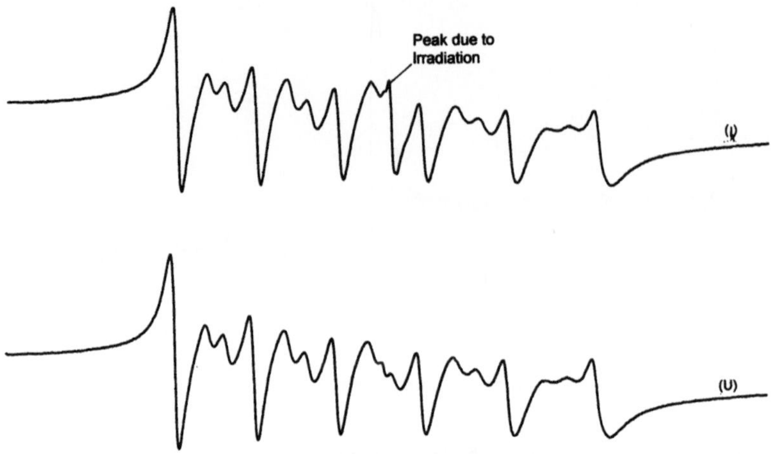

Fig. 4.7 ESR spectra from irradiated (I) and unirradiated (U) Norway Lobster cuticle (wide scan).

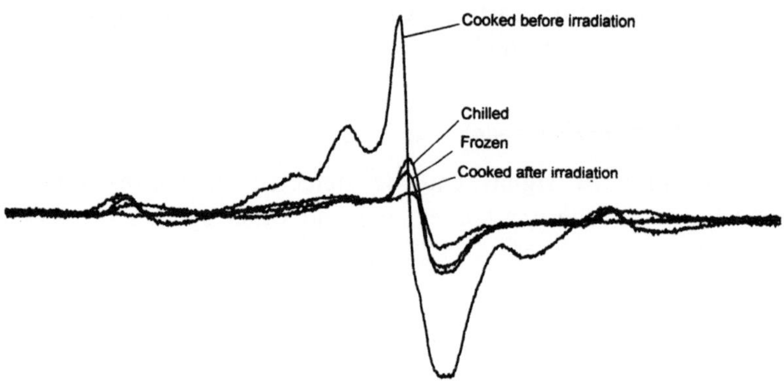

Fig. 4.8 Irradiation specific ESR spectra from irradiated Norway Lobster cuticle showing the effect of different processing procedures.

exoskeleton of Norway lobster was similar, but that the intensity of the peak varied (Stewart *et al.*, 1993). Consequently, it seems that the part of the cuticle used for ESR analysis will not affect identification of irradiation treatment, but it could influence the estimation of dose in samples of unknown processing history.

(b) Other Crustacea. Work with other species of prawn and shrimp has indicated that the situation with *Crustacea* is more complex than for foods containing bone. There is evidence that the shape of the radiation-induced ESR signal is species dependent (Fig. 4.9). Even the same

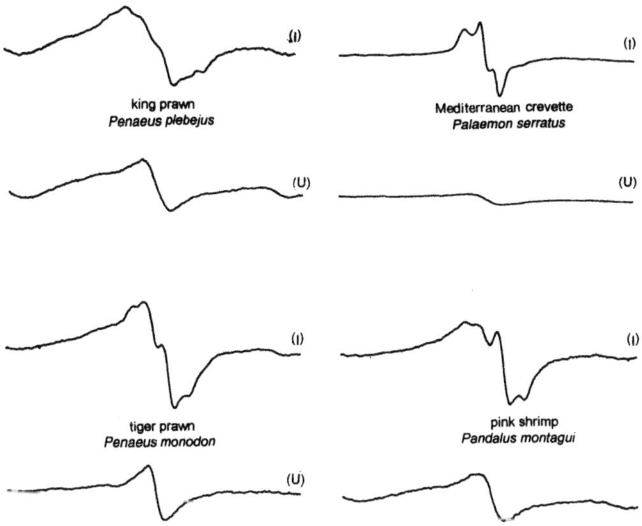

Fig. 4.9 ESR spectra from the irradiated (I) and unirradiated (U) cuticle of various species of *Crustacea* (Stewart, 1993).

species, when examined by several workers, has given radiation-induced signals which are quite different in shape (Desrosiers, 1989; Stewart *et al.*, 1994). The reason for these conflicting results is unknown, but it is possible that some species may not have been correctly identified. Therefore, in order to avoid any discrepancies in reported results, it has been recommended that in all future work each species should be given its scientific name (Anon., 1992).

The origin of the free radicals responsible for the radiation-induced signals in the exoskeleton of *Crustacea* has not been conclusively identified. It has been suggested that the complex signal observed in the cuticle of pink shrimp may be derived from chitin, which is a major component of the exoskeleton (Desrosiers, 1989). However, other components of the shell matrix, such as calcium, may also contribute to the radiation-induced signal (Dodd *et al.*, 1985). A knowledge of the origin of the radical (or radicals) responsible for the ESR signal may help to explain the varying responses which have been observed.

(c) Snails and other molluscs. Initial research with this type of product has indicated the presence of a weak Mn^{2+} signal in both the irradiated and unirradiated samples. A singlet due to irradiation can be isolated from the irradiated sample, and this has been shown to be stable for a period of weeks. However, the signal does not appear to respond linearly to dose beyond 2 kGy.

4.6.3 Fresh fruits

The water content of fresh fruits and vegetables varies from approximately 80 to 95%, and so radicals induced in the pulp by irradiation are not stable. However, the seeds, shells or skins can trap free radicals, and so could in principle be used to monitor radiation exposure. Amongst fruit, the strawberry is a likely candidate for irradiation, since the process can be used to extend shelf-life and inhibit mould growth. A multicomponent signal has been detected in strawberry achenes. In both irradiated and non-irradiated samples, the spectrum consists of a six-line signal due to Mn^{2+}, and a single central line, the intensity of which varies with water content and is thought to be related to a quinone radical. In irradiated achenes there is a novel radiation-induced signal, and it has been suggested that it originates from cellulose. Unlike the singlet, this signal is not affected by water content. The fact that this radiation-induced peak originates from cellulose suggests that it should form the basis of a test for all parts of fruits and vegetables which contain considerable amounts of cellulose in high dry matter environments.

Further research is required in order to examine the stability of the signals induced in irradiated fruits and vegetables since there are indications that stability varies with the food being examined.

4.6.4 Spices and herbs

The ESR signal derived from irradiated cellulose has also been observed in paprika and pepper. However, the detection of the cellulose lines can be difficult or impossible in the presence of the Mn^{2+} signal, especially if the spices have a low cellulose content (Helle and Linke, 1992).

4.6.5 Dried fruits

The ESR spectra of irradiated dried fruits have also been examined. They are complex and quite different to unirradiated samples which generally show no ESR signal, so making identification of irradiated samples relatively easy. The spectra are similar in shape to those derived from irradiated component sugars which make up 60–70% of the composition of the fruits. Irradiated dried pineapples, figs, papayas and apricots show spectra similar to irradiated saccharose which is present in high concentrations in these fruits, while irradiated raisins, dates, currants and sultanas contain less saccharose and they display spectra of a different shape (Helle and Linke, 1992).

4.6.6 Nuts

The ESR spectra of irradiated shells from walnuts, peanuts, hazelnuts and pistachio nuts are similar to those from irradiated strawberries, and the radiation-induced signal, which is thought to be due to cellulose, has potential for the identification of these products (Helle and Linke, 1992; Raffi *et al.*, 1992).

4.7 Packaging materials

Irradiated paper-based packaging shows the characteristic cellulose signal which exhibits a rapid decay in signal intensity during the first 24 h after irradiation, but thereafter is relatively stable showing only a gradual decline over a prolonged period of time (Fig. 4.10). ESR examination of samples of packaging has the advantage of being relatively quick and simple, but it should be remembered that the presence of the characteristic signal due to irradiation treatment only indicates that the packaging has been treated. It is not evidence that the contents have been irradiated, but it would suggest that the contents at least merit further investigation.

4.8 Multilaboratory intercomparisons

The ultimate proof that the ESR method is useful for the identification of irradiated food is its performance in blind trials. A number of studies have already been completed, and the results have been very encouraging (Stevenson and Stewart, 1995). In the majority of cases it has been possible to identify foods treated with irradiation at doses well below those likely to be used commercially. Following the success of these collaborative trials, standard methodologies for the application of the ESR method for the identification of irradiated food containing bone and some fruits (containing cellulose) have been prepared and submitted

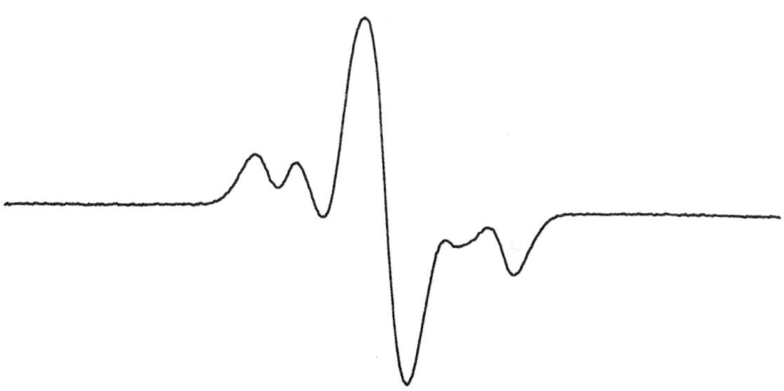

Fig. 4.10 ESR spectrum from irradiated paper based packaging.

to the European Committee of Normalization (CEN). At the time when the trials were carried out, there was not sufficient information available to permit inclusion of *Crustacea* among the products being examined, and thus validation of the method for the identification of these irradiated foods is only now being undertaken. As well as qualitative identification of irradiation, doses received by bones have also been estimated in three of the collaborative trials using the re-irradiation technique. Although the differences between the applied and estimated doses were sometimes quite large, and there were differences between laboratories, it was possible to distinguish between samples given low, medium and high doses of irradiation.

4.9 The future of ESR detection

ESR spectroscopy has proved to be a non-destructive technique which is capable of the quick and easy identification of a number of irradiated foods. However, further research is required into products such as *Crustacea*, exotic fruits, dried fruits and spices, in order to establish the origin of the radiation-induced signals, their stability, specificity and the minimum detectable doses. This fundamental research, together with collaborative testing of the procedures developed, should result in a greater number of standard methods. This research on detection methodology should help to reassure the consumer and ensure that the process of food irradiation is not abused.

References

Anon. (1981) Joint FAO/IAEA/WHO Expert Committee on Wholesomeness of Irradiated Food (JECFI). Report of the Working Party on Irradiation of Food, *WHO Technical Report Series 659*, WHO, Geneva.

Anon. (1984) Codex General Standards for Irradiated Foods and Recommended International Code of Practice for the Operation of Radiation Facilities used for the Treatment of Foods, *Codex Alimentarius Commission*, XV, 1st edn., Rome.

Anon. (1986) *Advisory Committee on Irradiated and Novel Foods. Report on the Safety and Wholesomeness of Irradiated Food*, HMSO, London.

Anon. (1988) Test market for irradiated strawberries in France. *Food Irradiation Newsletter*, **11**, 45.

Anon. (1989) International document on food irradiation, in *Acceptance, Control of and Trade in Irradiated Food*, IAEA, Vienna, p. 135.

Anon. (1990a) *The Food (Control of Irradiation) Regulations 1990*, No. 2490, HMSO, London.

Anon. (1990b) *Consumer Report: Food irradiation—The Consumer's View*, Survey Research Group, The Association for Consumer Research, London.

Anon. (1992) *Coordinated Research Programme on Analytical Detection Methods for Irradiation Treatment of Foods (ADMIT)*, Second Research Co-ordination Meeting, Hungary, IAEA, Vienna, p. 15.

Bordi, F., Fattibene, P., Onori, S. and Pantaloni, M. (1993) An alternative procedure for ESR identification of irradiated chicken drumsticks. *Appl. Radiat. Isotopes*, **44**, 443–447.
Bruhn, C.M. and Noell, J.W. (1987) Consumer in-store response to irradiated papayas. *Food Technol.*, **41**, 83–85.
Davidson, I.G. (1988) ESR studies on gamma-irradiated foods. *Ph.D. Thesis*, University of Aberdeen, Scotland.
Desrosiers, M.F. (1989) Gamma-irradiated seafoods: identification and dosimetry by ESR spectroscopy. *J. Agric. Food Chem.*, **37**, 96–100.
Desrosiers, M.F., McLaughlin, W.L., Sheahen, L.A., Dodd, N.J.F., Lea, J.S., Evans, J.C., Rowlands, C.C., Raffi, J.J. and Agnel, J-P.L. (1990) Co-trial on ESR identification and estimates of gamma-ray and electron absorbed doses given to meat and bones. *Int. J. Food Sci. Technol.*, **25**, 682–691.
Desrosiers, M.F., Wilson, G.L., Hunter, C.R. and Hutton, D.R. (1991) Estimation of the absorbed dose in radiation-processed food—1: Test of the EPR function by a linear regression analysis. *Appl. Radiat. Isotopes*, **42**, 613–616.
Dodd, N.J.F., Swallow, A.J. and Ley, F.J. (1985) Use of ESR to identify irradiated food. *Radiat. Phys. Chem.*, **26**, 451–453.
Dodd, N.J.F., Lea, J.S. and Swallow, A.J. (1988) ESR detection of irradiated food. *Nature (London)*, **334**, 387.
Dodd, N.J.F., Haishun, J., Lea, J.S. and Swallow, A.J. (1992) Factors influencing the yield of free radicals in irradiated chicken. *Int. J. Food Sci. Technol.*, **27**, 371–383.
Gray, R. and Stevenson, M.H. (1989a) The effect of post-irradiation cooking on the ESR signal in irradiated chicken drumsticks. *Int. J. Food Sci. Technol.*, **24**, 447–450.
Gray, R. and Stevenson, M.H. (1989b) Detection of irradiated deboned turkey meat using ESR spectroscopy. *Radiat. Phys. Chem.*, **34**, 899–902.
Gray, R. and Stevenson, M.H. (1990) Effect of length of storage on the ESR signal from various bones in irradiated chicken carcasses. *Int. J. Food Sci. Technol.*, **25**, 506–511.
Gray, R. and Stevenson, M.H. (1991) Effect of dose rate and length of storage on the ESR signal strength in irradiated chicken bone. *Int. J. Food Sci. Technol.*, **26**, 669–672.
Gray, R., Stevenson, M.H. and Kilpatrick, D.J. (1990) The effect of irradiation dose and age of bird on the ESR signal in irradiated chicken drumsticks. *Radiat. Phys. Chem.*, **35**, 284–287.
Helle, N. and Linke, B. (1992) ESR for detecting gamma-irradiated foodstuffs. *Brüker Report* 91/92, pp. 8–9.
Ikeya, M. (1993) *New Applications of Electron Spin Resonance. Dating, Dosimetry and Microscopy*, World Scientific, Singapore.
Marcotte, M. (1992) Irradiated strawberries enter the U.S. market. *Food Technol.*, **46**, 80–86.
Ostrowski, K., Dziedzic-Goclawska, A. and Stachowicz, W. (1980) Stable radiation-induced paramagnetic entities in tissue mineral and their use in calcified tissue research, in *Free Radicals in Biology* (ed. W.A. Pryor), Academic Press, London, pp. 321–344.
Ostrowski, K., Dziedzic-Goclawska, A., Stachowicz, W. and Michalik, J. (1981) Crystallinity of tissue mineral as evaluated by electron spin resonance spectrometry. *Basic Appl. Histochem.*, **25**, 79–83.
Raffi, J., Stevenson, M.H., Kent, M., Thiery, J.M. and Belliardo, J.-J. (1992) European intercomparison on electron spin resonance identification of irradiated foodstuffs. *Int. J. Food Sci. Technol.*, **27**, 111–124.
Stevenson, M.H. and Gray, R. (1989a) An investigation into the effect of sample preparation methods on the resulting ESR signal from irradiated chicken bone. *J. Sci. Food Agric.*, **48**, 261–267.
Stevenson, M.H. and Gray, R. (1989b) The effect of irradiation dose, storage time and temperature on the ESR signal in irradiated chicken drum-sticks. *J. Sci. Food Agric.*, **48**, 269–274.
Stevenson, M.H. and Gray, R. (1990) Can ESR spectroscopy be used to detect irradiated food? in *Food Irradiation and the Chemist* (eds. D.E. Johnston and M.H. Stevenson), Royal Society of Chemistry, Cambridge, pp. 80–96.
Stevenson, M.H. and Stewart, E.M. (1995) Identification of irradiated food: the current status. *Radiat. Phys. Chem.*, in press.

Stewart, E.M. (1993) The use of ESR spectroscopy for the detection of irradiated *Crustacea* with particular reference to *Nephrops norvegicus* (Norway lobster), *Ph.D. Thesis*, The Queen's University of Belfast, N. Ireland.

Stewart, E.M., Stevenson, M.H. and Gray, R. (1991) Use of ESR spectroscopy for the detection of irradiated whiting (*Merlangius merlangus*). *J. Sci. Food Agric.*, **55**, 653–660.

Stewart, E.M., Stevenson, M.H. and Gray, R. (1992) Detection of irradiation in scampi tails—effects of sample preparation, irradiation dose and storage on ESR response in the cuticle. *Int. J. Food Sci. Technol.*, **27**, 125–132.

Stewart, E.M., Stevenson, M.H. and Gray, R. (1993) The effect of irradiation dose and storage time on the ESR signal in the cuticle of different components of the exoskeleton of Norway lobster (*Nephrops norvegicus*). *Appl. Radiat. Isotopes*, **4**, 433–437.

Stewart, E.M., Stevenson, M.H. and Gray, R. (1994) Use of ESR spectroscopy for the detection of irradiated *Crustacea*. *J. Sci. Food Agric.*, **65**, 191–197.

Chapter 5

Ultracentrifugation of Food Biopolymers

S.E. Harding
Department of Applied Biochemistry and Food Science,
University of Nottingham, UK

5.1 Introduction and historical comment

Ultracentrifugation – whether it be preparative or analytical – is hardly a new technique. It was in the early 1920s that T. Svedberg, on sabbatical leave from the University of Uppsala, working with J.B. Nichols at Madison, Wisconsin, produced the first ultracentrifuge, known as the optical centrifuge. Svedberg then designed and produced the oil-turbine ultracentrifuge (the basis of his Nobel lecture in 1926) capable of speeds of 40 000 rev./min (producing gravitational fields close to 100 000 that of gravity). At these speeds it was possible to sediment *biopolymers* and two classes of experiment rapidly developed: preparative ultracentrifugation for the separation and purification of biopolymers (including density gradient methods), and analytical ultracentrifugation, where, by replacing the plastic rotor tubes with precision cells with windows at top and bottom, and constructing an optical system around these, the distribution of solute in these cells could be assayed, and from this information on the sizes, shapes and interaction properties of biopolymers could be obtained. Details of all this earlier history can be found in Svedberg's early papers such as Svedberg and Nichols (1923), Svedberg and Rinde (1923), and Svedberg and Fåhraeus (1926) – describing the first use of sedimentation equilibrium analytical ultracentrifugation to measure the molecular weight of a protein – and his Nobel Prize lecture (Svedberg, 1927).

Insofar as food biopolymers are concerned, Svedberg himself published several papers on using both sedimentation velocity and sedimentation equilibrium. This includes studies on egg albumin (Sjörgen and Svedberg, 1930; Svedberg and Erikkson, 1934), and gelatin solutions (Krishnamurti and Svedberg, 1930). Although the bulk of Svedberg's polysaccharide work was on cellulose and its derivatives (Jullander, 1985), work was published on polyuronides (i.e. polysaccharides with carboxyl groups) such as pectins, alginates and gum arabic. In these papers the necessity of working in salt solutions to suppress electric charge effects (polyelectrolyte behaviour) was clearly demonstrated. In another important contribution to our understanding of the behaviour of food biopolymers, Siholta and Svedberg (1948) gave a new ultracentrifuge method for determining the solvation (an important functional property) of food macromolecules. It would be, of course, wrong to assume that it was only Svedberg and his co-workers who provided all the early input into our knowledge of the solution properties of food macromolecules, although it is true to say that, without his contribution, the field now would be still rather primitive.

The great boom period of the ultracentrifuge was 1950–1970, with almost every Biochemistry Department and most Chemistry Departments in the UK possessing *and running* the famous Beckman Model E analytical ultracentrifuge or its MSE (Crawley, UK) equivalent. Although the vast bulk of the work during this period was on macromolecules (in particular proteins) other than from food, this period was not devoid of food activity. It was during this period, for example, that Johnson and co-workers, using sedimentation velocity, developed our understanding of the structure of gelatin gels (Johnson, 1964; Johnson and Metcalfe, 1967) and that the self-assembly properties of milk proteins became largely understood (McKenzie, 1970).

So why is a chapter on the ultracentrifuge present in a volume on *new* physico-chemical techniques? The reason is that there has been, over the last five years, a rebirth of interest in analytical ultracentrifugation, coupled with significant advances in instrumentation and analysis, all of which can, and indeed is, being applied to food polymer research. Preparative ultracentrifugation has always remained an important separation tool, but interest in *analytical* ultracentrifugation, after the boom period 1950–1970, had dramatically declined so that, up to 1987, only a handful of workers were active in this field worldwide. Protein biochemists had largely switched to gel filtration and gel electrophoresis for their purity and molecular weight work (together with sequence analysis and mass spectrometry for the latter) and to NMR and X-ray crystallography for their conformation work. Polysaccharide chemists and the synthetic polymer community also switched to gel filtration, and also to light scattering methods for absolute molecular weight and size determinations. Towards the end of the 1980s, however, scientists began to realize the limitations of these other procedures; and this coincided with a growing demand for physico-chemical types of measurement, not only from food scientists, but also from molecular biologists desperate for relatively inexpensive ways (in terms of quantities of material required) of characterizing their gene products.

The result of this rebirth has been twofold: the emergence of national centres (in the UK and USA) for this type of measurement, and the launch of the first commercially available analytical ultracentrifuge (with full on-line data capture and analysis facilities) for almost two decades.

The aim of this chapter is to show just why modern analytical ultracentrifugation is worthy of consideration by food scientists and technologists interested in characterizing and understanding the behaviour of the macromolecular components of food products. The main focus of attention will be on food proteins and food polysaccharides (the latter proving quite a challenge for most physical methods because of their heterogeneity and high non-ideality in solution), in both dilute and concentrated solution form, and how the modern analytical ultracentrifuge can provide information on food biopolymers about in terms of:

(i) the size (in terms of either the molecular weight or dimensions);
(ii) the size distribution (for a polydisperse solution of macromolecules);
(iii) the solution conformation (in terms of the overall or gross conformation, and flexibility);
(iv) the purity of a macromolecular preparation (in terms of a sedimentation velocity diagram or density spectrum);
(v) the interactions between macromolecules (including self-association behaviour, complex formation);
(vi) the behaviour of gels; and
(vii) interfacial diffusion phenomena, and diffusion through gels and matrices.

The four analytical ultracentrifuge techniques used to provide this information are: sedimentation velocity analysis; sedimentation equilibrium analysis; density gradient equilibrium analysis; and diffusion analysis. It should be borne in mind, however, at the outset that ultracentrifugation is at its most powerful when used with complementary techniques such as (for conformation analysis) X-ray scattering, viscometry, rotational diffusional techniques and electron microscopy, or (for protein subunit composition work) gel electrophoresis, or (for molecular weight distribution studies) gel filtration and light scattering.

5.2 Types of analytical ultracentrifuge measurement

What sort of information can we get from sedimentation analysis in the analytical ultracentrifuge? It depends on the type of ultracentrifuge technique we apply – all are possible with the same instrumentation. *Sedimentation velocity* can provide us with information on the sample heterogeneity, shape information – in some cases to surprising detail – and also interaction information by, for example, assaying for what we call co-sedimentation phenomena (i.e. species sedimenting at the same rate). At lower rotor speeds, *sedimentation equilibrium* can provide absolute size and size distribution information (in terms of molecular weight averages and molecular weight distributions). There are two other important types of analytical ultracentrifuge measurement – namely isopycnic (= constant density) *density gradient analysis*, important for assaying the composition (and hence purity) of a macromolecular preparation, and finally *diffusion analysis*. Although dynamic light scattering is now the method of choice for the measurement of translational diffusion coefficients, the optical system on the analytical ultracentrifuge is proving very useful for investigating the diffusion of molecules through matrices, as well as towards and through interfaces in two-phase systems.

5.3 Instrumentation/optical systems

An analytical ultracentrifuge is simply a high-speed ultracentrifuge (up to 60 000 rev./min) with an appropriate optical system for observing and recording solute distributions during and at the end of the sedimentation process. There are three types of analytical ultracentrifuge now in common use. Firstly there are increasing numbers of the new Beckman Optima XLA (Schachman, 1989; Giebeler, 1992) analytical ultracentrifuge (Fig. 5.1(a)) being installed, the initial model with just scanning absorption optics, and also a more recent version with scanning Rayleigh interference optics. Although the Beckman Model E ultracentrifuge (with refractive index optics and absorption optics, although the latter is very noisy) has not been commercially available now for almost 20 years, there are still many in use (many adapted and modified, Fig. 5.1(b)) – a testimony to the reliability of this instrument. In East European countries, analytical ultracentrifuges made by the MOM optical works at Budapest (with refractive index and absorption optics) are also still in use (Görnitz and Linow, 1992).

Concentration distributions of a macromolecular solution (whether the macromolecule be a food protein, food polysaccharide or whatever) during ultracentrifugation can be recorded using refractive index optics. There are two types: Interference and Schlieren.

Interference: This gives a record of solute concentration as a function of radial distance, and is popular for sedimentation equilibrium measurements. Extraction of molecular weight parameters is quite laborious, but it gives the most accurate results for weight-average molecular weight data. Although in the past the fringe patterns were unwieldy to analyse, computers have now made this a realistic possibility, particularly for the analysis of distributions of sedimentation coefficient and molecular weight (Stafford, 1992).

Schlieren: This is another name for 'refractive index gradient' optics. It gives a direct record of the concentration gradient as a function of radial position, and is ideal for sedimentation velocity (where a sedimenting boundary appears as a peak) or isopycnic density gradient equilibrium (where a band appears as a positive peak followed by a negative peak). It can also be applied to solute distributions at sedimentation equilibrium, where it gives the *z*-average molecular weight directly.

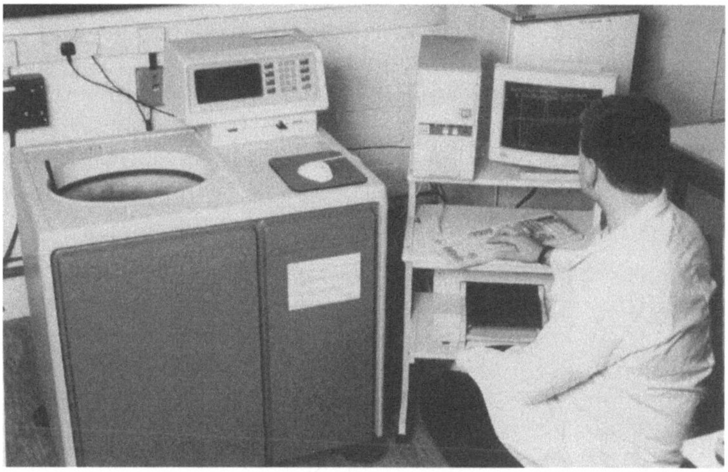

(a)

(b)

Fig. 5.1 (a) Beckman Optima XLA analytical ultracentrifuge. (b) Beckman Model E analytical ultracentrifuge (with modifications).

Interference optics requires a minimum solute concentration of ~0.2 mg/ml, whereas Schlieren optics normally requires at least ~ 0.8 mg/ml, although with modern data capture procedures lower concentrations are possible (Rowe *et al.*, 1992). Both types of refractive index optics can be applied to food polysaccharides, food proteins or any food macromolecules, since all have a refractive index considerably different from that of water or any aqueous solvent.

There are also two other types of optical system, both depending on whether or not a suitable chromophore is present on the macromolecule (either naturally occurring or synthetically attached).

Absorption optics: Insofar as food macromolecules are concerned, until recently this technique was limited to food proteins using UV absorption at 278 nm, or for coloured proteins like haemoglobin and myoglobin, absorption in the visible region. With the new Beckman Optima XLA ultracentrifuge and its stable light source down to 220 nm, some polysaccharides can also be detected. For proteins, concentrations down to ~0.2 mg/ml can be detected and fluorescein labelled macromolecules (λ_{max} ~ 480 nm) can be detected to lower concentrations. Details can be found in Lewis (1992).

Fluorescence optics: This is a variant of the above, in which a fluorescing wavelength of fluorescein or another suitable chromophore is used. It needs a specially adapted instrument (Schmidt and Riesner, 1992) although work is currently underway in producing a commercially available variant of the XLA ultracentrifuge with this facility (T. Laue, private communication)

A preparative ultracentrifuge can also be used as an analytical tool for sedimentation equilibrium measurements, if the macromolecule is labelled with an appropriate chromophore (absorption, fluorescent or radio-label) and the final equilibrium distribution is pooled and assayed immediately at the end of the ultracentrifuge run (Howlett, 1992), although this is only a substitute for proper analytical ultracentrifugation.

5.4 Sedimentation velocity: shape analysis and homogeneity

A typical analytical ultracentrifuge cell contains two sector-shape channels (one for the solution, one for the reference solvent) which can take between 0.1 and 0.8 ml. With an *analytical* ultracentrifuge, using the appropriate optical system one can record the position and rate of movement of the sedimenting boundary within the solution channel. (For a simple description of the method and what it can do, see Harding (1994a).) Figure 5.2 compares patterns for a monodisperse food protein (β-lactoglobulin) recorded using absorption optics at 278 nm, a food polysaccharide (a heat degraded alginate) recorded using Schlieren op-

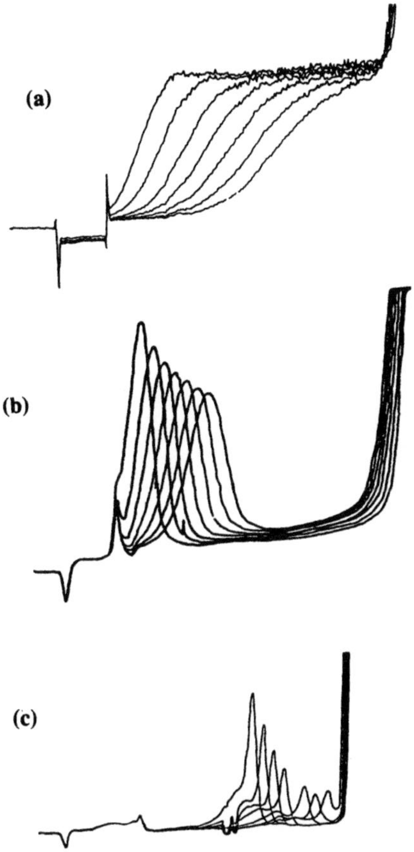

Fig. 5.2 Optical records of sedimentation velocity of food biopolymers in the analytical ultracentrifuge. The direction of sedimentation in each case is from left to right. (a) Sedimenting boundary for a monodisperse β-lactoglobulin B preparation recorded using scanning absorption optics in Optima XLA ultracentrifuge. Monochromator wavelength = 279 nm; scan interval = 20 min; rotor speed = 50 000 rev./min; temperature = 25.0°C; loading concentration, $c^o \sim 1.0$ mg/ml; $s_{25} = (2.97 \pm .03)$ S (from Harding and Advani, 1995). (b) Sedimenting boundary for a food polysaccharide (heat-treated sodium alginate) recorded using scanning Schlieren optics on an MSE Centriscan ultracentrifuge. Monochromator wavelength = 546 nm; scan interval = 30 min; rotor speed = 49 000 rev./min; temperature = 20.0°C, $c^o \sim 5.0$ mg/ml; $s_{20} = (1.22 \pm 0.05)$ S. (c) Sedimenting boundaries for *Chromobacter viscosum* lipase using scanning Schlieren optics on the MSE Centriscan. Monochromator wavelength = 546 nm; scan interval = 10 min; rotor speed = 49 000 rev./min; temperature = 20.0°C; two clear sedimenting boundaries are $s_{20} = (3.1 \pm 0.1)$ S (faster peak) and $s_{20} \sim 1.0$ S (slower peak) (from Simpkin *et al.*, 1991).

tics, and another food protein (lipase), containing some contaminating material, again recorded using Schlieren optics. The rate of movement of a sedimenting boundary from these patterns per unit centrifugal field gives the sedimentation coefficient s. By adjusting this (using simple formulae) to standard conditions (water as solvent at 20°C), and extrapolating it

(a)

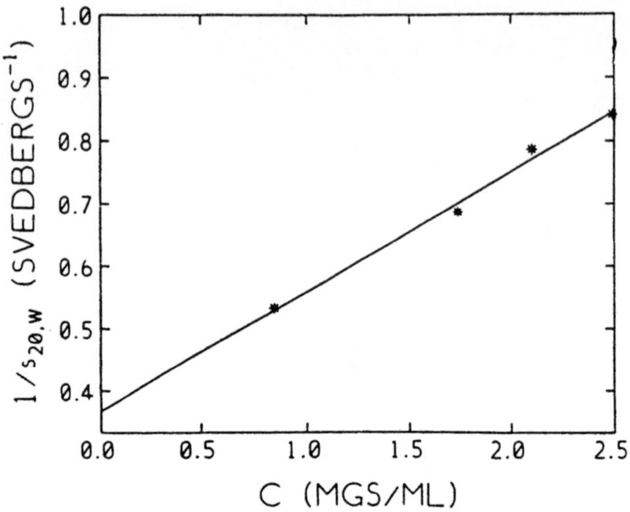

(b)

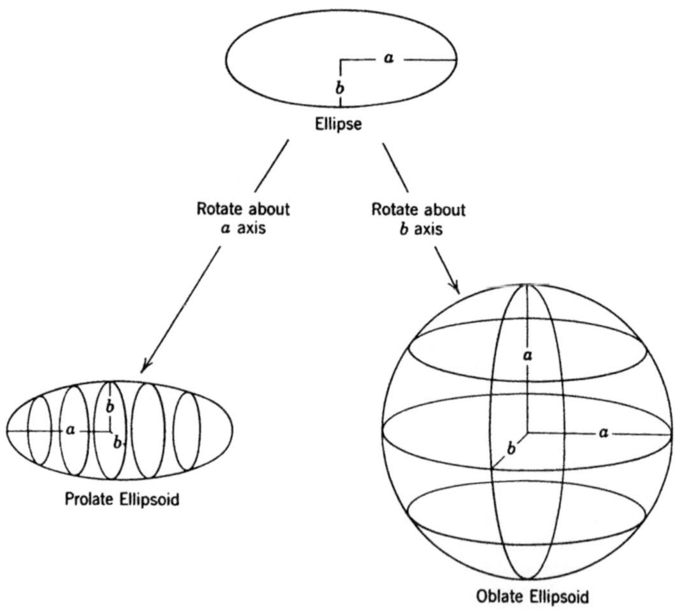

Fig. 5.3 (a) Plot of the reciprocal of the sedimentation coefficient $s_{20,w}$ versus concentration c for sodium carboxymethyl cellulose in phosphate-chloride buffer (pH 6.8, $I = 0.10$). $s^{\circ}_{20,w} = (2.7 \pm 0.2)$ S; $k_s = (521 \pm 34)$ ml/g (from Anderson, 1991). N.B. For food polysaccharides the reciprocal plot usually gives the best linear fit; for food proteins, a plot of $s_{20,w}$ against c is usually better (Rowe, 1977). (b) Ellipsoids of revolution models (from Tanford, 1961). The shape is given by the axial ratio a/b in both cases.

(or the reciprocal thereof) to zero concentration, we can get from the intercept the corrected sedimentation coefficient, $s^o_{20,w}$ (unit = seconds or Svedbergs, S, where 1 S = 10^{-13} s), and from the slope we can get the sedimentation concentration regression coefficient, k_s (unit = ml/g). Then from both parameters, shape information can be extracted (Fig. 5.3a). Table 5.1 gives some s and k_s values for a range of food biopolymers.

Besides providing shape information, for each scan or sedimentation diagram, the form and number of the sedimenting boundaries can provide a rapid indication of the heterogeneity of a sample, with impurities, etc., showing up as separate boundaries. Figure 5.2(c) shows such a case for a preparation of lipase (the pattern scanned seven times at 10-min intervals) containing a slower moving impurity.

For shape analysis there are two general stategies. One strategy, which is applied to *relatively* rigid structures (for example, many food protein systems), is to use ellipsoid or bead models to represent the overall structure of these molecules in *solution*. The sort of information that can reasonably be sought with this approach is the overall or gross conformation of the food protein in solution: its axial dimensions, how its

Table 5.1 Sedimentation velocity parameters for some food macromolecules

Macromolecule	M^a	$s^o_{20,w}$ (S)	k_s (ml/g)	Reference
Ovalbumin	45000	3.42[b]	6.2	Miller and Golder (1952); Holt (1970); Harding (1981b)
Bovine serum albumin	67000	4.31	5.4	Shulman (1953); Baldwin (1957)
β-Lactoglobulin (B) (Dimer)	36000	2.83	4.6	Cecil and Ogston (1949); Advani and Harding (1995)
Myosin	490000	5.6	94	Emes and Rowe (1978)
Collagen α-1 chain		3.16	56.9	Nishihara and Doty (1958)
Lipase (*Chromobacter viscosun*)	35000	3.17	18.7	Simpkin *et al.* (1991); Simpkin (1994)
Lipase (*Pseudomonas* spp.)	38000	2.99	7.0	Simpkin *et al.* (1991); Simpkin (1994)
Lipase (*Aspergillus* spp.)		3.01	9.5	Simpkin *et al.* (1991); Simpkin (1994).
Alginate	130000	2.4	449	Anderson (1991)
Carboxymethylcellulose		2.7	522	Anderson (1991)
Xylan	177000	2.4	270	Elizabeth *et al.* (1995)
Xylinan	275000	9.4	273	Berth *et al.* (1995)
κ-Carrageenan	284000	4.2	591	Day *et al.* (1995)
Pectin (citrus)	90000	1.9	128	Harding *et al.* (1991b)
Pectin (lime)	130000	2.3	196	Harding *et al.* (1995)

$s^o_{20,w}$ values for proteins normally correct to ±1% or better; for polysaccharides to ±5%. k_s values normally correct to ±2% for proteins, ±20% for polysaccharides. Some k_s values have been corrected for 'radial dilution' and to 'solution density' (Rowe, 1977). Most determinations done in (aqueous) solvents of ionic strength 0.1 M for proteins, 0.1–0.3 M for polysaccharides, and proteins and most polysaccharides buffered to pH between 6 and 7.

[a]Weight-average values for polydisperse polysaccharide samples.

[b]After correction from 25°C to 20°C.

subunits are arranged etc. The second strategy is for those molecules like food polysaccharides where it is unreasonable to assume, at least *a priori*, a rigid conformation; here it is necessary to perform more general modelling. The approach here is firstly to identify the conformation type (random coil, rigid rod or compact sphere, or something in between these extremes), and once this is established then to make further deductions about the conformation. For example, if the conformation is that between a random coil and rigid rod, we ask what is the flexibility of the molecule in terms of parameters like the persistence length a and the contour length L? If the conformation is between that of rigid rod and compact sphere, then, and only then, can ellipsoid or bead modelling be applied. It has to be borne in mind that, whatever strategy is actually followed, sedimentation data are best used in conjunction with other hydrodynamic data.

5.4.1 Ellipsoid and bead modelling

This can be done via calculation of the so-called frictional ratio, the ratio of the friction coefficient, f, of the macromolecule to the frictional coefficient of a spherical particle, f_o, with the same mass and anhydrous volume. The ratio f/f_o is related to the sedimentation coefficient by

$$(f/f_o) = \{M(1 - \bar{v}\rho)]/[N_A \times (6\pi\eta_o s^o_{20,w})]\}[(4\pi N_A/3\bar{v}M)]^{1/3}, \quad (5.1)$$

where M is the molar mass (g/mol), $\bar{v}$ the partial specific volume (ml/g) (typically ~0.73 ml/g for proteins, 0.45–0.65 ml/g for polysaccharides), ρ is the density (g/ml) of the solution (which can be approximated as the density of the solvent (Harding and Johnson, 1985)), η_o (poise) is the solvent viscosity (in this case, after the correction referred to above, the density of water at 20.0°C) and N_A is Avogadro's number (mol^{-1}). The quantity f/f_o, in turn, is a function of the hydration of the macromolecule, w (mass H_2O bound/mass of macromolecule) and the conformation via a particle shape factor known as the 'Perrin shape factor' P (also known as the frictional ratio due to shape), in recognition of Perrin (1936) who developed the theory for the frictional coefficients of ellipsoids:

$$P = (f/f_o)[(w/\bar{v}\rho) + 1]^{-1/3}. \quad (5.2)$$

The factor $[(w/\bar{v}\rho) + 1]$ is sometimes referred to as the swelling of the macromolecule due to hydration, i.e. $v_s/\bar{v}$ with v_s the swollen specific volume (ml/g). For a food protein the value of $v_s/\bar{v}$ is typically ~1.4. For a food polysaccharide it can be as high as ~100.

If the macromolecule is fairly rigid, P can be related directly to the axial dimensions of the macromolecule using ellipsoid of revolution models. Ellipsoids of revolution are so-called because they are the

Table 5.2 Axial dimensions of some proteins from X-ray crystallography

Protein	Dimensions (Å)	Reference
Myoglobin	43 × 35 × 23	Kendrew *et al.* (1958)
Haemoglobin	64 × 55 × 50	Perutz *et al.* (1960)
Lysosyme	45 × 30 × 30	Blake *et al.* (1965)
Cytochrome *c*	35 × 25 × 25	Dickerson and Geiss (1969)
Ribonuclease	38 × 28 × 22	Kartha *et al.* (1967)
Carboxypeptidase	50 × 42 × 38	Lipscomb *et al.* (1971)
Pre-albumin	70 × 55 × 50	Blake *et al.* (1978)

structures formed by rotating an ellipse, of semi-axes a and b ($a > b$), about either its major or a-axis (to give a three-dimensional ellipsoid of semi-axes a, b and b, known as a prolate ellipsoid) or rotated about its minor axis b (to give a three-dimensional ellipsoid of semi axes a, a and b (Fig. 5.3(b)). Table 5.2 gives the dimensions of some protein molecules from their X-ray crystal structures, and it can be seen that some of these can be reasonably approximated by two equal axes (especially two equal minor axes, namely the prolate case). The axial ratio (a/b) characterizing such ellipsoids is related to P by

$$P = \frac{(1 - b^2/a^2)^{1/2}}{(b/a)^{2/3} \ln \dfrac{1 + (1 - b^2/a^2)^{1/2}}{b/a}} \tag{5.3a}$$

(prolate ellipsoid), or for an oblate ellipsoid from

$$P = \frac{(a^2/b^2 - 1)^{1/2}}{(a/b)^{2/3} \tan^{-1}(a^2/b^2 - 1)^{1/2}}. \tag{5.3b}$$

Table 5.3 gives a/b as a function of P in the range $1 \leqslant (a/b) \leqslant 100$. A simple analytical inversion of these formulae giving a/b directly in terms of P is not possible. However, simple polynomial expansions are possible (Laue *et al.*, 1992) giving a/b values correct to better than ~0.1%:

$$(a/b) = 1 + 2.346X^{1/2} + 8.297X + 8.4X^2 - 0.4589X^3 + 0.0314X^4 \tag{5.4a}$$

for the prolate case, and

$$(a/b) = 1 + 2.975X^{1/2} + 8.48X + 13.295X^2 - 3.219X^2 + 0.0996X^4 \tag{5.4b}$$

for the oblate case, and for both cases $X = P - 1$.

A problem with this treatment is that f/f_0 (and hence P and the sedimentation coefficient) is a relatively insensitive function of shape. A more serious problem is that an assumption has to be made concerning the hydration w (or equivalently $v_s/\bar{v}$). A value of 0.35 has in the past been popularly taken for proteins (Tanford, 1961) although a survey of 21 proteins by Squire and Himmel (1979) has suggested w values in the range (0.53 ± 0.26) (see also Pessen and Kumosinski (1985) and Laue

Table 5.3 Values of the ellipsoid of revolution shape parameters *P*, *R*, Π as a function of axial ratio for prolate and oblate ellipsoids of revolution

Axial ratio (*a*/*b*)	*P* (prolate)	*R* (prolate)	Π (prolate)	*P* (oblate)	*R* (oblate)	Π (oblate)
1	1.000	1.600	3.200	1.000	1.600	3.200
2	1.044	1.471	3.122	1.042	1.494	3.180
3	1.113	1.291	2.960	1.105	1.369	3.179
4	1.183	1.138	2.778	1.166	1.274	3.192
5	1.250	1.017	2.601	1.224	1.203	3.208
6	1.314	0.921	2.438	1.277	1.148	3.225
7	1.375	0.844	2.291	1.327	1.105	3.241
8	1.434	0.782	2.159	1.373	1.071	3.255
9	1.490	0.730	2.041	1.417	1.043	3.268
10	1.543	0.686	1.935	1.458	1.019	3.280
20	1.996	0.465	1.288	1.783	0.901	3.352
30	2.359	0.379	0.980	2.020	0.856	3.384
40	2.671	0.332	0.799	2.212	0.833	3.403
50	2.950	0.302	0.679	2.375	0.819	3.415
60	3.205	0.280	0.593	2.519	0.810	3.423
70	3.442	0.264	0.527	2.648	0.803	3.429
80	3.664	0.251	0.476	2.765	0.798	3.434
90	3.874	0.240	0.435	2.873	0.793	3.437
100	4.074	0.231	0.400	2.974	0.790	3.441

To measure *P*, the sedimentation coefficient and an assumed value for the hydration of the molecule are required. To measure *R*, the sedimentation concentration dependence parameter and the intrinsic viscosity [η] are required. To measure Π, the second virial coefficient *B* (from sedimentation equilibrium measurements) and [η] are required.

et al. (1992) for further discussions on this). For polysaccharides, hydration levels are usually much higher and it is almost impossible to give a reasonable estimate for them (although it should be borne in mind that most polysaccharides cannot be represented by these rigid models in any case).

An alternative approach is to use the alternative shape function known as the Wales–van Holde ratio $k_s/[\eta]$ (Wales and van Holde, 1954), which is often referred to by the symbol *R* (Harding and Rowe, 1982). This combination of the concentration dependence parameter k_s with the intrinsic viscosity [η] gives, after certain assumptions and approximations, a shape function which is not dependent on any knowledge of *w* or v_s, and is a more sensitive function of axial ratio *a*/*b* (Table 5.3). The full functional dependence of *R* on *a*/*b* can be found in Rowe (1977). Table 5.4 gives estimates for the shapes of some food proteins determined in this way. Another shape function not requiring knowledge of the hydration level is the Π function (Harding, 1981a). This comes from measurements of the thermodynamic second virial coefficient *B*, [η], and the molar mass *M*:

$$\Pi = \{2BM/[\eta]\} - \{f(Z)/[\eta]M\}. \qquad (5.5)$$

Here $f(Z)$ is a function of the net charge *Z* on the macromolecule. This function is zero at the isoelectric point of proteins and tends to zero at

Table 5.4 Axial ratios of food proteins from k_S and intrinsic viscosity $[\eta]$ measurements

Protein	k_S (ml/g)	$[\eta]$ (ml/g)	R $(= k_S/[\eta])$	axial ratio $(a/b)^a$	Reference
Ovalbumin	5.45	3.49	1.56	1.5	Miller and Golder (1952); Holt (1970); Harding (1981b)
Bovine serum albumin	5.4	3.9	1.38	2.3	Baldwin (1957); Tanford and Buzzell (1956)
β-Lactoglobulin (B) (dimer)	4.6	2.86	1.61	1.0	Advani and Harding (1995); Townend (1960)
Collagen (seven fractions sonically degraded to varying extents)	250–142	1075–245	0.23–0.58	100–15	Nishihara and Doty (1958); see also Creeth and Knight (1965)

[a]Value for the equivalent prolate ellipsoid. The k_s values have been corrected for radial dilution and to solution density (Rowe, 1977).

Table 5.5 Axial ratios of food proteins from the Π function

Protein	Π	Axial ratio $(a/b)^a$	Reference
Haemoglobin	3.20	1.0	Harding (1981a)
Ovalbumin	3.18	1.0–2.0	Harding (1981b)
Myosin[b]	0.47	80	Harding (1987)

[a]Value of the equivalent prolate ellipsoid.
[b]An ellipsoid is a poor approximation here because of the flexibility of the molecule; only a very approximate estimate of a/b is possible.

high ionic strengths. If not negligible, it can be measured by, for example, titration (see Jeffrey *et al.* (1977) who have done this for ovalbumin). The second thermodynamic virial coefficient B (A_2 in some texts) can be measured by sedimentation equilibrium (see below) or from osmotic pressure or light scattering measurements. Table 5.3 also includes Π as a function of a/b and Table 5.5 gives an example of its application to three well-known food proteins.

Another combined function involving s and $[\eta]$ (referred to as the Scheraga and Mandelkern (1953) β-function), the first published hydration independent shape parameter, is also independent of w or v_s but is extremely insensitive to a/b, and it is really only of use as a quasi-constant parameter for evaluating the molar mass from $s^{o}_{20,w}$ and $[\eta]$ measurements.

In many cases a crude ellipsoid of revolution, with two axes necessarily equal, gives a poor approximation to the true overall (or gross) conformation of proteins in solution. Furthermore, a decision has to be made *a priori* as to which ellipsoid of revolution model to apply: oblate or prolate? (Usually it is the latter which gives the better representation.) For cases like these the much more general tri-axial ellipsoid, with three unequal semi-axes ($a \geqslant b \geqslant c$) and hence two characteristic axial ratios (a/b, b/c), is much more appropriate. Although this model is considerably more complicated to apply (it involves the use of *two* hydration independent shape functions and computer-graphical intersection procedures), the necessary theory and computational procedures have been developed. By using a combination of Π with a radius of gyration function (from light scattering), the overall rod conformation and axial ratio of myosin is successfully predicted $(a/b,\ b/c) = (80,\ 1)$ *without assuming* a priori *a prolate ellipsoid* (Harding, 1987).

Even with the extra degree of freedom the general tri-axial ellipsoid gives, there are many structures which cannot be reasonably represented by a symmetric ellipsoidal structure. The classic example from the non-food protein world is the antibody molecule. The basic idea is that the frictional ratio f/f_o (and hence P and $s^o_{20,w}$) can be calculated for a given array of spherical beads which do not have to be equal in size. This is done for various possible models for a given macromolecule, and the one which gives the predicted f/f_o (or P or $s_{20,w}$) in closest agreement with the experimentally measured value is chosen as the best model. Because of uniqueness problems (i.e. the possibility of other equally complex but quite different models giving similar agreement), it is necessary to rely on (i) other hydrodynamic or scattering data (such as $[\eta]$, rotational diffusion coefficients or the radius of gyration) and (ii) a close starting estimate (from electron microscopy, X-ray crystallography, etc.) to the true conformation. This method has had extensive application to several protein systems (Garcia de la Torre, 1989), especially immunological ones like immunoglobulins (Byron, 1992) and complement (Perkins, 1989), and has successfully been applied to model the segmental flexibility of myosin (Iniesta *et al.*, 1988; Garcia de la Torre, 1989). Surprisingly, so far as the author is aware, it has not been applied to the study of seed or milk proteins, where it would appear to have potential application in terms of modelling the arrays of subunits, although similar approaches using light or low-angle X-ray scattering have been applied (Plietz *et al.*, 1983).

5.4.2 General modelling

For polydisperse and not-so-rigid macromolecules, like many polysaccharides, we have to use shape analysis by sedimentation velocity in a general way, first of all to indicate what is the conformation *type* of the macromolecule (random coil, rigid rod or compact sphere, or whatever). For this purpose relations, known as the Mark–Houwink–Kuhn–Sakurada (MHKS) relations and their corresponding (exponential) coefficients have proved vital. One of these is the b coefficient which comes from the relation between $s^o_{20,w}$ and the molecular weight M:

$$s^o_{20,w} \propto M^b. \tag{5.6}$$

(Similar coefficients exist for the relationships of the intrinsic viscosity, the diffusion coefficient, and radius of gyration with M (Yamakawa, 1971; Smidsrød and Andresen, 1979). The MHKS-b coefficient is usually

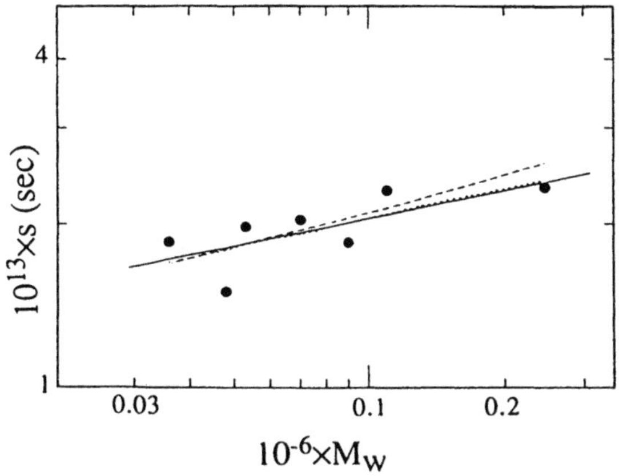

Fig. 5.4 Double-logarithmic plot of $s^{o}_{20,w}$ versus weight-average molar mass M^{o}_{w} (from low-speed sedimentation equilibrium) for citrus pectin fractions. Solid line corresponds to a fit of the data to the MHKS equation. Dashed line corresponds to the fit to equations representing a worm-like coil with persistence length 60 nm and mass per unit length 420 g mol^{-1} nm^{-1}, values which are only consistent with a rigid rod model (from Harding *et al.*, 1991b).

obtained by preparing a homologous series (i.e. the same polymer but different molecular weights) of the polymer (by chromatographic separation, ultrasonication or heat degradation) and then taking the slope of a double logarithmic plot of log $s^{o}_{20,w}$ *versus* log M (Fig. 5.4). The characteristic values of b for spheres, random coils and rods are respectively 0.667, 0.4–0.5 and 0.15 (Smidsrød and Andresen, 1979). The Wales–van Holde parameter $k_s/[\eta]$ ($\equiv R$) is also useful in this context, having a values of ~1.6 for spheres and random coils, and <<1 for rods. Some examples of applications of both these parameters to food polysaccharides are given in Table 5.6.

The MHKS-b coefficient can also be used to model the flexibility of a polymer in terms of wormlike-coil models (Hearst and Stockmayer, 1962; Yamakawa, 1971) from the ratio of the contour length L to the persistence length a (see Harding *et al.*, 1991a).

Table 5.6 Polysaccharide shapes from sedimentation velocity

Polysaccharides[a]	$k_s/[\eta]$	b	Conformation
Dextran fractions		0.44	Random coil
DIT-dextrans		0.56	Semi-flexible coil
Pullulans	1.4	0.45	Random coil
β-Glucans	0.4		Extended
Amylopectin	1.45		Spheroidal
Alginates	0.6		Extended
Citrus pectins	0.2	0.17	Rigid rod

From Harding (1992) and references cited therein.
[a]All in aqueous solvents apart from amylopectin (90% DMSO, 10% H_2O).

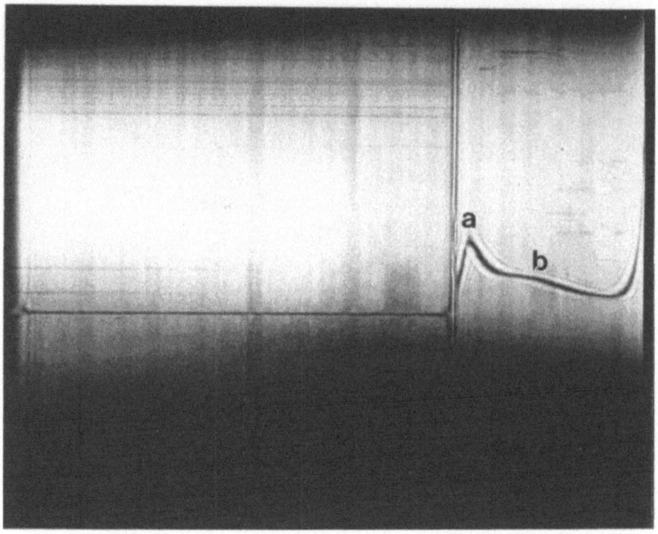

Fig. 5.5 Sedimentation velocity diagram for a solution containing 1 mg/ml bovine serum albumin and 1 mg/ml of an alginate (Pronova, Protan inc., Drammen, Norway) in a phosphate-chloride buffer (pH 6.8, $I = 0.10$ M). Rotor speed = 34 790 rev./min; temperature = 25.0°C. Peak (a) is unbound alginate ($s_{20} = 1.5$ S); peak (b) is BSA–alginate complex sedimenting at ~ 10.6 S. The BSA control sediments at ~ 3.9 S (from Harding *et al.*, 1993).

Sedimentation velocity can also be used to assay for interaction in a mixed solute system, by using the principle of co-sedimentation. Figure 5.5 shows such an example, showing interaction between bovine serum albumin and alginate. If the species in a mixture have chromophores absorbing in different regions of the UV or visible spectrum, the absorption optical system can be used to optimal effect (Marsh and Harding, 1993).

5.5 Sedimentation equilibrium: molecular weight distribution analysis

Whereas in a sedimentation velocity experiment at relatively high rotor speeds – for a globular protein or linear polysaccharide, say, 40 000–50 000 rev./min – the sedimentation rate and hence the sedimentation coefficient is a measure of the size and shape of the molecule, at much lower speeds (say 10 000 rev./min or less) in a sedimentation *equilibrium* experiment the forces of sedimentation and diffusion on the macromolecule become comparable. Now, instead of getting a sedimenting boundary, one gets, after a period of time (from a few hours to a few days depending on the nature of the solute) a steady state equilibrium distribution of macromolecules with a low concentration at the meniscus building up to a high concentration at the cell base. This final steady state pattern is a function *only* of molecular weight and related parameters (virial coefficients and association constants where appropriate) and not on molecular shape, since, at equilibrium, there is no further movement of the macromolecules; thus frictional effects through shape variation are not involved. (For a simple description of this method and what it can do, see Harding (1994b).) So, like 'static' (as opposed to 'dynamic') light scattering, sedimentation equilibrium is an absolute way of obtaining the molecular weight or 'molar mass' M (g/mol) of a food biopolymer, and Table 5.7 gives examples for a range of polysaccharides and seed proteins. Although a sedimentation equilibrium experiment is longer to perform compared to a light scattering measurement (it can take up to 3 days), it is usually less prone to problems (such as clarification, etc.) With the ability to run samples multiply (up to 9 at a time in the Beckman Optima XLA), it may become the main method of choice once again.

Table 5.7 Molecular weights of food polysaccharides and some 12S seed proteins (weight averages from sedimentation equilibrium analysis)

Biopolymer	M_w	Reference
Pullulan P100	95000	Kawahawa *et al.* (1984)
Pullulan P800	76000	Kawahawa *et al.* (1984)
Pullulan P1200	124000	Kawahawa *et al.* (1984)
Dextran T500	500000	Edmond *et al.* (1968)
Guar	800000	Gaisford *et al.* 1986
Locust bean gum (cold water soluble fraction)	340000	Gaisford *et al.* (1986)
Locust bean gum (hot water soluble fraction)	330000	Gaisford *et al.* (1986)
Glucomannan	280000	Winwood (1986)
β-Glucan (barley)	160000	Woodward *et al.* 1983
β-Glucan (wheat)	230000	Harding *et al.* (1991a)
Chitosan (Protan Sea Cure)	162000	Errington *et al.* (1993)
Chitosan (Trondheim KN50-1)	64100	Errington *et al.* (1993)
Chitosan (Trondheim KN50-4)	4300	Errington *et al.* (1993)
Xanthan (Kelco RD)	5900000	Dhami *et al.* (1995)
Alginate (Kelco 'Pro-nova')	210000	Horton *et al.* (1991)
Pectin (Koch-Light, citrus)	90000	Harding *et al.* (1991b)
Pectin (green tomato)	160000	Seymour and Harding (1987)
Pectin (red tomato)	96000	Seymour and Harding (1987)
Pectin (red tomato, antisense polygalacturonase)	135000	Smith *et al.* (1990)
Carmin (from safflower seed)	260000	Prakash (1992) and refs. therein
α-Globulin (from sesame seed)	270000	Prakash (1992) and refs. therein
Arachin (from peanut)	330000[a]	Prakash (1992) and refs. therein
Brassin (from rapeseed)	300000	Harding *et al.* (1987)
Helianthin (from sunflower seed)	300000[a]	Prakash (1992) and refs. therein

These values are normally precise to ± 5–15%.
[a]From combining the sedimentation coefficient with the diffusion coefficient via the Svedberg equation (Harding, 1994a).

(a)

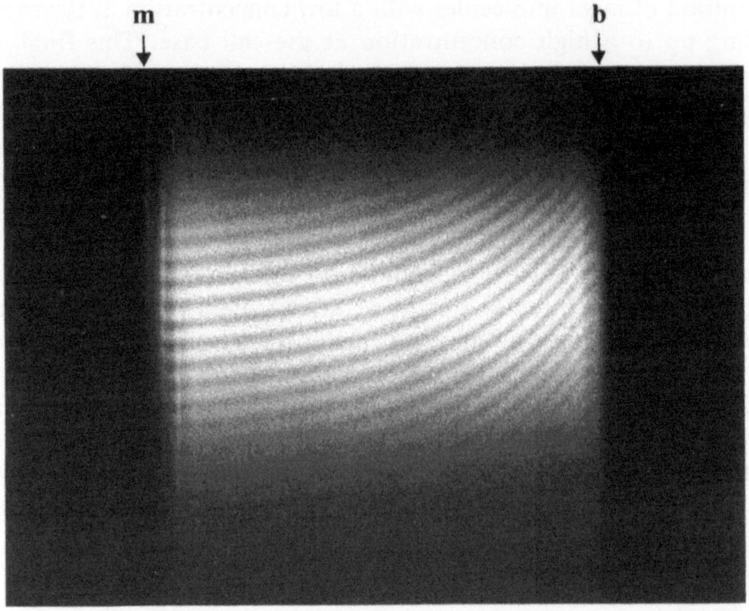
m
b

(b)

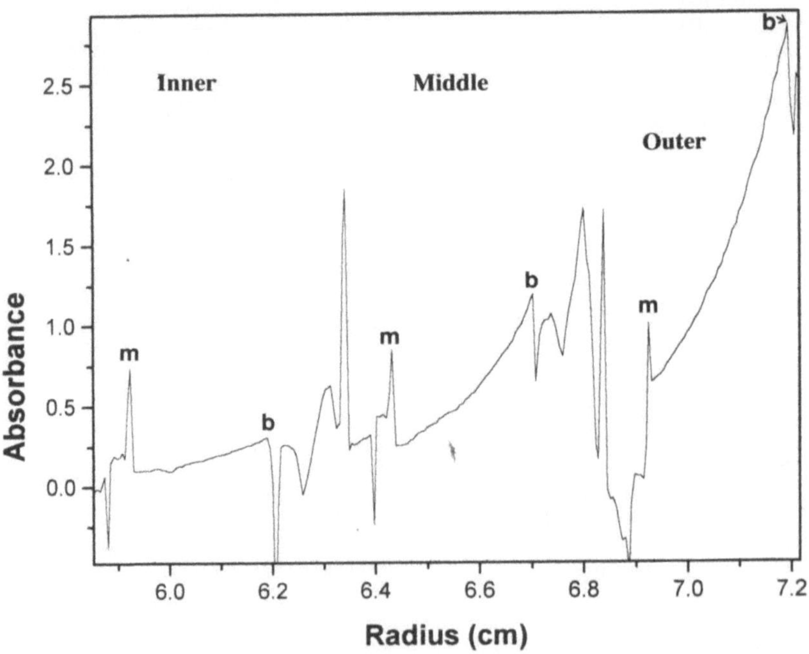
Inner
Middle
Outer
m
b
m
b
m
b
Absorbance
Radius (cm)
0.0
0.5
1.0
1.5
2.0
2.5
6.0
6.2
6.4
6.6
6.8
7.0
7.2

(c)

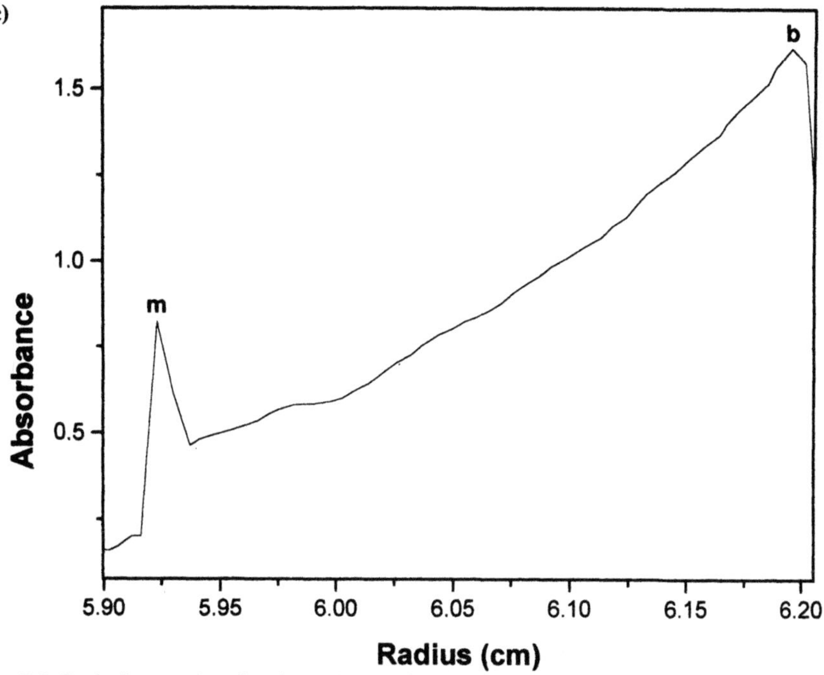

Fig. 5.6 Optical records of solute distributions of food biopolymers at sedimentation equilibrium. The direction of the centrifugal field is from left to right. (a) Rayleigh interference profiles for ovalbumin (M = (45 000 ± 2000) g/mol). Phosphate-chloride buffer, pH 6.8, $I = 0.10$ M). Each fringe profile represents a plot of concentration of the protein (relative to the solution/air meniscus 'm') against distance from the centre of the ultracentrifuge rotor; 'b' represents the base of the cell. Rotor speed = 17 250 rev./min; temperature = 20.0°C. Loading concentration $c^{o} \sim 0.5$ mg/ml. (b) Absorption optics profiles (recorded at a wavelength λ of 280 nm) for β-lactoglobulin B. A special multichannel cell was used permitting three samples: inner ($c^{o} = 0.1$ mg/ml); middle ($c^{o} = 0.2$ mg/ml) outer: ($c^{o} = 0.3$ mg/ml). The same buffer as (a) was used. Rotor speed = 15 000 rev./min. Because of restrictions of the Lambert–Beer law, with the outer channel only absorbances < 1.5 could be used. (c) Rescan of the inner channel of (b) but using a wavelength of 230 nm to enhance the signal; this picks up the peptide bond in the protein which usually gives a higher absorbance than the aromatic amino acid absorbances at 280 nm.

The most accurate way of recording these final steady state concentration distributions is using Rayleigh interference optics (Fig 5.6a). More appropriately for proteins, we can take advantage of absorption of the aromatic amino acids or the peptide, bond to use the much more convenient absorption optics set for the ultraviolet (Fig. 5.6(b,c)): if absorption optics are used patterns can be routinely captured automatically on-line and analysed (Giebler, 1992; Lewis, 1992). With the Rayleigh interference system, it is much more complicated, and indeed expensive, to perform on-line data capture (Laue, 1992). Whatever the system, these patterns are usually now captured and read automatically, either directly into a computer, or via photography or chart recorder output first (Rowe *et al.*, 1992).

How do we get molecular weight (or molar mass) information? The interfaced computer converts the digitized information of Fig. 5.6 into an accurate record of log concentration (expressed in terms of absorbance units, A, or fringe displacement units, J) versus radial distance squared. For fairly ideal monodisperse systems (e.g. dilute solutions of some small proteins) such plots are linear, but for non-ideal or heterogeneous systems they are not (Fig. 5.7). The average slope of these plots, linear or otherwise, gives what we call the whole distribution weight average molar mass, or whole cell weight average molar mass, $M^{o}_{w,\,app}$. (We use a procedure known as 'MSTAR' for this purpose (Harding *et al.*, 1992).) These are apparent values, hence the subscript 'app'. This is because of the thermodynamic non-ideality of the system arising from excluded volume effects (related to the effective volume, after hydration, of the macromolecular species) and from charge effects if the macromolecule is highly charged. These effects are usually represented by the virial term B (A_2 in some texts) in the relation

$$(1/M^{o}_{w,\,app}) = (1/M^{o}_{w})(1 + 2BM^{o}_{w}c). \tag{5.7}$$

So the quantity $(1 + 2BM^{o}_{w}c)$ is the factor by which, at a given concentration c, the measured molar mass, $M^{o}_{w,\,app}$, differs from the true or

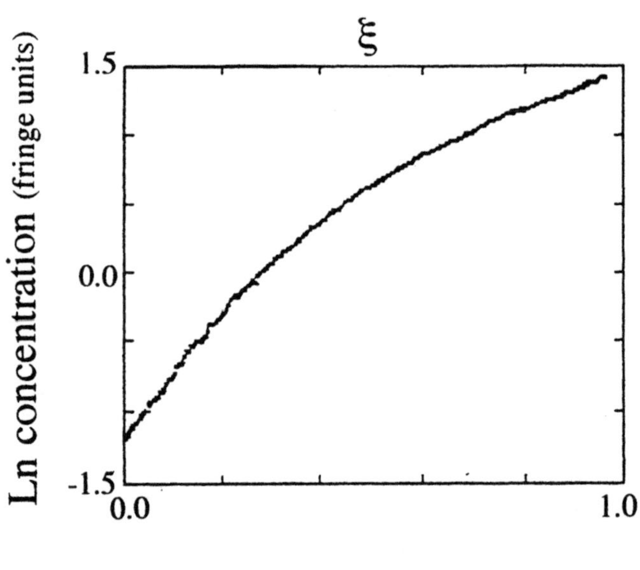

Fig. 5.7 Plot of log concentration (expressed in Rayleigh fringe units J) versus the normalized radial displacement squared parameter ξ for κ-carrageenan, where $\xi = (r^2 - a^2)/(b^2 - a^2)$, r is the radial displacement at a given point in the distribution, and a and b are the corresponding positions of the meniscus and cell base, respectively. Rotor speed = 6419 rev./min; temperature = 21.0°C (pH 6.5, I = 0.10 M). The downward curvature (negative $d^2 \ln J/d\xi^2$) is caused by the large degree of non-ideality of this macromolecule at the loading concentration employed (~ 1.4 mg/ml). $M^{o}_{w,\,app}$ = 77 000 g/mol at this concentration compared with the ideal M^{o}_{w} of 284 000 g/mol. ($B \sim 3.4 \times 10^{-3}$ ml mol g^{-2}) (from Day *et al.*, 1995).

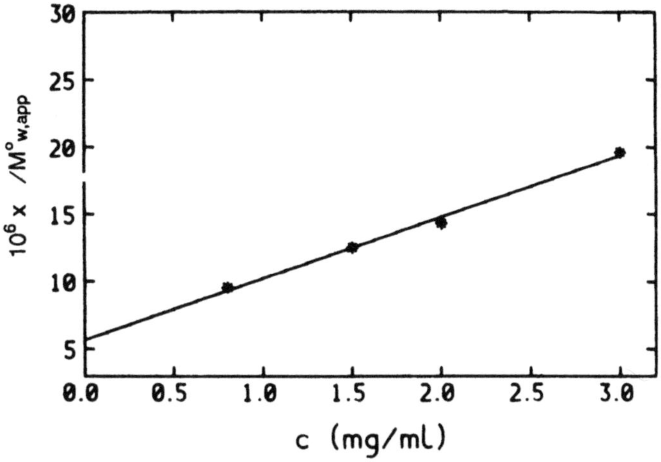

Fig. 5.8 Plot of $1/M^{\circ}_{w,app}$, versus loading concentration c for the fibre polysaccharide xylan. $M^{\circ}_{w,} = 177\,000$ g/mol; $B = 2.3 \times 10^{-3}$ ml mol g^{-2} (from Elizabeth *et al.*, 1995).

'ideal' molar mass, M°_{w}. The minimum concentration required for a sedimentation equilibrium run is in the range ~0.1–0.3 mg/ml, if 'long' (i.e. 20–30 mm) path length cells are used. At these concentrations, for food proteins and many food polysaccharides, this factor is relatively small (5% or less) and so it is reasonable to approximate M°_{w} as $M^{\circ}_{w,app}$ (Harding *et al.*, 1991a). For other food polysaccharides – especially the polyanions such as alginates and xanthan – where $(1 + 2BM^{\circ}_{w}c)$ is > 1.05, or where working at low concentrations is not possible or may lead to loss of structural integrity of the molecule (possible dissociation of a multi-subunit protein), a range of loading concentrations must be used and an extrapolation to zero concentration is necessary (Fig. 5.8). But a word of warning is required: although the fitting of reciprocal $M^{\circ}_{w,app}$ as a function of concentration gives reliable B values, depending on the form of the extrapolation an extrapolation of $1/M^{\circ}_{w,app}$ can give poor estimates of M°_{w} and often it is safer to extrapolate $M^{\circ}_{w,app}$ values directly (and manually). This is specially true for highly concentration dependent systems where other virial terms (C, D or A_3, A_4, etc.) may be necessary (Horton *et al.*, 1991).

Local slopes along the curves of Fig. 5.7 can also be taken to give point or 'local' average molecular weights ($M_{w,app}$) at a given radial position in the ultracentrifuge cell, and these can be particularly useful for assaying interacting systems, especially when given as a function of the local concentration in the cell. One of the classic examples of this, which has been extensively studied (see Visser *et al.*, 1972), is the effect of temperature on the association/dissociation equilibrium of β-lactoglobulin

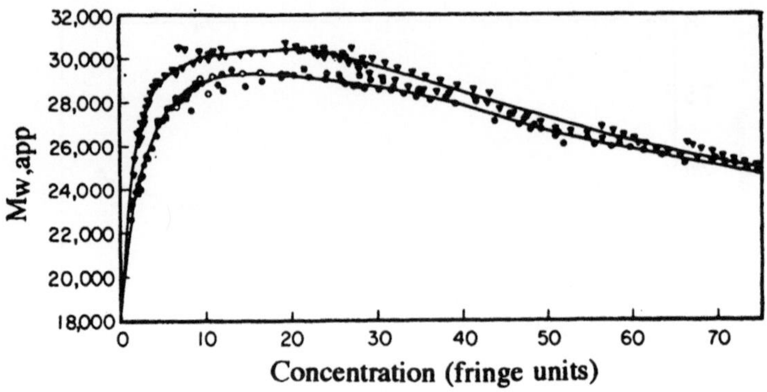

Fig. 5.9 Association–dissociation plot of apparent weight average molecular weight as a function of concentration (expressed in fringe numbers) for β-lactoglobulin B at low pH (pH 2.64, $I = 0.16$ M). Upper curve, temperature = 15°C, lower curve, temperature = 25°C (from Visser *et al.*, 1972). The several kinds of points which provide the information for the two interpolation curves (open and closed circles and triangles) represent different cell loading concentrations. The fact that for a given temperature they overlap is diagnostic of genuine self-association behaviour (Roark and Yphantis, 1969). The fits give values for the association (dimerization) constant 'k_2' of (3900 ± 1010) ml/g at 15°C and (2310 ± 50) ml/g at 25°C.

B at low pH (Fig. 5.9). Further, if $M^{\circ}_{\mathrm{w,app}}$ is given as a function of the square of the radial displacement, then the value of $M^{\circ}_{\mathrm{z,app}}$ can be estimated from.

$$M^{\circ}_{\mathrm{z,app}} = \{M_{\mathrm{w,app}}(b)\cdot c(b) - M_{\mathrm{w,app}}(a)\cdot c(a)\}/\{c(b) - c(a)\}, \qquad (5.8)$$

where a is the radial position of the cell meniscus and b the cell base. (In some special cases the number average molecular weight can also be obtained from Rayleigh optical records.) $M_{\mathrm{z,app}}$ values can also be obtained directly by using the Schlieren optical system mentioned earlier in the context of sedimentation velocity (see Rowe *et al.*, 1992). The ratio of the z-average to the weight-average molecular weight is often used as an 'index of polydispersity' of polymeric samples, as it can be related to the breadth of the distribution. For example, for a preparation of the polysaccharide xylan (which provides fibre in some food products), a value of $M^{\circ}_{\mathrm{z,app}}/M^{\circ}_{\mathrm{w,app}}$ of ~1.2 has been obtained (Elizabeth *et al.*, 1995) using this procedure.

There are three other ways of using sedimentation equilibrium data to get molecular weight distributions (Harding, 1994c). Perhaps the simplest in principle and most useful in practice is to combine it with gel filtration, giving the latter an 'absolute' basis (i.e. not subject to assumptions concerning the conformation of calibration standards). Gel filtration is a very simple but powerful way of separating macromolecules of different sizes in a polydisperse solution. It can also give molecular

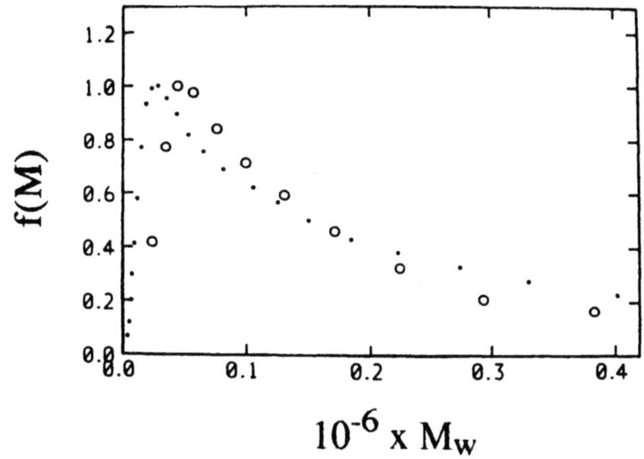

Fig. 5.10 Distribution of molecular weight $f(M)$ for a citrus pectin determined by gel permeation chromatography with columns calibrated by absolute molecular weight measurements on pectin fractions by low speed sedimentation equilibrium (open circles). The results agree well with a similar calibration procedure using classical light scattering measurements (small filled circles). Reproduced, with permission, from Harding *et al.* (1991).

weight distributions by comparing elution profiles with those of standard molecules of known molecular weight and the same conformation as the molecules being studied. Although the latter requirement is usually satisfied if globular proteins are being studied and globular protein standards are used, it fails for synthetic polymer and polysaccharide systems where so-called 'polystyrene' or 'pullulan' standards are often used. These problems can be avoided by taking fractions from the column and separately determining their molecular weights by sedimentation equilibrium (Fig. 5.10).

5.6 Analytical density gradient sedimentation equilibrium

5.6.1 Composition analysis

Density gradient sedimentation equilibrium procedures using *preparative* ultracentrifuges are routinely used in biochemical science for purifying macromolecular systems on the basis of their density; the method is sometimes called *isopycnic* density gradient equilibrium. The analogous situation in an analytical ultracentrifuge permits us to assay for purity. (This was the classical method used by Meselson *et al.* (1957) to show that the replication of DNA was semi-conservative.) The idea is to have as a solvent a dense material which redistributes so that there is a distribution or 'gradient' of density throughout the sedimentation cell. Any dissolved macromolecule will move until at equilibrium it reaches its isodensity point, i.e., the position in the cell where its own density matches that of the solvent. In aqueous solvents caesium salts are often used to produce the gradient (in the range 1.2–1.7 g/ml). Proteins, polysaccharides and nucleic acids have their isodensity points at ~1.3 g/ml, ~1.6 g/ml and ~1.7 g/ml, respectively, and hence this method is very useful for checking the purity of food polysaccharide or food protein preparations (see Harding, 1992).

5.6.2 Gels

The technique of sedimentation equilibrium can also be applied to the study of the structure of gels, especially if the Schlieren optical system is used, and it can provide complementary information to conventional rheological methods. For example, because the network concentration will vary in the gel as a function of radial position, it is possible to monitor the swelling pressure and other thermodynamic properties of the gel as a function of concentration. These procedures are outside the scope of this introductory text; for further details the interested reader should consult Borchard (1991), Cölfen and Borchard (1991) and Holtus *et al.* (1991). Using the absorption optical system, and after selection of an appropriate wavelength (i.e. in which the gel matrix is invisible), it is possible in principle to follow the diffusion of small molecules – including small proteins – through the gel, as a probe into gel structure.

5.6.3 Diffusion analysis

Historically, the ultracentrifuge was used as a tool for measuring the diffusion coefficients of macromolecules in solution using a procedure known as 'boundary spreading'. That is, how fast a sedimenting boundary broadened out was a function of the diffusion coefficient; then by combining the diffusion coefficient with the sedimentation coefficient, the molecular weight could be obtained via a relation known as the 'Svedberg equation'. Nowadays, the more rapid technique of dynamic light scattering is the method of choice for diffusion measurements, but for certain systems the analytical ultracentrifuge has clear advantages. One of these is for the analysis of the diffusion of small molecules or macromolecules through concentrated media, including gels and two-phase systems. With the latter, it is possible to model membrane phenomena and the diffusion of small molecules through food gels/matrices (a low rotor speed is chosen simply to minimize convection effects). Applied in this way, the ultracentrifuge is not being used as a sedimentation tool as such, but rather the optical system alone is being

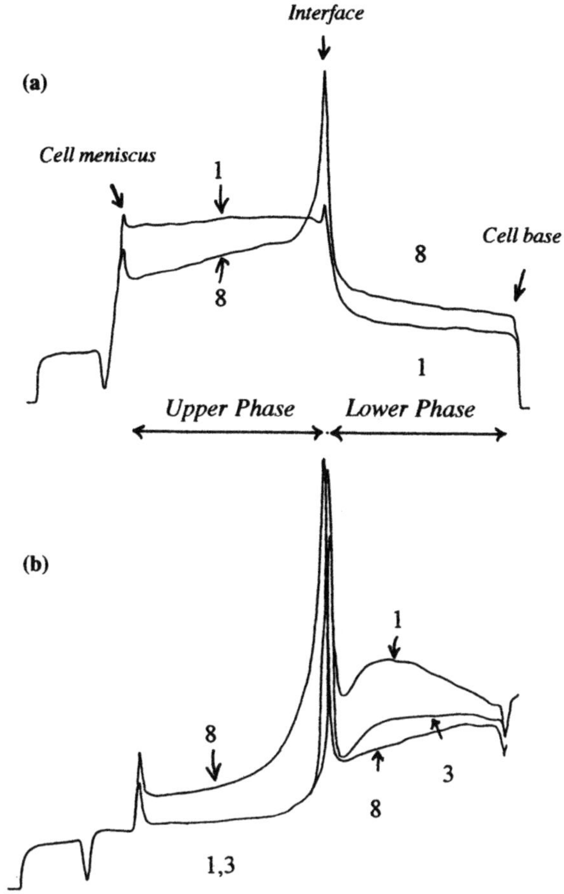

Fig. 5.11 Diffusion of lipase (*Chromobacter viscosum*) through an interface separating two aqueous phases, monitored using the absorption optics (at 280 nm) on an analytical ultracentrifuge. The phases are formed by mixing dextran and polyethylene glycol in the appropriate ratio. The numbers (1,3,8) correspond to various scan times. (a) Lipase initially in the top (i.e. less dense) phase in the ultracentrifuge cell. (b) Lipase initially in the 'bottom' phase. Note in both cases the massive accumulation of protein at the interface, and also (in (b) only) the irregularities near the interface on the lower phase side (from Simpkin, 1994; see also Harding and Tombs, 1989).

used (Fig. 5.11). Nonetheless relatively detailed information can be drawn concerning the diffusion through and towards interfaces between two phase systems (Tombs and Harding, 1988; Harding and Tombs, 1989). This has proved highly valuable in the study of lipases which are being increasingly considered for use in oil and fat processing (Macrae and Hammond, 1985), and their structure and interfacial behaviour has recently been studied in some depth by this technique (Simpkin, 1994).

References

Advani, M. and Harding, S.E. (1995) Unpublished results.

Anderson, M.T. (1991) *The Interaction of Mucous Glycoproteins with Polymeric Materials*, Ph.D. Dissertation, University of Nottingham, UK.

Baldwin, R.L. (1957) Boundary spreading in sedimentation velocity experiments. *Biochem. J.*, **65**, 503–512.

Berth, G., Jumel, K. and Harding, S.E. (1995) Unpublished results.

Blake, C.C.F., Loenig, D.F., Mair, G.A., North, A.C.T., Phillips, D.C. and Sarma, V.R. (1965) Structure of hen egg-white lysosyme. *Nature (London)*, **206**, 757–761.

Blake, C.C.F., Geisow, M.J. and Oatley, S.J. (1978) Structure of pre-albumin: secondary, tertiary and quaternary interactions determined by Fourier refinement at 1–8 Å. *J. Mol. Biol.*, **121**, 335–356.

Borchard, W. (1991) Swelling pressure equilibrium of swollen crosslinked systems in an external field. I: Theory. *Prog. Colloid Polym. Sci.*, **86**, 84–91.

Byron, O.D. (1992) *Solution Studies on the Conformation and Assembly of the Monoclonal Antibody B72.3*, Ph.D. Dissertation, University of Nottingham, UK.

Cecil, R. and Ogston, A.G. (1949) The sedimentation constant, diffusion constant and molecular weight of β-lactoglobulin. *Biochem. J.*, **44**, 33–35.

Cölfen, H. and Borchard, W. (1991) Swelling pressure equilibrium of swollen crosslinked systems in an external field. III: Unsolved problems concerning the systems gelatin/water and κ-carageenan/water. *Prog. Colloid Polym. Sci.*, **86**, 102–110.

Creeth, J.M. and Knight, C.G. (1965) On the estimation of the shape of macromolecules from sedimentation and viscosity measurements. *Biochim. Biophys. Acta*, **102**, 549–558.

Day, K.M., Dhami, R. and Harding, S.E. (1995) Unpublished results.

Dhami, R., Harding, S.E., Mitchell, J.R., Hunt, T. and Jones, T. (1995) Unpublished results.

Dickerson, R. and Geiss, I. (1969) *The Structure and Action of Proteins*, Benjamin, California.

Edmond, E., Farquhar, S., Dunstone, J.R. and Ogston, A.G. (1968) The osmotic behaviour of Sephadex and its effects on chromatography. *Biochem. J.*, **108**, 755–763.

Elizabeth, N.J., Dhami, R. and Harding, S.E. (1995) Unpublished results.

Emes, C.H. and Rowe, A.J. (1978) Hydrodynamic studies on the self association of vertebrate skeletal muscle. *Biochim. Biophys. Acta*, **537**, 110–124.

Errington, N., Harding, S.E., Vårum, K.M. and Illum, L. (1993) Hydrodynamic characterization of chitosans of differing degrees of acetylation. *Int. J. Biol. Macromol.*, **15**, 113–117.

Gaisford, S.E., Harding, S.E., Mitchell, J.R. and Bradley, T.D. (1986) A comparison between the hot and cold water soluble fractions of two locust bean gum samples. *Carbohydr. Polym.*, **6**, 423–442.

Garcia de la Torre, J. (1989) Hydrodynamic properties of macromolecular assemblies, in *Dynamic Properties of Biomolecular Assemblies*, Royal Society of Chemistry Special Publication no. 74, Cambridge, U.K. pp. 3–31.

Giebeler, R. (1992) The Optima XL-A: a new analytical ultracentrifuge with a novel precision absorption optical system, in *Analytical Ultracentrifugation in Biochemistry and Polymer Science*, (eds. S.E. Harding, A.J. Rowe and J.C. Horton) Royal Society of Chemistry, Cambridge, chap. 2.

Görnitz, E. and Linow, K.-J. (1992) Experience with the MOM 3180 analytical ultracentrifuge, in *Analytical Ultracentrifugation in Biochemistry and Polymer Science*, (eds. S.E. Harding, A.J. Rowe and J.C. Horton) Royal Society of Chemistry, Cambridge, chap. 3.

Harding, S.E. (1981a) A compound hydrodynamic shape function derived from viscosity and molecular covolume measurements. *Int. J. Biol. Macromol.*, **3**, 340–341.

Harding, S.E. (1981b) Could egg albumin be egg shaped? *Int. J. Biol. Macromol.*, **3**, 398–399.

Harding, S.E. (1987) A general method for modelling macromolecular shape in solution. A graphic (P–G) intersection procedure for triaxial ellipsoids. *Biophys. J.*, **51**, 673–680.

Harding, S.E. (1989) Modelling the gross conformation of assemblies using hydrodynamics: the whole body approach, in *Dynamic Properties of Biomolecular Assemblies*, Royal Society of Chemistry, Special Publication No. 74. Cambridge, pp. 32–56.

Harding, S.E. (1992) Sedimentation analysis of polysaccharides, in *Analytical Ultracentrifugation in Biochemistry and Polymer Science*, (eds. S.E. Harding, A.J. Rowe and J.C. Horton) Royal Society of Chemistry, Cambridge, chap. 27.

Harding S.E. (1994a) Determination of macromolecular homogeneity, shape and interactions using sedimentation velocity analytical ultracentrifugation, *Methods Mol. Biol.*, **22**, 61–73.

Harding, S.E. (1994b) Determination of absolute molecular weights using sedimentation equilibrium analytical ultracentrifugation. *Methods Mol. Biol.*, **22**, 75–84.

Harding S.E. (1994c) The analytical ultracentrifuge spins again. *Trends Anal. Chem.*, **99**, 439–446.

Harding S.E. and Johnson, P. (1985) The concentration dependence of macromolecular parameters. *Biochem. J.*, **231**, 543–547.

Harding S.E. and Rowe, A.J. (1982) Modelling biological macromolecules in solution. I. The ellipsoid of revolution, *Int. J. Biol. Macromol.*, **4**, 160–164.

Harding S.E. and Tombs, M.P. (1989) Protein diffusion through interfaces, *Interactions in Protein Systems* (eds. K.D. Schewnke and B. Raab), Abhandlungen der Akademie der Wissenschaften der DDR, Akademi-Verlag, Berlin, pp. 65–72.

Harding S.E., Kyles, P., West, G. and Norton, G. (1987) The relative molecular mass, heterogeneity and subunit composition of the 12S globulin from oil seed rape. *Biochem. Soc. Trans.*, **15**, 684.

Harding S.E., Vårum, K.-M., Stokke, B.-T. and Smidrsød, O. (1991a) Molecular weight determination of polysaccharides. *Adv. Carbohydr. Anal.*, **1**, 63–144.

Harding S.E., Berth, G., Ball, A., Mitchell, J.R. and Garcia de la Torre, J. (1991b) The molecular weight distribution and conformation of citrus pectins in solution studied by hydrodynamics. *Carbohydr. Polym.*, **16**, 1–15.

Harding S.E., Horton, J.C. and Morgan, P.J. (1992) MSTAR: A FORTRAN program for the model independent molecular weight analysis of macromolecules using low speed or high speed sedimentation equilibrium, in *Analytical Ultracentrifugation in Biochemistry and Polymer Science*, (eds. Harding S.E., A.J. Rowe and J.C. Horton), Royal Society of Chemistry, Cambridge, chap. 15.

Harding S.E., Jumel, K., Kelly, R., Gudo, E., Horton, J.C. and Mitchell, J.R. (1993) The structure and nature of protein-polysaccharide complexes, in *Food Proteins: Structure and Functionality*, (eds. K.-D. Schwenke and R. Mothes), VCH Verlagsgesellschaft, Weinheim, Germany, pp. 216–226.

Harding S.E., Horton, J.C. and Berth, G. (1995) *Ultracentrifuge Studies on Lime Pectin*. Manuscript in preparation.

Hearst, J.E. and Stockmayer, W.H. (1962) Sedimentation constants of broken chains and wormlike coils. *J. Chem. Phys.*, **37**, 1425–1433.

Holt, J.C. (1970) *A Study of the Reversibility of Denaturation of Ovalbumin*, Ph.D. Dissertation, University of London, UK.

Holtus, G., Cölfen, H. and Borchard, W. (1991) Swelling pressure equilibrium of swollen crosslinked systems in an external field. II: The determination of molecular parameters of gelatin/water gels from the swelling pressure-concentration curves. *Prog. Colloid Polym. Sci.*, **86**, 92–101.

Horton, J.C., Harding S.E., Mitchell, J.R. and Morton-Holmes, D.F. (1991) Thermodynamic non-ideality of dilute solutions of sodium alginate studied by sedimentation equilibrium ultracentrifugation. *Food Hydrocolloids*, **5**, 125–127.

Howlett, G.J. (1992) The preparative ultracentrifuge as an analytical tool, in *Analytical Ultracentrifugation in Biochemistry and Polymer Science*, (eds. S.E. Harding, A.J. Rowe and J.C. Horton), Royal Society of Chemistry, Cambridge, chap. 4.

Iniesta, A., Diaz, F.G. and Garcia de la Torre, J. (1988) Transport properties of rigid bent-rod macromolecules and of semi-flexible broken rods in the rigid body treatment. Analysis of the flexibility of myosin rod. *Biophys. J.*, **54**, 269–275.

Jeffrey, P.D., Nichol, L.W., Turner, D.R. and Winzor, D.J. (1977) The combination of molecular covolume and frictional coefficient to determine the shape and axial ratio of a rigid macromolecule. Studies on ovalbumin. *J. Phys. Chem.*, **81**, 776–781.

Johnson, P. (1964) A sedimentation study on gel systems. *Proc. R. Soc. London, Ser. A*, **278**, 527–542.

Johnson, P. and Metcalfe, J.C. (1967) Physico-chemical studies on gelatin gels from soluble and insoluble collagens. *Eur. Polym. J.*, **3**, 423–447.

Jullander, I. (1985) Svedberg and the Polysaccharides, in *Physical Chemistry of Colloids and Macromolecules (The Svedberg Symposium)*, (ed. B. Rånby), Blackwell Scientific, Oxford, UK, pp. 29–36.

Kartha, G., Bellow, J. and Harker, D. (1967) Tertiary structure of ribonuclease. *Nature, (London)*, **213**, 862–865.

Kawahawa, K., Ohta, K., Miyamoto, H. and Nakamura, S. (1984) Preparation and solution properties of pullulan fractions as standard samples for water-soluble polymers. *Carbohydr. Polym.*, **4**, 335–356.

Kendrew, J.C., Bodo, G., Dintzis, H.M., Parish, H.M., Wycoff, H. and Phillips, D.C. (1958) A three-dimensional model of the myoglobin molecule obtained by X-ray analysis. *Nature (London)*, **181**, 662–666.

Krishnamurti, K. and Svedberg, T. (1930) The ultracentrifuge study of gelatin solutions. *J. Am. Chem. Soc.*, **52**, 2897–2906.

Laue, T.M. (1992) On-line data aquisition and analysis from the Rayleigh interferometer, in *Analytical Ultracentrifugation in Biochemistry and Polymer Science*, (eds. S.E. Harding, A.J. Rowe and J.C. Horton), Royal Society of Chemistry, Cambridge, chap. 6.

Laue, T.M., Shah, B.D., Ridgeway, T.M. and Pelletier, S.L. (1992) Computer aided interpretation of analytical sedimentation data for proteins, in *Analytical Ultracentrifugation in Biochemistry and Polymer Science*, (eds. Harding S.E., A.J. Rowe and J.C. Horton), Royal Society of Chemistry, Cambridge, chap. 7.

Lewis, M.S. (1992) Data acquisition and analysis systems for the absorption optical system of the analytical ultracentrifuge, in *Analytical Ultracentrifugation in Biochemistry and Polymer Science*, (eds. Harding S.E., A.J. Rowe and J.C. Horton), Royal Society of Chemistry, Cambridge, chap. 8.

Macrae, A.R. and Hammond, R.C. (1985) Present and future applications of lipases. *Biotechnol. Gent. Eng. Rev.*, **3**, 193–217.

Marsh, E.N. and Harding, S.E. (1993) Methylmalonyl-CoA mutase from *Propionibacterium shermanii:* characterization of the cobalamin-inhibited form and subuniy-cofactor interactions using analytical ultracentrifugation. *Biochem. J.*, **290**, 551–555.

Meselson, M., Stahl, F.W. and Vinograd, J. (1957) Equilibrium sedimentation of macromolecules in density gradients, *Proc. Natl. Acad. Sci. USA*, **43**, 581–583.

McKenzie, H.A. (ed.) (1970) *Milk Proteins, Chemistry and Molecular Biology*, Academic Press, New York. Vol. I.

Miller, G.L. and Golder, R.H. (1952) Sedimentation studies with the Spinco ultracentrifuge. *Arch. Biochem. Biophys.*, **36**, 249–258.

Nishihara, T. and Doty, P. (1958) The sonic fragmentation of collagen macromolecules. *Proc. Natl. Acad. Sci. USA*, **44**, 411–417.

Pessen, H. and Kumosinski, T.F. (1985) Measurements of protein hydration by various techniques. *Methods Enzymol.*, **117**, 219–255.

Perkins, S.J. (1989) Hydrodynamic modelling of complement, in *Dynamic Properties of Biomolecular Assemblies*, Royal Society of Chemistry Special Publication no. 74, Cambridge, U.K. pp. 226–245.

Perrin, F. (1936) (Translated title) Brownian motion of an ellipsoid II. Free rotation and depolarization of fluorescence. Translation and diffusion of ellipsoidal molecules. *J. Phys. Radium*, **7**, 1–11.

Perutz, M.F., Rossmann, M.G., Cullis, A.F., Muirhead, H., Will, G. and North, A.C.T. (1960) Structure of haemoglobin-A. Three-dimensional Fourier synthesis at 5.5 Å resolution, obtained by X-ray analysis. *Nature (London)*, **185**, 416–422.

Plietz, P., Damaschun, G., Müller, J.J. and Schwenke, K.-D. (1983) The structure of the 11-S globulins from sunflower and rape seed. A small-angle X-ray scattering study. *Eur. J. Biochem.*, **130**, 315–320.

Prakash, V. (1992) Sedimentation analysis of seed proteins, in *Analytical Ultracentrifugation in Biochemistry and Polymer Science*, (eds. Harding S.E., A.J. Rowe and J.C. Horton), Royal Society of Chemistry, Cambridge, chap. 24.

Roark, D.E. and Yphantis, D.A. (1969) Studies of self-associating systems by equilibrium ultracentrifugation. *Ann. N.Y. Acad. Sci.*, **164**, 245–278.

Rowe, A.J. (1977) The concentration dependence of transport processes. A general description applicable to the sedimentation, translational diffusion and viscosity coefficients of macromolecular solutes. *Biopolymers*, **16**, 2595–2611.

Rowe, A.J., Wyne-Jones, S., Thomas, D.G. and Harding, S.E. (1992) Methods for off-line analysis of sedimentation velocity and sedimentation equilibrium patterns, in *Analytical Ultracentrifugation in Biochemistry and Polymer Science*, (eds. Harding S.E., A.J. Rowe and J.C. Horton), Royal Society of Chemistry, Cambridge, chap. 5.

Schachman, H.K. (1989) The analytical ultracentrifuge reborn. *Nature (London)*, **941**, 259–260.

Scheraga, H.A. and Mandlekern, L. (1953) Consideration of hydrodynamic properties of proteins. *J. Am. Chem. Soc.*, **75**, 179–184.

Schmidt, B. and Riesner, D. (1992) A fluorescence detection system for the analytical ultracentrifuge and its application to proteins nucleic acids, viroids and viruses, in *Analytical Ultracentrifugation in Biochemistry and Polymer Science*, (eds. S.E., Harding, A.J. Rowe and J.C. Horton), Royal Society of Chemistry, Cambridge, chap. 11.

Seymour, G.B. and Harding, S.E. (1987) Analysis of the molecular size of tomato (*Lycopersicon esculentum* Mill) fruit polyuronides by gel filtration and low-speed sedimentation equilibrium. *Biochem. J.*, **245**, 463–466.

Shulman, S. (1953) The determination of sedimentation constants with the oil-turbine and Spinco ultracentrifuges. *Arch. Biochem. Biophys*, **44**, 230–239.

Siholta, H. and Svedberg, T. (1948) A new method to determine the degree of solvation of macromolecules. *Acta Chem. Scand.*, **2**, 474–485.

Simpkin, N.J. (1991) *Solution Behaviour and Phase-Phase Diffusion of Microbial Lipases,* Ph.D. Dissertation, University of Nottingham, UK.

Simpkin, N.J., Harding, S.E. and Tombs, M.P. (1991) Solution behaviour of *Chromobacter viscosum* and *Pseudomonas* sp. lipases: no evidence of self-association. *Biochem. J.*, **273**, 611–613.

Smith, C.J., Watson, C.F., Morris, P.C., Bird, C.R., Seymour, G.B., Gray, J.E., Arnold, C., Tucker, G.A., Schuch, W., Harding, S.E. and Grierson, D. (1990) Inheritance and effect on ripening of antisense polygalacturonase genes in transgenic tomatoes. *Plant Mol. Biol.*, **14**, 369–379.

Squire, P.G. and Himmel, M. (1979) Hydrodynamics and protein hydration. *Arch. Biochem. Biophys.*, **196**, 165–177.

Sjögren, B. and Svedberg, T. (1930) The pH stability region of egg albumin. *J. Am. Chem. Soc.*, **52**, 5187–5192.

Smidsrød, O. and Andresen, I.L. (1979) *Biopolymerkjemi*, Tapir, Trondheim, Norway.

Stafford, W.F. (1992) Methods for obtaining sedimentation coefficient distributions, in *Analytical Ultracentrifugation in Biochemistry and Polymer Science*, (eds. S.E. Harding, A.J. Rowe and J.C. Horton), Royal Society of Chemistry, Cambridge, chap. 20.

Svedberg, T. (1927) *Prix Nobel 1926*, Stockholm; see also Farber, E. (ed.) (1953) *Nobel Prize Winners in Chemistry 1901–1961*, Abelard-Schuman, London, pp. 99–102.

Svedberg, T. and Erikkson. I.-B. (1934) Ultracentrifugal study of the action of papain on ovalbumin. *Biol. Bull.*, **66**, 191–223.

Svedberg, T. and Fåhraeus, R. (1926) A new method for determination of the molecular weight of the protein. *J. Am. Chem. Soc.*, **48**, 430–438.

Svedberg, T. and Nichols, J.B. (1923) Determination of size and distribution of size of particles by centrifugal methods. *J. Am. Chem. Soc.*, **45**, 2910–2917.

Svedberg, T. and Rinde, H. (1923) The determination of the distribution of size of particles in disperse systems. *J. Am. Chem. Soc.*, **45**, 943–954.

Tanford, C. (1961) *Physical Chemistry of Macromolecules*, Wiley, New York.

Tanford, C. and Buzzell, J. G. (1956) The viscosity of aqueous solutions of bovine serum albumin between pH 4.3 and 10.5. *J. Phys. Chem.*, **60**, 225–231.

Tombs, M.P. and Harding, S.E. (1988) Protein transport processes in the water–water interface in incompatible two-phase systems, in *Advances in Separations using Aqueous Phase Systems in Cell Biology and Biotechnology*, (eds. D. Fisher and I.A. Sutherland) Plenum, New York, pp. 229–232.

Townend, R., Weinberger, L. and Timasheff, S.N. (1960) Molecular interactions in β-lactoglobulin. IV. The dissociation of β-lactoglobulin below pH 3.5. *J. Am. Chem. Soc.*, **82**, 3175–3179.

Visser, J., Deonier, R.C., Adams, E.T. Jr. and Williams, J.W. (1972) Self-association of β-lactoglobulin B in acid solution and its variation with temperature. *Biochemistry*, **11**, 2634–2643.

Wales, M. and van Holde, K.E. (1954) The concentration dependence of the sedimentation constants of flexible macromolecules. *J. Polym. Sci.*, **14**, 81–86.

Winwood, R. (1986) *Functional Properties of a Novel Food Gelling System*, Ph.D. Dissertation, University of Nottingham, UK

Woodward, J.R., Phillips, D.R. and Fincher, G.B. (1983) Water-soluble (1–3),(1–4)-β-D-glucans from barley (*Hordeum vulgare*) endosperm. I. Physicochemical properties. *Carbohydr. Polym.*, **3**, 143–156.

Yamakawa, H. (1971) *Modern Theory of Polymer Solutions*, Harper and Row, New York.

Chapter 6

Rheology of Biopolymer Solutions and Gels

S.B. Ross-Murphy
Division of Life Sciences,
King's College London, UK

6.1 Introduction

One of the most familiar and characteristic properties of high polymers is their ability to alter quite markedly, and even in the case of gels to immobilize, the flow of many times their weight of solvent. This effective change in viscosity has been known for many years to be dependent on the size of the macromolecule in solution, and so on its molecular weight M_r (Morris, 1984; Ross-Murphy, 1984, 1994a; Launay *et al.*, 1986). Since gels have a nominally infinite M_r, the viscosity of a gel 'solution' is also infinite, and so such systems will show elasticity but not flow. In this chapter we consider the rheology of biopolymer solutions and gels. To do this, we first have to introduce the concept of biopolymer size and shape and from these introduce the idea of concentration regimes, and how they affect rheological properties.

6.2 Biopolymer size and shape

Few of the functional properties of food macromolecules are directly governed by their primary sequence structure. Much more significant are the space filling conformations adopted by the biopolymers. The main biopolymers of interest to the food industry are proteins and polysaccharides (Ross-Murphy, 1994a). For polysaccharides, the bulkiness of residue substituents offers more of a steric restriction to rotation than for protein (peptide) bonds. For the latter the normal entropic drive towards randomness of overall configurations is countered by enthalpic intermolecular or intramolecular interactions between non-adjacent residues, such as α-helices and parallel and anti-parallel β-pleated sheet structures.

The secondary structure described above leads quite naturally to the tertiary structure, i.e. the shape of the whole macromolecule. For example, many non-structural proteins have a compact globular shape, while fibrous proteins such as tropocollagen consist of a triple super-helix. Marine polysaccharides such as carrageenan and agarose involve quite extensive double helical structure. Groups of macromolecules can form large quaternary structures, from the octahedral arrangement of globular sub-units in soy proteins, and the granule structure of starches, through to the macroscopic forms, both in native animal and vegetable matter, and in 'denatured' food materials.

Most polymers behave as flexible coils in dilute solution, and their solution properties may be described by the random flight model. In this model, the coil (root-mean-square) size, the radius of gyration R_g, is proportional to the square root of their molecular weight (relative molecular mass) M_r. For spherical particles, an approximation to the shape of globular proteins, the volume of the sphere $\propto R_g^3$, and thus $R_g \propto M_r^{1/3}$, while for perfectly rigid (but infinitely thin) rods $R_g \propto M_r$. Typical sizes, in terms of radius of gyration, are 5 nm for globular proteins, 25 nm for flexible polysaccharides, and >100 nm for very stiff (worm-like or rod) proteins and polysaccharides.

6.3 Concentration regimes

We now introduce the effect of concentration (or volume fraction) on the material properties of biopolymer solutions and dispersions, in terms of a 'penetrable sphere' model representing the time average shape of the biopolymer (Morris, 1984; Ross-Murphy, 1984, 1994a; Launay *et al.*, 1986). The volume of this equivalent sphere is $(4/3)\pi R_g^3$. For a solution of C kg of particles per m^3 (= C grams per litre), there will be CN_A/M_r particles, where N_A is Avogadro's number. The total volume of particles is then $(4/3)\pi R_g^3 CN_A/M_r$. If this volume is greater than the total volume occupied by the number of particles (or macromolecules), we have complete occupancy of space – the volume fraction of particles is 1. What actually occurs as we approach this concentration (packing fraction) depends crucially upon the penetrability (or permeability) of our spherical particle. If this is essentially non-existent, as in a 'rigid' protein molecule, there will be a dramatic increase in viscosity as we approach some critical packing fraction some way below unit volume fraction. For spheres this critical fraction is around 60%. For rigid anisotropic particles it is lower still, since, as we have discussed before, each of these will sweep out a larger volume than the rigid sphere (Fig. 6.1(a)). However, for penetrable spheres such as these, the effect is different, since we can exceed 'total occupancy' by interpenetration, and the apparent packing fraction can exceed 100%.

It has become conventional to denote the total occupancy concentration by C^* (where C^* is proportional to M_r/R_g^3). (Strictly, however, there is a transition regime rather than a sharp discontinuity, and other

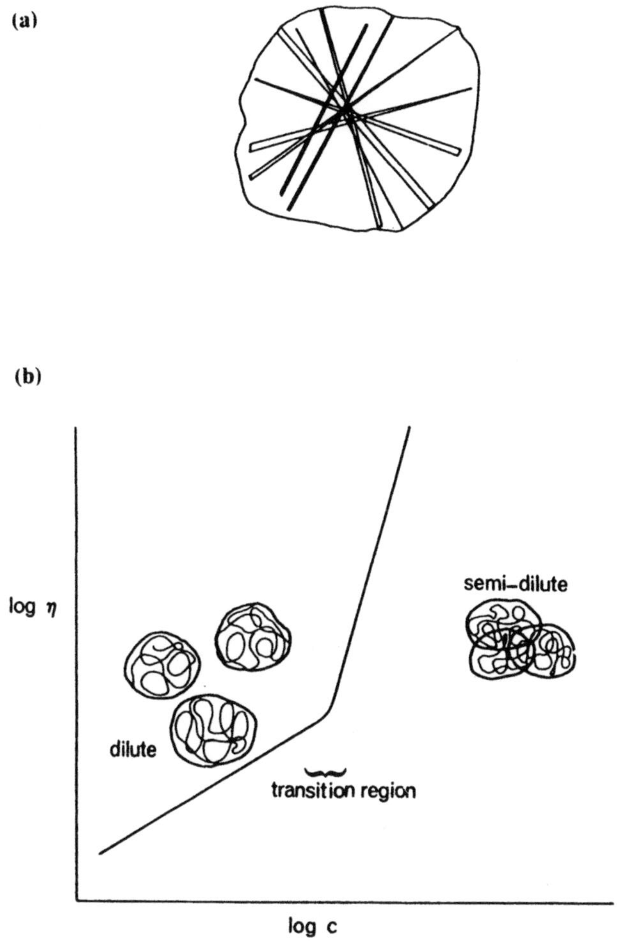

Fig. 6.1 (a) Time averaged sphere representing the hydrodynamic volume swept out by a rod macromolecule. (b) Increase in solution viscosity with concentration in passing from the dilute to the semi-dilute regime. Reproduced with permission from Ross-Murphy (1994).

'critical concentrations' (C^{**}, C_{cr}) are defined in the literature.) Concentrations below C^* are denoted 'dilute', and those well above C^* are termed 'semi-dilute' (Fig. 6.1(b)). Here 'dilute' has a specific meaning, in terms of the space occupancy, so that whether or not a solution is regarded as dilute does *not* depend on the concentration say in weight percent. In other words, a 10% solution of small particles could be dilute, and a 1% solution of larger particles could be semi-dilute. Higher concentrations still (say $>50\%$) can be called concentrated, but these scarcely if ever occur for biopolymer systems.

The important part of this discussion is that, for dilute solutions, we expect to see the behaviour of isolated macromolecules. For semi-dilute solutions, however, we have a more homogeneous view of a collection of polymer residues (segments), without knowing which segment belongs to which individual chain. It also follows that if a viscosity increment is required at low concentration, polymers that have extended coil or rod-like chain profiles will be much more effective than hard sphere or globular macromolecules.

6.3.1 Dilute solutions and intrinsic viscosity

In the dilute regime, the 'viscosity increment' over the solvent depends upon the size and shape of small numbers of macromolecules. For solutions only just more viscous than the solvent, we can estimate the macromolecular properties of an individual polymer. The first 'viscosity increment' rule was that of Einstein, who wrote

$$\eta = \eta_s(1 + 2.5\phi + \ldots), \qquad (6.1)$$

where ϕ is the 'volume fraction' of spherical particles, η the solution viscosity and η_s the solvent viscosity. Rearranging this equation, and if higher terms can be neglected (as $\phi \to 0$), we have

$$(\eta - \eta_s)/\eta_s = 2.5\phi + \ldots = [\eta]_{spheres}. \qquad (6.2)$$

If the disperse phase and fluid densities are close to unity, the volume fraction is equal to the concentration in units of g ml^{-1} (10^3 kg m^{-3}), and $[\eta]$ is known as the intrinsic viscosity. The quantity $[\eta]$ is mostly defined in units of reciprocal percentage concentrations (g dl^{-1}), so that from equation (6.2) the intrinsic viscosity of small spherical particles is 2.5×10^{-2} dl g^{-1}.

For spheres, according to the Einstein rule, $[\eta]$ is independent of size and therefore M_r. (Note that $[\eta]$ is not a 'viscosity' at all, as it has units of reciprocal concentration, rather than Pa s or Poise.) In practice this value is very much a lower bound, and when $[\eta]$ has been calculated for other molecular shapes, significantly larger values are obtained. For example, $[\eta]$ has been evaluated for solid prolate (rod shaped) ellipsoids as a function of axial ratio (L/d). When this ratio becomes very large, we have $[\eta] \propto (L/d)^{9/5}$. For a rod molecule it is reasonable to assume that $M_r \propto L$, so that the two extreme cases of a rod and a sphere give, respectively,

$$[\eta]_{rod} \propto M_r^{9/5} \quad \text{and} \quad [\eta]_{sphere} \propto M_r^0.$$

According to the Mark–Houwink equation, equation (6.3), the intrinsic viscosity depends on M_r in the following way:

$$[\eta] = K' M_r^{\alpha}, \qquad (6.3)$$

where $\alpha = 0$ for a sphere, 1.8 for a rod, and 0.5–0.8 for flexible polymers. Previously C^* was defined to be $M_r/((4/3)\pi R_g^3 N_A)$, but because $[\eta]$ is defined as the volume occupied by a gram of polymer molecules, we can see that $C^* \approx 1/[\eta]$. This is the definition we shall employ in the rest of this chapter.

6.3.2 Dilute solution viscometers

The most widely used viscometer for very dilute solutions is of the glass capillary type. This has two etched (upper and lower) marks, and the time for the solvent, t_s and the solution, t to flow between these, is noted. The relative viscosity for the solution is then the ratio t/t_s. Automatic timing and dilution viscometers can be obtained, and the whole glassware can be immersed in a finely controlled temperature bath (< 0.05°C is really required).

For exceptionally stiff and elongated macromolecules (like the microbial polysaccharide xanthan), the measured $[\eta]$ will be lowered by the shear rate in a capillary system. Consequently a concentric cylinder dilute solution viscometer, where the maximum shear rate is much lower, is preferred. The solution is poured into a narrow gap between two concentric cylinders (Couette geometry). One of these (e.g. the outer cylinder) turns at a constant rate (the rotor), while the inner cylinder (the stator) is attached, for example, to a thin torsion wire. In a typical modern version the speed of the rotor ($\propto$ shear rate) is controlled by a voltage ramp using a microcomputer, which can then correlate the sample response torque voltage ($\propto$ shear stress), and confirm the shear stress/shear rate linearity.

6.3.3 Determination of $[\eta]$

The ratio of the viscosities for the solvent and the solution gives the relative viscosity η_r for the solution, i.e. $\eta_r = \eta/\eta_s$. If η_r lies in the range say 1.2–2.2, the intrinsic viscosity may be obtained as follows. The specific viscosity η_{sp} is defined to be $(\eta_r - 1)$ and so we have

$$[\eta] = \lim_{C \to 0} \{\eta_{sp}/C\} \approx \lim_{C \to 0} \{\ln(\eta_r)/C\}, \tag{6.4}$$

where the plots of η_{sp}/C and $\ln(\eta_r)/C$ versus C are due to Huggins and Kraemer, respectively. Both plots are usually made, and the mean intercept obtained (Morris and Ross-Murphy, 1981).

6.3.4 Semi-dilute solutions: shear rate effects

At some concentration above C^*, individual macromolecules so interpenetrate each others' domains that there is a rather sudden change in flow properties, and the viscosity becomes much more shear rate dependent. Until now we have mentioned the effect of shear rate only briefly, because for dilute solutions of biopolymers it is only rarely significant. However, most biopolymer solutions are pseudoplastic ('shear thinning'), so that viscosity decreases with the shear rate, and this effect becomes increasingly noticeable as the concentration increases. Consequently, we digress briefly to explain this effect in more detail.

For fluids shear stress depends upon the rate of change of strain with time, and so the usual (shear) viscosity is defined by

$$\eta = \tau/(\mathrm{d}\gamma/\mathrm{d}t) = \tau/\dot{\gamma}, \tag{6.5}$$

where $\dot{\gamma}$ is the shear strain rate (or shear rate), and τ is the shear stress. The most usual trace of viscosity against shear rate for semi-dilute polymer solutions shows a progressive decrease, i.e., so-called pseudoplastic behaviour. At very low shear rates the viscosity shows a shear rate independent Newtonian plateau, sometimes called the zero shear (rate) viscosity η_0. The opposite to pseudoplastic behaviour, dilatancy ('shear thickening') but with a zero shear viscosity, is sometimes seen for more dilute swollen particulate systems, such as starch dispersions. For all shear rate dependent solutions, it is useful to measure η over as wide a range as possible. Measurement over >5 decades of strain rate is possible with certain sophisticated viscometers, where typical low shear rates would be 10^{-2} to 1 reciprocal second (written s^{-1}), and high shear rates greater than say $10^3\,\mathrm{s}^{-1}$. Such instruments are sometimes called 'rheogoniometers' or 'mechanical spectrometers'.

The essential features of such a controlled strain rheometer are a vertically mounted motor (which can drive either steadily in one direction or can oscillate), and opposed to this a transducer housing a stress (torque) transducer (Ferry, 1980). The sample geometry, one half attached to the motor and the other to the transducer, can be changed from say Couette to cone/plate or disc/plate, and the sample temperature controlled. In recent years controlled stress rheometers have become very popular, since they are intrinsically cheaper to construct, and they can provide quite good specifications at comparatively low cost. In these, a stress is produced, for example by employing a two-phase AC motor, and this drives the upper member. The strain induced in the sample can be measured on the same fixture using an optical encoder or radial position transducers. This arrangement has the advantage that a constant low stress can be applied and η_0 determined.

The most convenient steady shear geometry is a cone and plate; the shear rate in this design is constant across the gap, and is given by $\dot{\gamma} = \Omega/\Theta$, where Ω is the radial velocity of rotation (radian s^{-1}) and the cone angle Θ is measured in radians. This constancy of shear rate is a special feature of the cone-plate geometry compared to say Couette or parallel plate designs. In the latter, the edge of the rotating plate gives a maximum shear rate, whereas at the centre the shear rate is zero.

Well above C^*, the simplest explanation of shear rate effects is to say that at the lowest shear rates, intermolecular entanglements disrupted by the imposed deformation are replaced by new interactions between different partners, with no overall change in the extent of entanglement, and therefore in solution properties. This situation corresponds to the Newtonian region of the flow curve. The onset of more pronounced shear thinning occurs when the rate of externally imposed motion becomes progressively greater than the rate of formation of new entanglements.

Several workers have suggested that empirical master curves of viscosity at all shear rates and concentrations can be constructed by plotting $\log(\eta/\eta_0)$ against $\log(\dot{\gamma}/\dot{\gamma}_x)$ where x is a fraction, say 0.5, and $\dot{\gamma}_x$ is the shear rate required to reduce η to $x\eta_0$ for different concentrations (and M_r) (Launay *et al.*, 1986). Other values of the fraction x, say 0.9 or 0.1, have also been suggested, although the precise choice appears relatively unimportant. Usually the viscosity versus shear rate profile shows a so-called power law region at medium to high shear rates where the plot of $\log \eta$ versus $\log \dot{\gamma}$ has a maximum slope of *ca.* − 0.8.

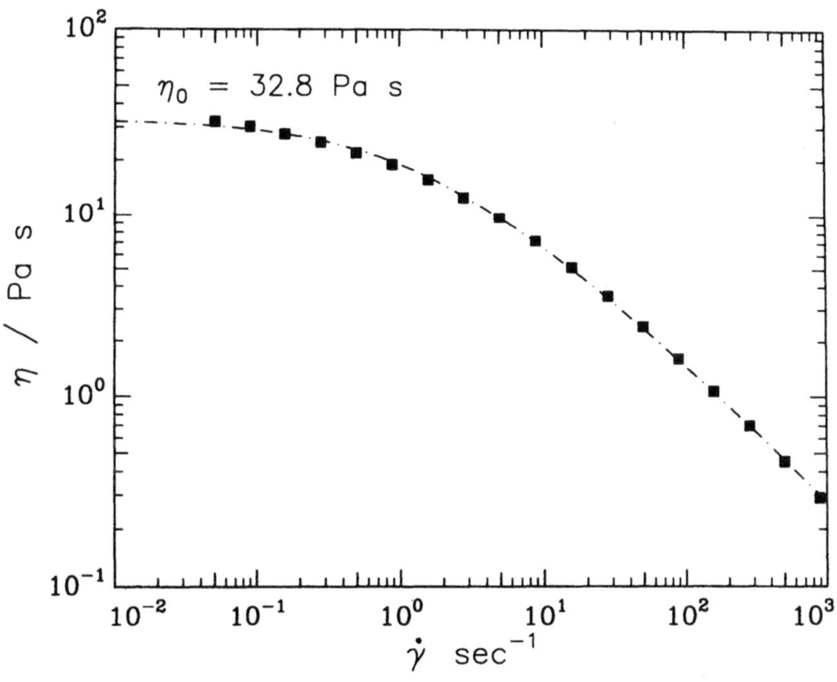

Fig. 6.2 Double log (viscosity versus shear-rate) plot for a guar solution (squares) and the corresponding fit to the Cross equation (line).

Over a range of shear rates, and despite the value of x, the above method of rescaling the data is quite satisfactory, although it may fail at high shear rates since then there is a tendency to approach a high shear rate (so-called upper) Newtonian viscosity plateau. For this reason it is widely accepted that the Cross equation (1965), equation (6.6), should be used instead. This has three independent parameters, η_∞, the high shear rate viscosity (which is not generally related to η_s), m, a power-law exponent, and λ, a relaxation time:

$$\eta = \eta_\infty + (\eta_0 - \eta_\infty)/[1 + (\lambda\dot{\gamma})^m] \qquad (6.6)$$

This is one of many empirical equations, but it is probably the most successful and widely used of these. Figure 6.2 shows a typical series of measurements for a guar solution, and the corresponding Cross fit; it also shows how useful the equation can be in estimating η_0, particularly when this lies below the instrument shear rate range.

6.3.5 Semi-dilute solutions: zero shear rate viscosity and entanglement systems

Well above C^*, the polymer segment concentration becomes uniform, and the zero shear viscosity, whether measured directly or extrapolated using equation (6.6), changes from approximately first power dependence of C to a proportionality of, say, $C^{3\text{–}4.5}$ at high concentrations. In some works this change is shown as a sharp transition, even when plotted in a (double logarithmic) viscosity versus C plot, but in practice the change is less sharp (and depends on molecular weight polydispersity). Figure 6.3 illustrates measurements for guar solutions of different concentration (Robinson *et al.*, 1982). The effect of chain shape and size is removed by plotting against $C[\eta]$ rather than against C itself. We now refer to the 'critical' breakpoint concentration as C_{cr}, although confusingly in some earlier work it was referred to as C^*. (In practice C_{cr} is *ca.* 4 times C^*.) C_{cr} can only be located precisely if this upper slope regime is reached, and this line extrapolated to meet that drawn through the low concentration data. In some publications, the upper slope is shown to begin at a concentration denoted C^{**} (around 10 times C^*). Data for many relatively flexible (so-called random coil) polysaccharides such as guar gum, locust bean gum, carboxymethylcellulose, etc., can be superposed onto a single 'master plot' of viscosity, either η_0 or zero shear rate specific viscosity η_{sp} (= η_0 divided by solvent viscosity) against this parameter $C[\eta]$ (Morris *et al.*, 1981). Polymer solutions above C_{cr} are often known as *entanglement networks* or *entangled solutions.*

What should be clear from the preceding discussions is that for a biopolymer solution to be an effective thickener, $[\eta]$ has to be large for

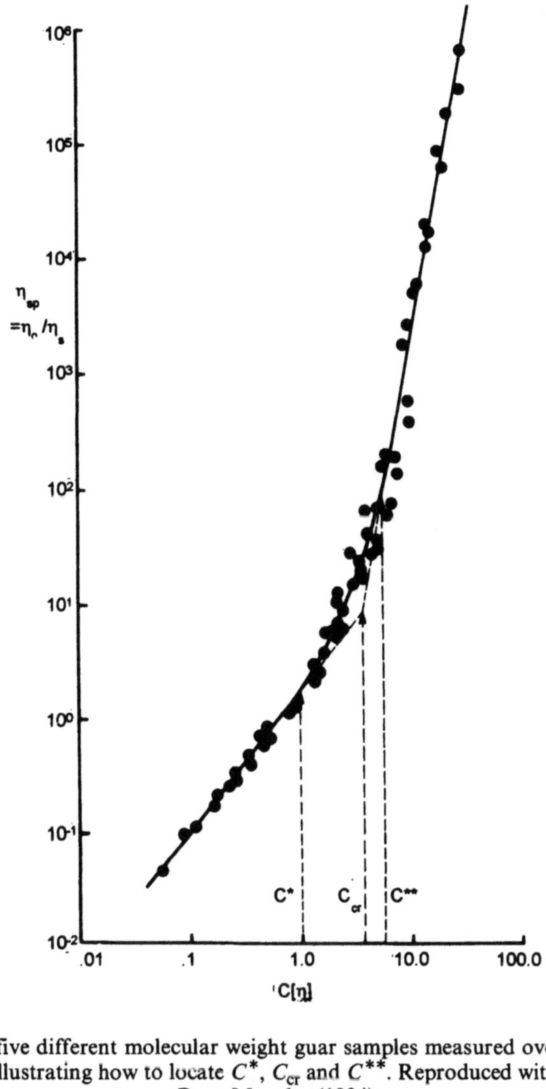

Fig. 6.3 η_{sp} for five different molecular weight guar samples measured over a wide range of concentrations, illustrating how to locate C^*, C_{cr} and C^{**}. Reproduced with permission from Ross-Murphy (1994).

a given M_r. This means that semi-flexible polymers will thicken much more effectively than compact or flexible chain systems, since the same viscosity will be found at much lower concentrations. Consequently, because of their compact shape, globular proteins are hardly used as thickeners. The intrinsic viscosity [η] of such proteins is *ca.* 3 ml/g (not much greater than for Einstein spheres), so any significant viscosity effects would only be seen at concentrations greater than $C^* \sim 1/[\eta] = 33$ wt%. As implied above, effective polysaccharide thickeners include the galactomannans (guar gum) and certain cellulose

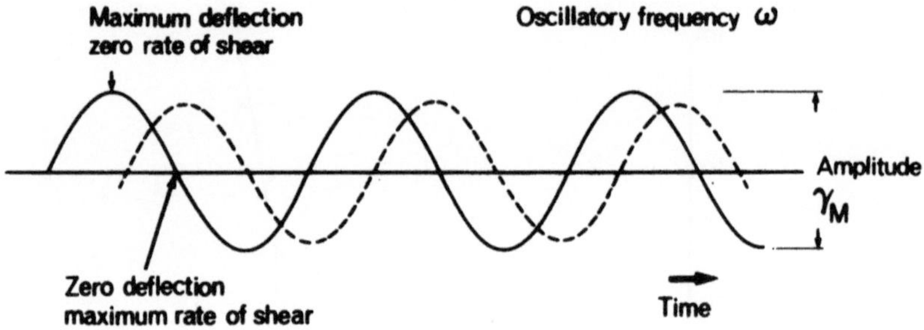

Fig. 6.4 Sinusoidal oscillatory shear. At the extremes of the oscillatory cycle shear strain γ (solid) is maximum, but shear rate, $\dot{\gamma}$ (dotted) is zero, while the converse is the case at the null strain position. Reproduced with permission from Ross-Murphy (1994).

derivatives. Structural evidence suggests that these systems do not undergo any (coil–helix) transition on heating or cooling. Since they are of quite high molecular weight (typically $> 10^6$), and not so flexible, they have a high $[\eta]$ for a given M_r and can often be used at concentrations at/or below 1 wt%.

6.4 Oscillatory measurements: mechanical spectroscopy

As already mentioned above, more expensive rheometers can apply an oscillatory strain to a biopolymer sample, and measure the corresponding oscillatory stress (Ferry, 1980; Ross-Murphy, 1994). Figure 6.4 illustrates the principle – a sinusoidal oscillation of maximum strain γ_M and oscillatory frequency ω is applied to a sample, here using parallel plate geometry. If the material is perfectly elastic, then the resultant stress wave is exactly in phase with the strain wave. On the other hand, if the material is a purely viscous fluid, since the rate of change of the sinusoidal oscillation is a maximum when the strain is zero, the resultant stress wave will be exactly 90° out-of-phase with the imposed deformation. In general, the stress wave will have a phase difference δ $(0 < \delta < 90°)$, so that δ, or more usually $\tan \delta$, is a measure of the viscous/elastic ratio for the material at frequency ω. The elastic (in-phase) and viscous (out-of-phase) components of the stress wave are separable, and they define the in-phase, shear storage modulus G' and the out-of-phase, shear loss modulus G'', given by $G^* = \tau^*/\gamma_0 = \sqrt{G'^2 + G''^2}$. Here, τ^* is the complex shear stress, γ_0 the maximum strain, their ratio G^* is the complex shear modulus, η^* is the dynamic viscosity, equal to G^*/ω, and $\tan \delta$ is the ratio G''/G'.

For many polymer solutions, the frequency dependence of η^* and the shear rate dependence of η are observed to be very similar at the same corresponding values of ω (measured in radians s^{-1}) and $\dot{\gamma}(s^{-1})$. This observation, often called the Cox–Merz rule (Ferry, 1980), has been observed for many polysaccharide entanglement solutions such as our model system guar. However, there are also exceptions, especially for so-called *weak gels* (see below). In these cases useful information can still be extracted from the comparison of (η, $\dot{\gamma}$) and (η^*, ω) profiles.

6.4.1 Strain dependence

The two moduli G' and G'' refer to measurements made in the range where G^* is independent of the applied oscillatory strain. At higher strains G^* tends to decrease with increasing ω (although some ordered protein systems, e.g. fibrin gels, show an increase). The maximum linear strain, γ_0, is very dependent on the nature of the system being investigated. For some systems including strong gels and entanglement solutions, we have $\gamma_0 \geqslant 20\%$, whereas for weak gels it can be very much (say up to 1000 times) smaller. Colloidal and particulate networks are often of this type (Ross-Murphy and Shatwell, 1993). More significantly, such systems do not obey the Cox–Merz rule, and $\eta^*(\omega)$, measured in the small strain limit, lies above $\eta(\dot{\gamma})$, except at high frequencies, when the curves may converge. Such behaviour is usually associated with a tendency to form aggregated structures or dispersions, which are then broken down under the applied strain, and biopolymer solutions (dispersions) showing this behaviour have been called *weak gels*.

6.4.2 Entanglement solutions, weak gels and strong gels

Some years ago we adopted a strategy to distinguish between biopolymer solutions and gels falling into the categories denoted by the terms entanglement solutions, weak gels and strong gels (Ross-Murphy, 1984). Although this involves a wide range of measurements, including both steady and oscillatory rheological techniques, the design below can be regarded as the minimum set.

1. Measure G^*, the complex shear modulus *versus* (maximum) oscillatory strain deformation γ_0 from about 0.1% up to 100% in 5% increments, and at a radial frequency ω of 10 rad s^{-1}. A double log plot of G^* *versus* γ_0 will be generated.
2. Using a strain within the linear viscoelastic region established above, measure G', G'' and η^* ($= G^*/\omega$) from $\omega = 0.1$ to 100 rad s^{-1}, at 1/3 decade intervals of ω. This gives the so-called mechanical spectrum, where log G', log G'' and log η^* are plotted *versus* log ω.

The results from these experiments are then examined. All of the dependencies discussed below refer to double logarithmic presentations.

The strain dependence from experiment 1 will be very similar for entanglement solutions and strong gels, with the linear viscoelastic strain γ_0 extending to approximately 0.20 (20%), while for weak gels γ_0 is usually < 5%.

The result of experiment 2, the frequency dependence or mechanical spectrum, for a polymer solution, is that the low frequency dependence of G', G'' and η^* are as illustrated in Fig. 6.5(a) with $G'' >> G'$. Here we have $G'' \propto \omega$ and $G' \propto \omega^2$. From the definition of $\eta^* = G^*/\omega$, we have $\eta^*(\omega)$ independent of ω – and we should have a Cox–Merz superposition of η_0 and η_0^*. As the frequency is increased, the $G' \propto \omega^2$ behaviour begins

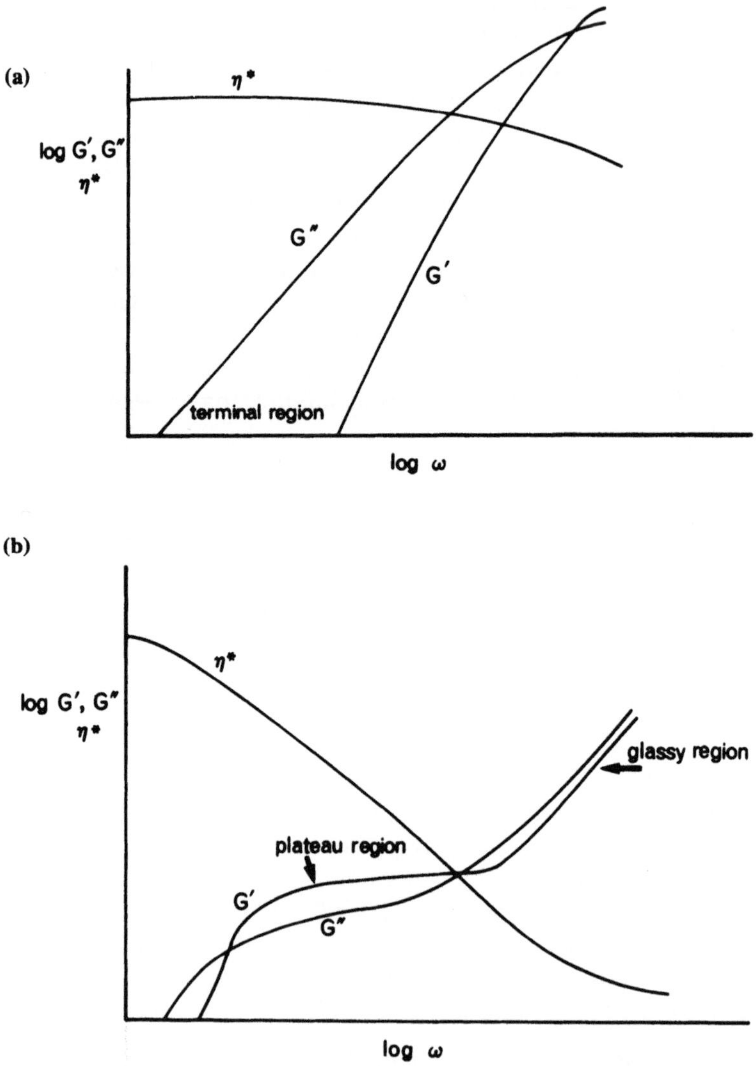

(c)

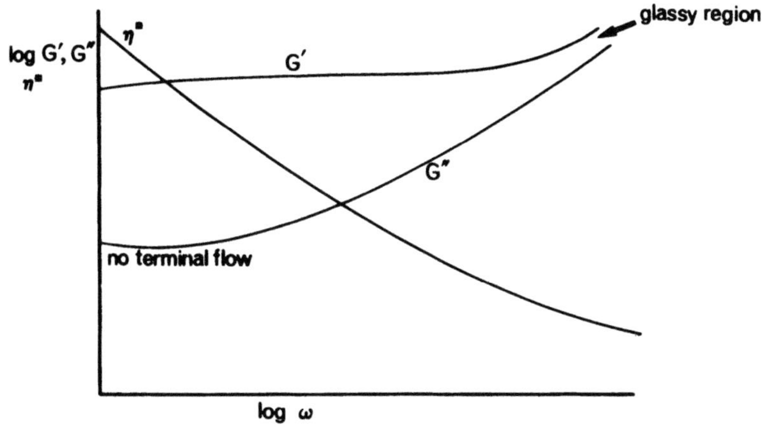

Fig. 6.5 Mechanical spectra, G', G'' and η^* plotted logarithmically against frequency for (a) a dilute solution, (b) a more entangled semi-dilute solution and (c) a gel. Reproduced with permission from Ross-Murphy (1994).

to dominate, so that there is a 'cross-over'. At higher frequencies G' and G'' become much less frequency dependent, and we have entanglement behaviour with a high frequency 'rubbery' plateau – the value of G' is the plateau modulus. Finally we pass through the transition region into the 'glassy' state. Figure 6.5(b) represents 10–15 decades of frequency, and no mechanical instrument can be expected to access more than a fraction of this range, at constant temperature. However, increasing the temperature can be equivalent to lowering the effective frequency of the experiment as we shall discuss below.

For both strong and weak gels the results are very different with $G'(\omega) > G''(\omega)$ (Fig. 6.5(c)) and both moduli being largely independent of frequency, so the system has an apparent equilibrium modulus (the modulus measured at zero frequency). This is exactly what we might have expected for an elastic material where G is independent of frequency. For real biopolymer systems, the behaviour is usually not so straightforward, so that both G' and G'' are slightly frequency dependent, both increasing with ω, and the exponent of $\eta(\dot{\gamma})$ is typically $\lessdot -1$. This behaviour is sustained even if the experiment is extended down to very low frequencies. However, using the above strategy we have established criteria for separating strong and weak gels, since they should give similar mechanical spectra, but have quite different strain dependencies.

(a) Entanglement solutions. The data for guar solutions illustrates the expected features very well. At low frequencies, G' does approach an ω^2 dependence, and we have $G'' \propto \omega$; and η^* approaches a Newtonian

plateau as discussed above. By measuring G' and G'' as a function of ω for different concentrations, it was observed that the cross-over frequency consistently decreases with increasing concentration; the system is a model Newtonian fluid at low concentrations, and shows a distinct plateau modulus at the highest frequency.

(b) Weak gels. The main properties of weak gel systems, well exemplified by xanthan solutions, have been discussed already. They produce a gel-type mechanical spectrum with $G' > G''$ and both of these are largely frequency independent, but the linear strain region is quite small. As we will see below, they also fail to obey the Cox–Merz steady and dynamic viscosity superposition rule. Weak gels tend not to show a zero shear viscosity plateau, either in steady or dynamic experiments. Instead the apparent power law exponents of $\log \eta$ *versus* $\log \dot{\gamma}$ are high ($\geqslant -0.8$), and the overall viscosity *versus* shear rate profile is more like that of a so-called viscoplastic material with $\tau = \tau_0 + k\dot{\gamma}^n$, where τ_0 is a finite yield stress, rather than a normal pseudoplastic fluid, where τ_0 is zero.There is considerable discussion (in the author's opinion much of it pedantic) as to whether a yield stress can really exist. For this reason the term 'apparent yield stress' is often preferred. As $\dot{\gamma} \rightarrow 0$, we have the apparent zero shear viscosity, defined as $\tau/\dot{\gamma}$ tends to infinity, rather than a finite (Newtonian) value. This range of properties is much more commonly seen with concentrated colloidal dispersions rather than with polymer solutions.

(*c*) *Strong gels.* The most widely studied strong gel system is almost certainly gelatin. Data for gelatin gels follow the expected frequency dependence provided sufficient time has been allowed for the gel to set, and that care has been taken in ensuring a reproducible temperature history. Kinetic gelation measurements (gel cure experiments) in which G' and G'' are monitored as a function of time have been carried out by many workers. Gelatin gels are, of course, thermoreversible, and gel melting profiles can also be monitored as a function of concentration and primary chain molecular weight in a rheological experiment. For substantially cured systems, G' is essentially independent of frequency, while G'', although essentially parallel and considerably lower in value that G' (by a factor of > 20), can show a small dip centred around say 1 rad s^{-1}.

More recently there have been many publications on polysaccharide gels including studies on carrageenans, pectin, agarose, amylose and amylopectin gels and gellan (a new microbial polysaccharide) gels. Some of these are extremely sensitive to slip in cone-plate or parallel plate geometries. A novel slotted Couette geometry (Richardson and Goycoolea, 1994) has proved very useful in such cases.

6.4.3 *Concentration dependence of modulus for strong gels*

Most data give G' (or more strictly the pseudo-equilibrium modulus G) approximately proportional to $C^{1.8-2.5}$, at high concentrations (greater than say $5C_0$), but with a higher exponent as C decreases and approaches C_0, the critical gel concentration, as illustrated in Fig. 6.6. This concentration C_0 is not related to C^* in any simple way, and C_0 is certainly not always equal to C^* as some workers have suggested. Several molecular models have been suggested to describe the concentration dependence of modulus for pectins, carrageenans, gelatin, agarose and amylose (Clark and Ross-Murphy, 1987). Theoretical fits to modulus/concentration data for different M_r fractions of synthetic amyloses are consistent with a mechanism involving interchain double helices.

In a typical gel cure experiment for heat-set protein gels, a solution of the protein is introduced into the rheometer and then 'ramped up' to close to, or above the protein denaturation temperature (as determined by, for example, DSC), and then maintained constant for say 3 h. The temperature can then be reduced, and G', G'', η^* measured as a function of ω and γ at different temperatures. There are, generally, no large relaxation processes (tan δ maxima) in the frequency window examined, but perhaps most interesting are the shapes of the gel cure traces, and

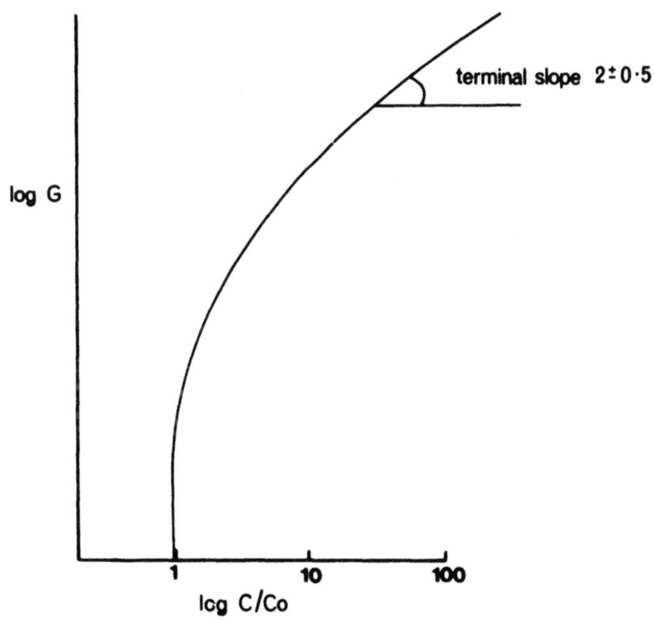

Fig. 6.6 Idealized representation of the modulus/concentration behaviour for a gel, showing the terminal slope regime (typically accessed when $C/C_0 > 10$). Reproduced with permission from Ross-Murphy (1994).

(a)

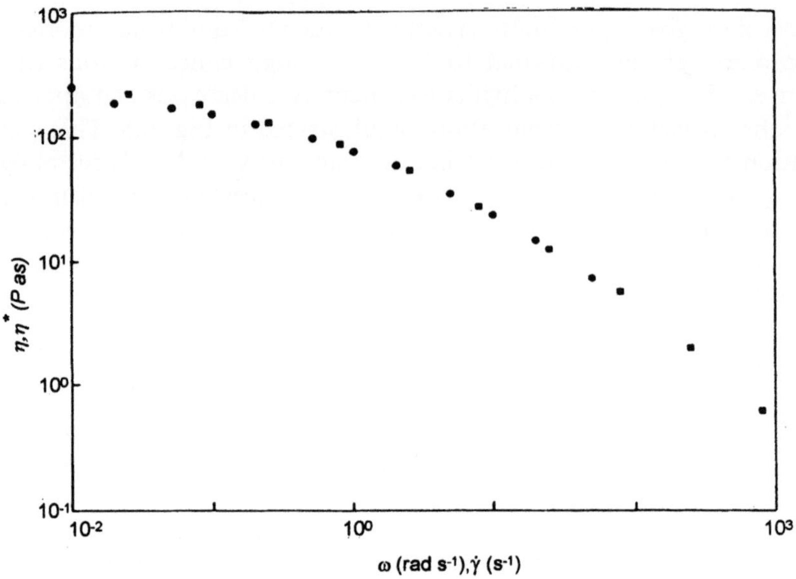

(b)

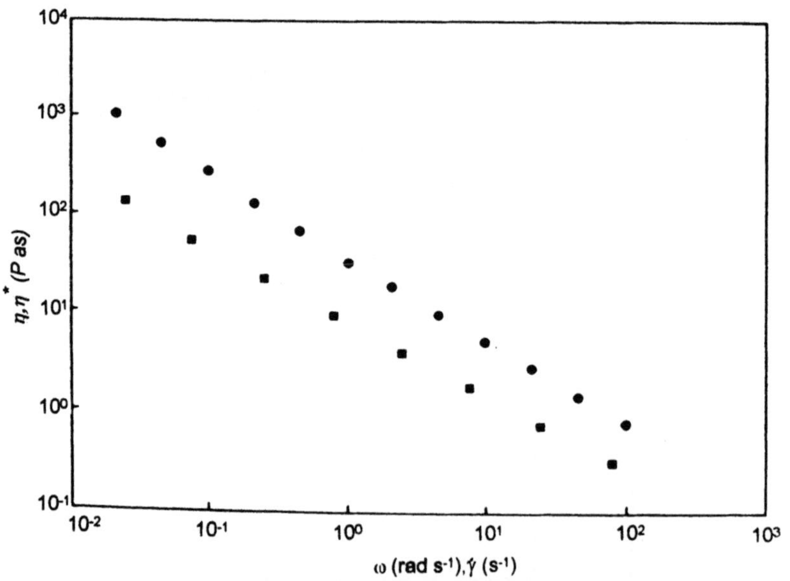

Fig. 6.7 'Cox–Merz plot' of log (steady shear viscosity (squares) and dynamic viscosity (circles)) plotted against log (shear rate and radial frequency), respectively for (a) a guar solution and (b) a xanthan solution. Reproduced from *International Journal of Biological Macromolecules* **9** (1987), pp. 250 and 257 by permission of the publishers, Butterworth Heinemann Ltd.©

the dependence of G' (less so G'') on concentration, pH and ionic strength.

6.4.4 Combined steady and oscillatory shear measurements

By combining oscillatory and steady shear measurements, we can derive still further information, since a strong gel will rupture completely, whereas a weak gel will flow, but without Cox–Merz superposition of η and η^*. In fact $\eta^*(\omega)$, measured in the small strain limit, will be greater than $\eta(\dot{\gamma})$, except at high frequencies, when the curves may converge. Such behaviour is usually associated with a tendency to form aggregated structures or dispersions, which are broken down under the applied strain.

The mechanical spectra of weak gels, such as xanthan dissolved in dilute salt solution, do not change significantly on dilution until quite low concentrations are reached, and then the behaviour becomes identical to that of an entangled polymer. This concentration dependence and the ability to flow without fracture – although without Cox–Merz superposition of $\eta(\dot{\gamma})$ and $\eta^*(\omega)$ – at higher concentrations, are the most important differences between a typical strong and weak gel system, as contrasted in Figs 6.7(a) and (b) (Richardson and Ross-Murphy, 1987a,b).

6.5 Time–temperature superposition

The final topic is that of time–temperature superposition. This technique is widely employed for synthetic polymer melts (Ferry, 1980), but its application to biopolymer solutions and gels has been less widespread. The principle is that lowering the temperature of a sample is equivalent to extending the frequency range to higher values and, correspondingly, measuring at higher temperatures is the same as measuring at lower frequencies. Consequently by making measurements at different temperatures, and rescaling all data to a single reference temperature, the effective frequency range of any oscillatory measurement can be extended.

The main caveat is that the sample should undergo no physical (or chemical) change when the temperature is altered. This is, of course, a very great (even absolute) limitation for some food biopolymers solutions and gels. Although the technique appears to work for some systems, it cannot be regarded as a reliable method. Nevertheless when applied successfully, as well demonstrated with recent data for pectins (Lopes da Silva *et al.*, 1994), it provides valuable extra information.

Acknowledgement

The author thanks Phillippa Rayment for helpful comments and for contributing the experimental data in Fig. 6.2.

References

Cross, M.M. (1965) Rheology of non-Newtonian fluids – a new flow equation for pseudoplastic systems, *J. Colloid Sci.*, **20**, 417–437.

Clark, A.H. and Ross-Murphy, S.B. (1987) Structural and mechanical properties of biopolymer gels. *Adv. Polym. Sci.*, **83**, 57–192.

Ferry, J.D. (1980) *Viscoelastic Properties of Polymers*, 3rd edn., Wiley, New York.

Launay, B., Doublier, J.L. and Cuvelier, G. (1986) Flow properties of aqueous solutions and dispersions of polysaccharides, in *Functional Properties of Food Macromolecules* (eds. J.R. Mitchell and D.A. Ledward), Elsevier Applied Science, Barking, UK, pp. 1–77.

Lopes da Silva, J.A., Gonçalves, M.P. and Rao, M.A. (1994) Influence of temperature on the dynamic and steady-shear rheology of pectin dispersions. *Carbohydr. Polym.*, **23**, 77–87.

Morris, E.R. (1984) Rheology of hydrocolloids, in *Gums and Stabilisers for the Food Industry* (eds. G.O. Phillips, D.J. Wedlock and P.A. Williams), Pergamon, Oxford, Vol. 2, pp. 57–78.

Morris, E.R. and Ross-Murphy, S.B. (1981) Chain flexibility of polysaccharides and glycoproteins from viscosity measurements. *Tech. Life Sci.*, **B310**, 1–46.

Richardson, R.K. and Goycoolea F.M. (1994) Rheological measurement of κ-carrageenan during gelation. *Carbohydr. Polym.*, **24**, 223–225.

Richardson, R.K. and Ross-Murphy, S.B. (1987) Non-linear viscoelasticity of polysaccharide solutions, 1. Guar galactomannan solutions. *Int. J. Biol. Macromol.*, **9**, 250–256.

Richardson, R.K. and Ross-Murphy, S.B. (1987) Non-linear viscoelasticity of polysaccharide solutions. 2. Xanthan polysaccharide solutions. *Int. J. Biol. Macromol.*, **9**, 257–264.

Robinson G., Ross-Murphy, S.B. and Morris, E.R. (1982) Viscosity–molecular weight relationships, intrinsic chain flexibility, and dynamic solution properties of guar galactomannan. *Carbohydr. Res.*, **107**, 17–32.

Ross-Murphy, S.B. (1984) Rheological methods, in *Biophysical Methods in Food Research, Critical Reports on Applied Chemistry*, (ed. H.W.-S. Chan), SCI Blackwell, Oxford, Vol. 5, pp. 138–199.

Ross-Murphy, S.B. (ed.) (1994a) *Physical Techniques for the Study of Food Biopolymers.* Blackie A & P, Glasgow.

Ross-Murphy, S.B. (1994b) Rheological methods, in *Physical Techniques for the Study of Food Biopolymers* (ed. S.B. Ross-Murphy), Blackie A & P, Glasgow, pp. 343–392.

Ross-Murphy, S.B. and Shatwell, K.P. (1993) Polysaccharide strong and weak gels. *Biorheology*, **30**, 217–227.

Chapter 7

Fracture Mechanics of Solid Foods

T. van Vliet and H. Luyten*
Department of Food Science,
Wageningen Agricultural University, The Netherlands

7.1 Introduction

Fracture properties are an important quality characteristic of solid and solid-like food products – with respect to eating quality, handling properties during storage (and/or further processing), and usage properties such as cutting and spreading. These properties pertain to fracture under widely differing conditions and speeds of deformation. During eating, for example, a solid or solid-like product is quickly deformed until it fractures or yields (the latter will not be dealt with specifically in this chapter, but see van Vliet *et al.* (1991)), whereas during storage the time-scale of the deformation, relevant for shape retention, may be of the order of days or much longer.

Fracture of foods is related to large deformations of the product. However, most of the fundamental studies on the mechanical properties of food have been done at small deformations. Often there is not a direct relation between small and large deformation properties (Atkins and Mai, 1985; Bremer *et al.*, 1990; van Vliet *et al.*, 1991, 1993). Fracture is considered to occur when all bonds break between the structural elements in a certain macroscopic plane, resulting in a breakdown of the structure of the material over length scales much larger than the size of the structural elements and ultimately in a falling apart of the material. Below we give first a short introduction to relevant parts of the theory of fracture mechanics and of the parameters required to describe fracture properties. Later, methods to evaluate fracture properties are dealt with.

7.2 Fracture mechanics

The starting point for the standard theory of fracture mechanics is that all materials are inhomogeneous and they contain weak points (inherent

* Present affiliation: Agrotechnological Research Institute, Wageningen, The Netherlands.

defects) which may be considered as small cracks, which leads to local higher stresses (higher forces per unit area). Fracture starts if the stress near such an inhomogeneity becomes larger than the local breaking stress. Fast and spontaneous growth of the crack may occur once this requirement is fulfilled, and if, due to this crack growth, more energy is released than is required to form the two new surfaces (Atkins and Mai, 1985). So, for understanding fracture processes, both stress and energy considerations have to be taken into account.

If a stress is applied to a material containing a tiny crack or notch, stress concentration will occur. The stress σ_l at the tip of a flat elliptical notch of length l with a radius r at the tip in an ideal, isotropic elastic material loaded perpendicular to the notch is (Gordon, 1978)

$$\sigma_l = \sigma[1 + 2\sqrt{(l/r)}], \tag{7.1}$$

where σ is the average stress remote from the notch. Equation (7.1) only applies to a two-dimensional sheet of material, but for the three-dimensional case no simple formula is available (Atkins and Mai, 1985). However, it indicates the approximate conditions under which local stresses may be higher than the average for materials where the stress carrying chains are connected by shear connections (e.g. in gel-like and glassy materials). If this is not the case, or if the connections are only very weak as in raw meat or in a rope, the resulting stress σ is approached by

$$\sigma = \sigma_i\left(\frac{w}{w - l}\right), \tag{7.2}$$

where σ_i is the stress is the absence of the notch, and w is the width of the piece of material. The resulting stress concentration is (much) lower (see also Fig. 7.1). So, regarding the effect of a crack perpendicular to the applied stress, one can distinguish notch-sensitive and notch-insensitive materials.

Fracture starts (is initiated) if the local stress in the material exceeds the breaking stress of the bonds between the structural elements giving the solid-like properties to the material. It propagates spontaneously with a certain speed if the differential energy released during crack growth surpasses the differential energy required to form two new surfaces.

On deformation of an ideal elastic material, the amount of strain (deformation) energy W' (J m^{-3}) elastically stored is

$$W' = \int_{\varepsilon=0}^{\varepsilon_f} \sigma(\varepsilon)\, d\varepsilon, \tag{7.3a}$$

where E is Young's modulus and ε the strain (relative deformation). For a linear elastic material, this gives

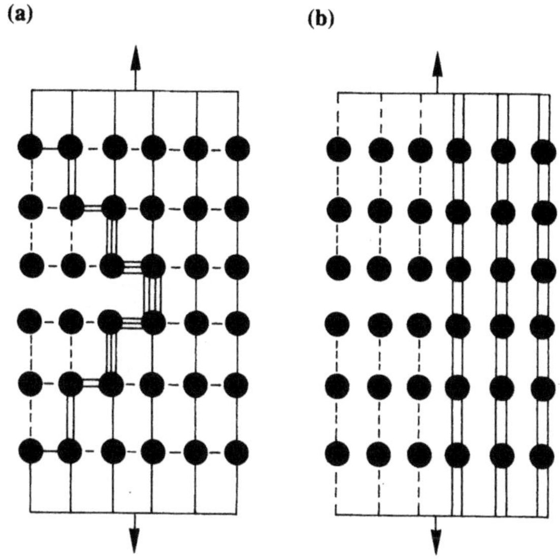

Fig. 7.1 Stress concentration in a material due to a tensile stress (very schematic representation). Connections between structural elements (black circles) carrying stress are indicated by full lines. The number of lines between two structural elements indicates the extent of stress concentration. (a) Material with shear connections between the stress carrying chains. (b) Material without shear connections.

$$W' = \tfrac{1}{2}\sigma\varepsilon = \frac{\sigma^2}{2E}. \tag{7.3b}$$

On growth of a crack or notch, the stress over the notch will become zero and will become relaxed in the region surrounding it (Fig. 7.2(a)). The volume in which the stress relaxes is proportional to $l^2 d$ where d is the width of the crack. The energy required for the formation of two new surfaces is proportional to ldR_s, where R_s (J m^{-2}) is the specific fracture energy. Fracture will propagate spontaneously if we have

$$W'(\tfrac{1}{2})\,\pi d[(l+\delta l)^2 - l^2] \geqslant R_s d\delta l. \tag{7.4}$$

Crack propagation starts if the depth of a notch or half the crack length is equal to a critical length l_c (Fig. 7.2b)). For an ideal, isotropic elastic material, l_c is given by

$$l_c = \frac{R_s}{\pi W'}. \tag{7.5a}$$

For a linear elastic material, this results in the relationship

$$l_c = \frac{2R_sE}{\pi\sigma_f^2}, \tag{7.5b}$$

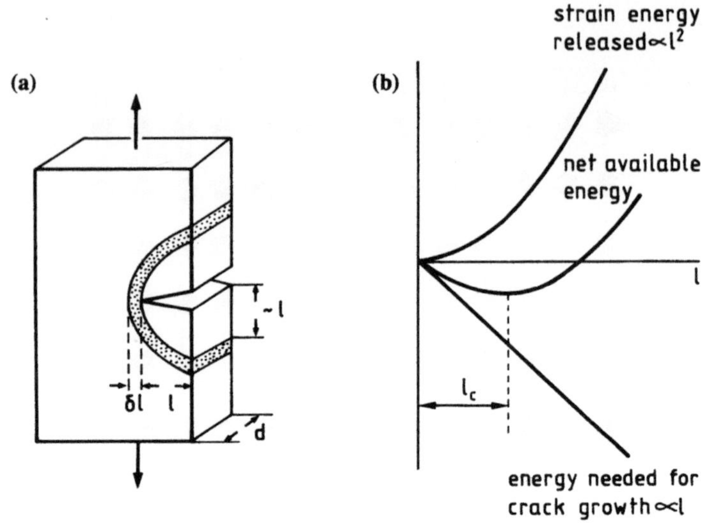

Fig. 7.2 (a) Volume in which strain energy is relaxed due to growth of a crack by a length δl. (b) Fracture energy needed and amount of stored energy available for crack growth as a function of crack length l.

where σ_f is the overall stress at the onset of fracture propagation. This stress is just the same as the stress for fracture initiation (equation (7.1)) in being proportional to $l^{-1/2}$.

The energy required for fracture propagation stems from the elastically stored strain energy. For non-ideal isotropic elastic materials, this energy will normally be lower than the amount of strain energy supplied to the material during deformation ($W' < \int \sigma(\varepsilon)\, d\varepsilon$). The latter involves also the extent of energy losses due to energy dissipation by viscous flow of the viscoelastic material and/or due to friction processes between different components of the material caused by inhomogeneous deformation of the material at large strains (van Vliet *et al.*, 1993). In general the energy balance can be written as

$$W = W' + W''_{\mathrm{m}} + W''_{\mathrm{c}} + W_{\mathrm{f}}, \tag{7.6}$$

where W is the amount of energy supplied to the material ($= \int \sigma(\varepsilon)\, d\varepsilon$), W' is the part of the deformation energy that is elastically stored, W''_{m} is the part dissipated due to flow of the network due to visco-elasticity of the material, W''_{c} is the part dissipated due to friction processes between different components of the material, and W_{f} is the net fracture energy (all in J m^{-3}). Energy dissipation may also occur during the transport to the tip of the crack of the strain energy which becomes available due to stress relaxation during crack growth. These energy dissipation processes cause the fracture behaviour of most foods (e.g. cheese, chocolate bars, bread) to depend on the rate of deformation. Moreover, it results in a slow down of crack growth speeds.

The relative importance of energy dissipation processes depends on the rate of deformation. It can be determined for large deformations by cycle experiments (Fig. 7.3) (Luyten *et al.*, 1991). Mostly W_m'' is more important at relatively low deformation speeds while the effect of W_c'' is more

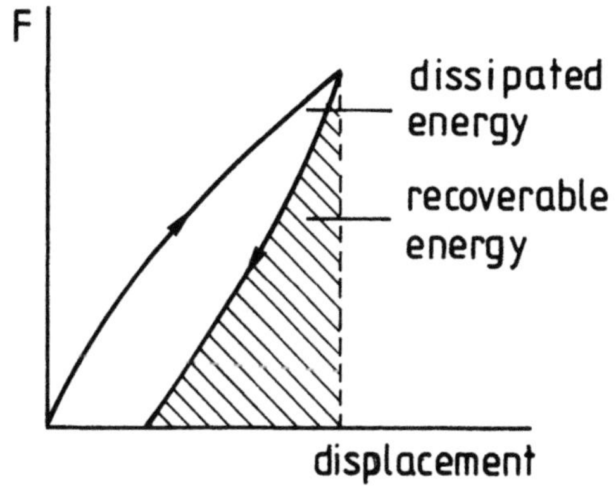

Fig. 7.3 Hypothetical force *F* *versus* displacement curve for loading and unloading (cycle) experiment of a test piece.

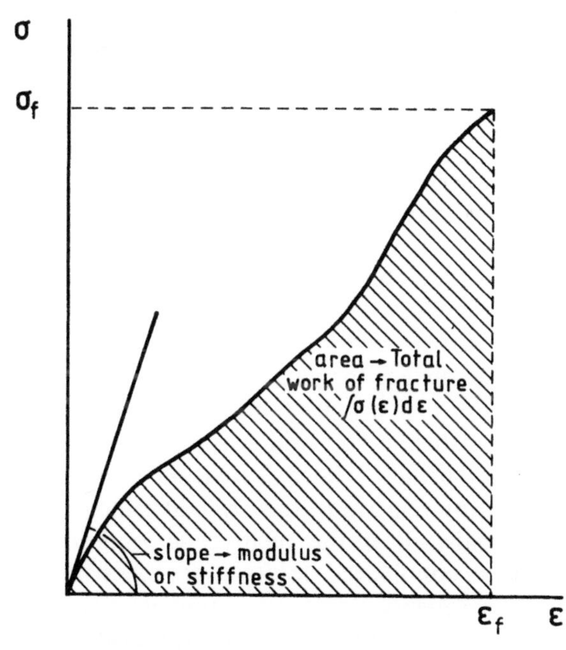

Fig. 7.4 Example of a stress σ *versus* strain ε curve for a non-linear deforming material. The material fractures at stress σ_f and strain ε_f.

important at higher speeds. This means that, for visco-elastic materials (e.g. Gouda cheese), the relative deformation at fracture may decrease with increasing deformation speed (less time for flow), while if W_c'' is important (e.g. starch gels) the reverse may be observed (van Vliet *et al.*, 1993). The occurrence of energy dissipation during deformation means that it is impossible to study fracture properties of food systems separately from large deformation properties.

The rather complicated character of the fracture process suggests that several parameters are required to describe it. An example of a so-called stress–strain curve up to fracture is given in Fig. 7.4. The most common characteristics used are fracture stress, fracture strain and work of fracture. However, for a complete description of the fracture behaviour of a material, other characteristics such as inherent defect length, crack speed and energy dissipation should be determined. The fracture parameters which can be determined experimentally depend strongly on the method used. In general, the methods that are easier to perform provide less information.

7.3 Methods to evaluate fracture properties

7.3.1 General description

Theoretically, three different types of simple deformation can be distinguished – namely uniaxial extension or compression, simple shear and all-sided compression. In many tests, the deformation is not just one of these simple types, but a combination, especially when large deformations (of non-brittle materials) are involved. This makes an exact description of the deformation very difficult or even impossible. Moreover, to be able to compare results obtained from different tests or various specimens, one has to correct for varying shape of the test pieces, to use an exact measure of the deformation and the applied forces, and to take into account possible effects of shear rates. The applied and determined forces are recalculated to stresses, which are given by the ratio of the force acting on a flat surface (or part thereof) divided by the area of that surface. A distinction can be made between shear (tangential) stresses acting in the plane of the surface and normal stresses acting perpendicular (normal) to the surface. The deformation is made quantitative by taking the deformation relative to the original dimension and expressing it as strain. (Formally, it is the relative deformation as the original distance between the condidered points approaches zero).

Most fracture experiments are done by applying a constant deformation rate or a constant stress. A constant deformation rate does not mean that also the strain rate is constant during the experiment (see below). Strain rate can be varied by varying the stress or the deformation rate applied.

7.3.2 Effects of experimental set-up

Results obtained in fracture experiments may be affected both by the particular test method chosen and by some general experimental factors as specimen dimensions, strain rate, temperature, stiffness of apparatus, etc. (Luyten *et al.*, 1992).

(a) Effects primarily related to the test piece

Size of test piece. As discussed above, fracture occurs due to growth of defects already present in the material. Thus, the size of a test piece has to be large compared to the defects that are controlling fracture. On the other hand, a test piece has to be homogeneous, which for most food systems sets an upper limit to the size.

The relative inaccuracy in the determination of the stress and strain increases with decreasing size of the specimen. This sets a minimum to the size of a test piece depending on the force required to deform it and the sensitivity of the apparatus.

As discussed above, the energy required for fracture propagation is roughly proportional to the width of the crack, while the stored strain energy available is to a first approximation proportional to the square of the width. This means that fracture propagation will not occur for specimens which are not large compared to the fracture-inducing inherent defects (Atkins and Mai, 1985). Therefore, it is very hard or even nearly impossible to fracture very small particles.

Indirect effects such as the dependence of heating or cooling rates on test piece size may be significant.

Shape of test piece. The homogeneity of the deformation may depend on test piece shape. For instance, during bending the test piece has to be long compared to its diameter, otherwise the deformation contains a large shear component. During uniaxial compression the height-to-diameter ratio should not be too high (e.g. for Gouda cheese not above 2 (Luyten, 1988)) in order to avoid buckling. In uniaxial compression, the friction between the specimen and the compression plates is, compared to other terms, relatively higher for a lower height-to-diameter ratio (Chu and Peleg, 1985). The kind of fracture (in shear or in tension) may also depend on this ratio (Luyten, 1988).

How the test piece is taken. Care must be taken that test piece preparation does not irreversibly affect its mechanical properties. After boring or cutting an elastic material, often the resulting test piece shape and dimensions may be different from that of the borer or mould used to produce it, like a piece of rubber cut out of a stopper by a cork borer

Fig. 7.5 Examples of shapes of a piece of young Gouda cheese (a) obtained with a cork borer and (b) by cutting with wires (after van Vliet and Peleg, 1991).

(Fig. 7.5). In the case of anisometric materials, the direction in which the specimen is taken from the product will affect results.

(b) Effects due to test conditions

Mode of deformation. The importance of a defect will depend on its direction with respect to the main direction of the strain. All deformations lead to a combination of strains in two (shearing deformation) or three directions in the test piece. For instance, during uniaxial compression, the test piece is compressed in one direction and extended in the two directions (in Cartesian coordinates) perpendicular to it, while in uniaxial tension the opposite is the case. So these modes of deformation differ in more aspects than just the sign of the direction of deformation.

Rate of deformation. For most food systems the large deformation and fracture properties depend on the strain rate. Normally, fracture stress increases with strain rate while fracture strain can decrease, increase or be independent of strain rate (van Vliet *et al.*, 1993).

Environmental conditions. Temperature and temperature history have a strong influence on the rheological and fracture properties of many food materials (e.g. cheese (Luyten, 1988), short cakes and margarines). Thus experiments must always be done at a given constant temperature. In fat-containing products, the temperature history will affect the amount of solid fat and hence the resulting mechanical properties. The picking up of moisture or the drying out of a test piece must also be prevented.

Apparatus. The deformation of the testing apparatus itself due to a certain load has to be negligible compared to the deformation of the specimen being tested. The same applies to the measuring bodies, e.g. the grips during tension tests, or any adhesive glue layers. In compression tests the nature of the material from which the plates are made will affect friction between the test piece and the plates.

7.3.3 *Uniaxial compression tests*

The most popular technique for determining large deformation and fracture properties is uniaxial compression of a test piece between two parallel plates. This type of test is easy to execute. One can apply a constant force (load) and measure the deformation as a function of time; or, what is normally done, one can apply a constant deformation rate and measure the required force.

Recalculation of a deformation in terms of strain is actually less straightforward than is often thought (Peleg, 1984). Frequently as a measure of strain, the so-called Cauchy or engineering strain is taken, which is given by $\Delta H/H_0$, where H_0 is the original height of the test piece and ΔH the change in height. An important disadvantage of this measure of strain is that, if one twice compresses a test piece x% and takes in the calculation of the second strain for H_0 the height of the already x% compressed test piece, one obtains a different strain from that obtained if one compresses the test piece immediately to $2x$%. Therefore, the natural or Hencky strain ε_H is to be preferred which gives at every deformation the actual strain. This measure of strain is defined as (Peleg, 1984; van Vliet, 1991)

$$\varepsilon_H = \int_{H_0}^{H} \left(\frac{dH}{H(t)}\right) = \ln\left(\frac{H}{H_0}\right). \tag{7.7}$$

At small deformations the Hencky and Cauchy strains are nearly the same (see Table 7.1 and Fig. 7.6).

Table 7.1 Comparison of Cauchy strain and Hencky strain

Cauchy ($\Delta H/H_0$)	−0.1	−0.2	−0.4	−0.6	−0.8
Hencky ($\ln(H(t)/H_0)$)	−0.105	−0.22	−0.51	−0.92	−1.61

In uniaxial compression tests, mostly cylindrical test pieces are used and occasionally cubes. In calculating the stress, a correction must be applied for the change in area on which the load acts, due to the compression of the test piece. As long as the volume and shape of the test piece do not change, the correction is easily made by taking for the actual surface area $A(t) = A(t = 0) \times (H_0/(H_0 - \Delta H))$.

There are several advantages of the uniaxial compression test method.

1. Preparation of the test pieces is mostly relatively simple.
2. If the test is executed in the right way (for example, with cylindrical test pieces having flat parallel ends), Young's modulus may be calculated from the stress–strain curve.
3. In principle, the fracture stress and strain and the total work of fracture can be calculated.

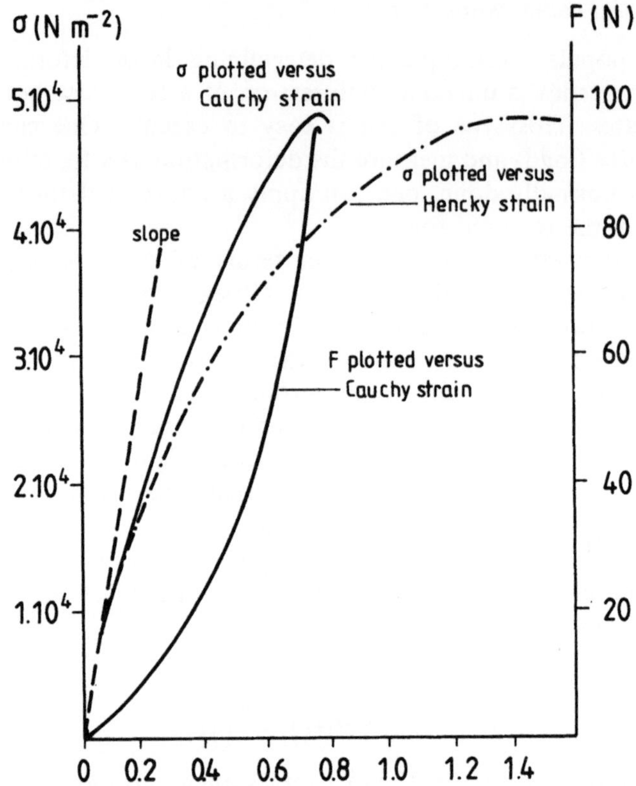

Fig. 7.6 Load F and stress σ *versus* relative deformation ε for an old Gouda cheese. The stress is plotted both as a function of the Cauchy strain and as a function of the Hencky strain.

4. The deformation rate can be easily varied.
5. Hysteresis or cycle experiments, giving data for the energy dissipation in the material due to large applied deformations, can be executed easily.

However, the uniaxial compression test method also has a few drawbacks, which are more or less serious, depending on the goal of the experiments.

1. It is not possible to determine various key fracture parameters such as crack speed, inherent defect length, notch sensitivity, and specific fracture energy (Luyten *et al.*, 1992).
2. Often fracture starts before it can be seen from the outside of the test piece or noticed by a change in the required stress (i.e. before the maximum stress is achieved) (Luyten *et al.*, 1991).
3. Friction between the test piece and the plates may affect the results and make calculation of the true stress and strain difficult (Casiraghi

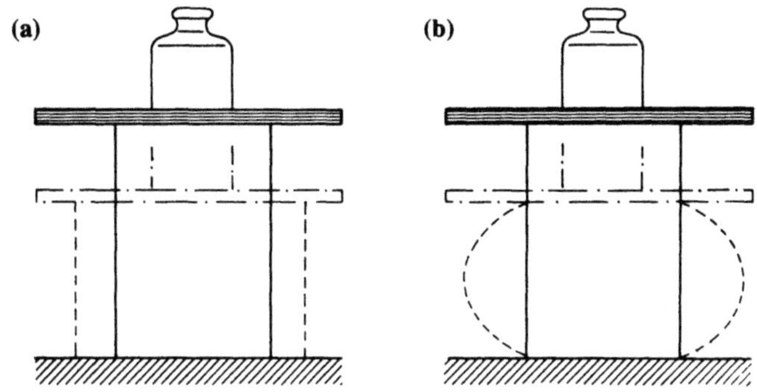

Fig. 7.7 Change in shape of a specimen during uniaxial compression. (a) No friction between test piece and the plates. (b) Strong friction between test piece and the plates.

et al., 1985; Chu and Peleg, 1985). Friction may lead to bulging of the middle part of the test piece (Fig. 7.7(b)). Assuming constant volume and a parabolic shape of the bulged part during compression, a correction procedure for bulging has been developed (Gent and Lindley, 1959). In general, friction will be more important for a low height-to-diameter ratio.

4. If, as is normally the case, the compression is performed at constant deformation rate, the strain rate will increase during the experiment. In terms of Hencky strain, the strain rate $\dot{\varepsilon}_H$ is

$$\dot{\varepsilon}_H = \frac{dH}{dt\,H(t)} = \frac{v}{H(t)} \tag{7.8}$$

where v is the compression speed. Equation (7.8) implies, for example, that, when $H(t)$ is 25% of H_0, $\dot{\varepsilon}_H$ is four times the initial value. Equation (7.8) also implies that the compression speed has to be adjusted if one wants to compare measurements on test pieces with various heights or if one wants to compare different types of tests, e.g. uniaxial compression with tension or bending tests.

5. For very brittle materials, or more generally, for products which fracture at low strains, fracture strain is hard to determine. This is especially the case if the surface of the food (e.g., in biscuits, chocolate bars, etc.) is not smooth. The product may fracture while the load is still only acting on the surface irregularities. Then the determination of fracture stress and strain in compression is not only badly reproducible, but in fact impossible (Fig. 7.8). In such circumstances, another measuring method such as three-point bending should be employed.

In summary, we can say that uniaxial compression is a useful method for routine determination of the large deformation and fracture properties

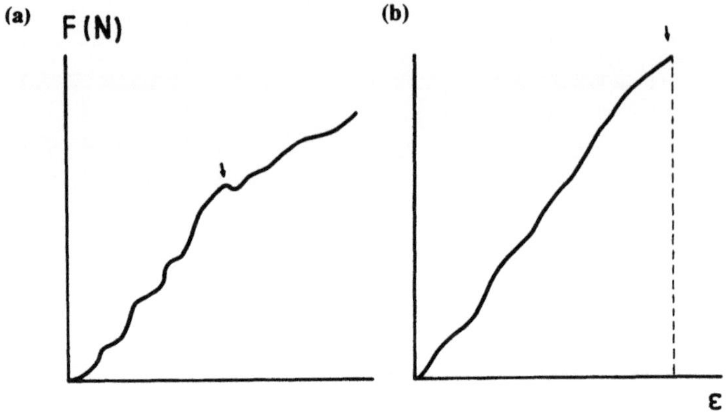

Fig. 7.8 Force F *versus* deformation ε curves for chocolate bars in (a) uniaxial compression and (b) three-point bending. Points where fracture was visible are indicated by the arrows.

of food systems (except for very brittle materials) provided one takes into account the drawbacks mentioned.

7.3.4 Uniaxial tension tests

This type of test largely parallels that of uniaxial compression. Tests are often performed on specially shaped test pieces in which an artificial notch may be made (Fig. 7.9). From the data obtained, the fracture stress, the fracture strain, the total work required for fracture, and Young's modulus may be calculated. Strain rate can be adjusted by varying the deformation rate. An important difference with uniaxial

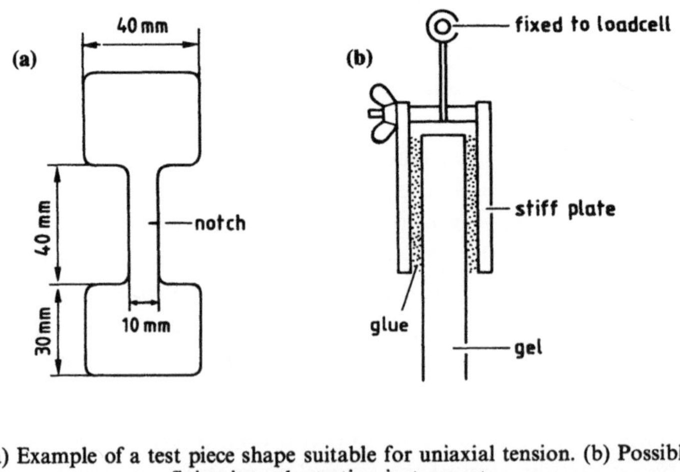

Fig. 7.9 (a) Example of a test piece shape suitable for uniaxial tension. (b) Possible way of fixing it to the testing instrument.

compression is that during an experiment at constant deformation rate the strain rate will decrease (equation (7.8)). There are also several other, more essential, differences which means that with tension tests, it is possible to determine all the more fundamental fracture parameters mentioned above. The main differences are described below.

The start of fracture in a uniaxial tension test can be noticed more easily and crack growth can be followed as a function of time. This allows the determination of the start of fracture initiation and of fracture propagation. Moreover, in theory, crack speed can be recorded. If an artificial notch is applied to the specimen, fracture normally starts at the tip of it, due to the stress concentration induced. By varying the notch length, notch sensitivity can be studied easily (Fig. 7.10) and the inherent defect length of the material can be determined from the notch length below which fracture stress is (nearly) independent of notch length (Luyten *et al.*, 1992). The specific fracture energy R_s can be determined in experiments on notched test pieces (Gordon, 1978) from the energy available in the volume of material around the notch per unit of fracture area, i.e.

$$R_s = \pi l W', \tag{7.9}$$

where l is the length of the notch and W' is the average elastic (stored) energy per unit volume in the whole test piece.

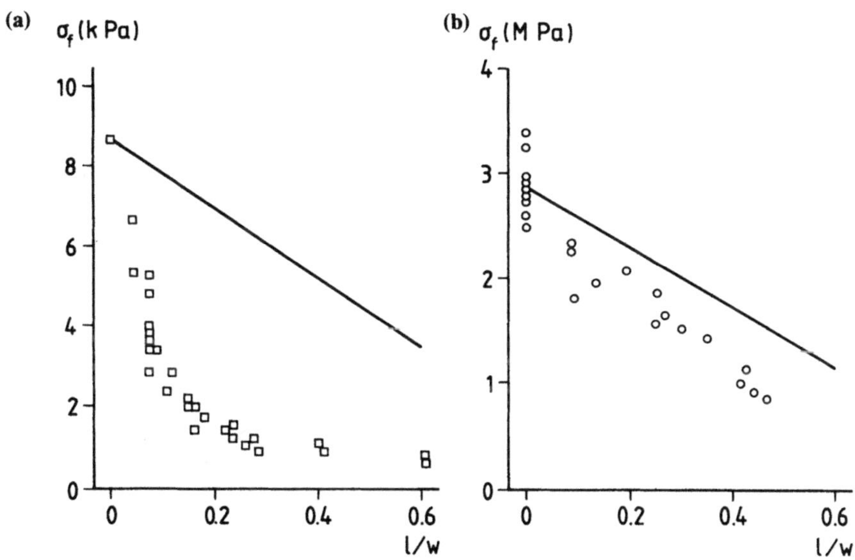

Fig. 7.10 Stress at fracture σ_f as a function of the relative length of the applied notch l for (a) 10 wt% potato starch gels in tension and (b) chocolate bars in three-point bending. The width of the starch gel is 10 mm, and the thickness of the chocolate bar is 4.5 mm. The lines give the theoretical dependence of σ_f on notch length for a notch-insensitive material.

The friction problem is completely avoided in a uniaxial tension test. However, this type of test also has a few drawbacks compared to uniaxial compression. Test piece preparation and the fixing of the sample to the instrument is much more difficult. Test pieces are required in general to be rather large, and therefore they have to be cut from fairly homogeneous large samples; this is a problem for many food products. Because they are often thin in the middle (e.g. as indicated in Fig. 7.9), fracture can occur readily during the fixing into the apparatus. A test piece may easily fracture at the grips. This can be largely avoided by using notched specimens of a dog-bone shape and special grips which do not damage the material. The glueing of the test piece to the measuring body is often a successful method. Because of the complicated shape of the test piece, however, it may be rather difficult to calculate the true stress and strain. Marking the test piece and filming the deformation of it during the test can be a good method to check the homogeneity of the deformation.

Summarizing, we note that uniaxial tension tests are good for determining fracture parameters, but, due to the fact that they are (rather) difficult to execute, they are not suited to routine testing.

7.3.5 Bending tests

Three-point bending tests are normally done (Fig. 7.11), but for very long (non-food) specimens four-point bending tests have also been described. In the latter case, the ratio of test piece length to diameter must be larger than is feasible for most food products. Bending tests are easy to perform, and it is not necessary to fix the specimen to the measuring body. Bending involves a combination of compression, tension and some shear. Fracture mostly starts at the outside of the tensed part of the test piece and therefore it can be observed. By applying artificial notches at that side of the test piece, notch sensitivity (Fig. 7.10), inherent defect length, and specific fracture energy can be determined.

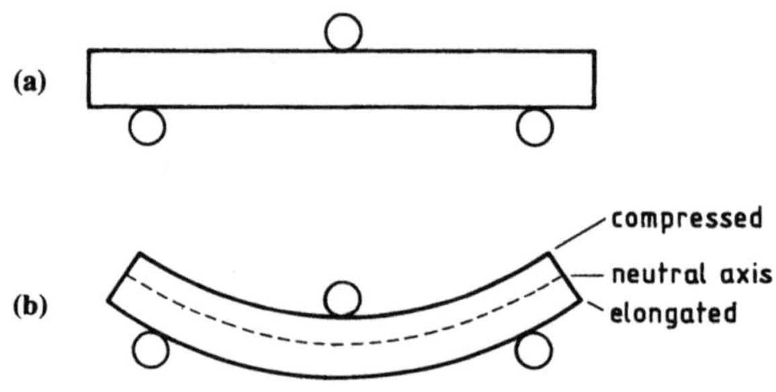

Fig. 7.11 Deformation during a three-point bending test: (a) before bending; (b) after bending.

For stiff food products, such as chocolate bars, it is possible to determine all the major fundamental fracture parameters with this type of test.

There are, however, several drawbacks of the method, as indicated below. Test pieces for bending have to be much longer than their diameter (thickness) which is often difficult to engineer. The required length (L) over diameter (d) ratio depends strongly on the type of material. Young (1989) gives for metal beams a minimum L/d ratio of 8 and for rectangular timber beams of 24. In practice, Luyten (1988) observed for cylindrical test pieces of Gouda cheese no effect of L/d on the observed fracture stress or strain for ratios of 3.3 or more.

The strain in the bent test piece is not nearly homogeneous (Fig. 7.11). The bottom side of the test piece is elongated while the upper side is compressed. In the middle is a 'neutral' axis. For large strains and for smaller span widths, the deformation contains a relatively important shear component.

Fracture strain must be low enough to allow calculation of the true fracture stress and strain (and to fracture the test piece at all). For the case of a rectangular shaped test piece with a height (thickness) b and a width d, the maximum stress σ_{max} and maximum strain ε_{max} can be calculated (Young, 1989):

$$\sigma_{max} = \frac{3FL}{2db^2}, \tag{7.10}$$

$$\varepsilon_{max} = \frac{6by}{L^2}. \tag{7.11}$$

Here, L is the span width between the supporting bars and y the displacement of the centre of the test piece in the direction of the load. Such stress and strain calculations are only accurate if (1) the strains involved are small, and (2) the stress and energy storage vary linearly with the distance from the neutral axis; this latter condition implies so-called linear elastic behaviour and an equal Young's modulus in compression and tension. This means, amongst other things, that the test material must be homogeneous.

There should not be a substantial bending of the test piece due to its own weight; so the modulus of the tested material should be rather high. In addition, the material must be rigid enough to prevent compression of the beam by the plunger.

Summarizing, we can say that three-point bending tests are very useful for studying rather brittle materials, and they are easily carried out.

7.3.6 Shearing tests

In these tests the applied deformations are clearly different from those in compression or tension. The shear stress and strain can be easily

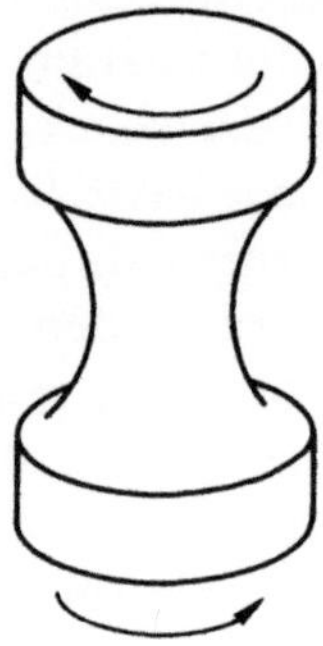

Fig. 7.12 Example of a dumbbell-shaped test piece suitable for use in a torsion experiment (after Montejano *et al.*, 1983).

calculated if a concentric cylinder geometry is used, but then the sample must be applied as a liquid. Zoon *et al.* (1989) studied in this way the large deformation and fracture behaviour of milk gels. In a torsion experiment the strain in the test piece is not homogeneous. Often special shaped test pieces are used (e.g. as shown in Fig. 7.12). Fixing of the specimen to the apparatus is rather difficult. It is often best in practice to glue it.

7.3.7 Controlled fracture tests

Several types of experiments are developed to obtain data for the fracture energy at an imposed and constant rate of fracture. Examples are wedge experiments, cutting experiments (with a knife or wire), the 'trouser-tear' test, the double cantilever beam test, and instrumented scissors (Vincent, 1992). Wedge experiments (Fig. 7.13) are successfully applied to several

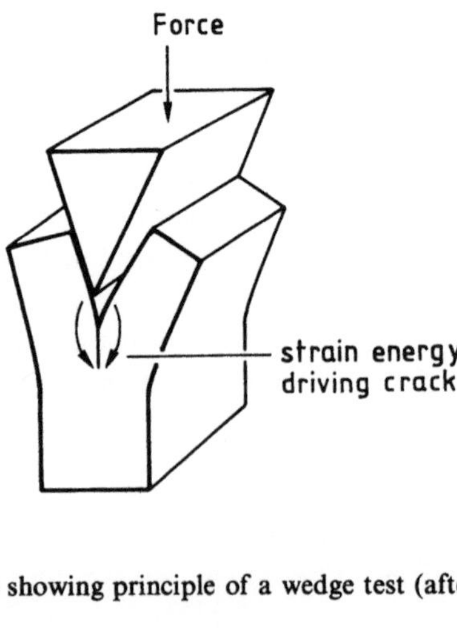

Fig. 7.13 Diagram showing principle of a wedge test (after Vincent, 1991).

food materials such as potato (Khan, 1989), apple (Vincent *et al.*, 1991), cheese (Luyten, 1988) and carrot (McGarry, 1993). Cutting experiments are very useful to obtain data for the fracture energy of soft materials like cheese (Luyten, 1988) or starch gels (Luyten *et al.*, 1992). In these experiments the materials have been cut with a wire. Cutting with a knife has been successfully applied to carrots (Atkins and Vincent, 1984).

By driving a wedge through a piece of material, the sides of the test piece are bent outwards (Fig. 7.13). The energy stored in the material by this process can be transported to the tip of the growing crack. The result is that the crack propagates ahead of the tip of the wedge. By adjusting the size and shape of the test piece, the rate of crack growth can be controlled, resulting in a crack speed equal to the driving speed of the wedge. If so, the force F_w required for this process is constant and the fracture energy can be calculated directly from F_w/d, where d is the width of the test piece (see below for cutting experiments). A typical force–time curve for a wedge or cutting experiment is shown in Fig. 7.14.

There are several advantages of wedge experiments: (1) The test piece can be very small. (2) The execution of the method and the calculation of the fracture energy is very simple, and so the test piece does not have to be clamped. (3) Fracture is in tension ahead of the wedge. Probably this is very similar to biting food with the teeth. Indeed, some good correlations with biting have been found (Vincent *et al.*, 1991). (4) Very small differences in fracture energy within a test piece, for example due to structural inhomogeneities, can be recognized.

Drawbacks of wedge experiments are as follows. (1) Only fracture energy can be determined; so additional experiments have to be done to obtain the other large deformation and fracture properties. (2) Side

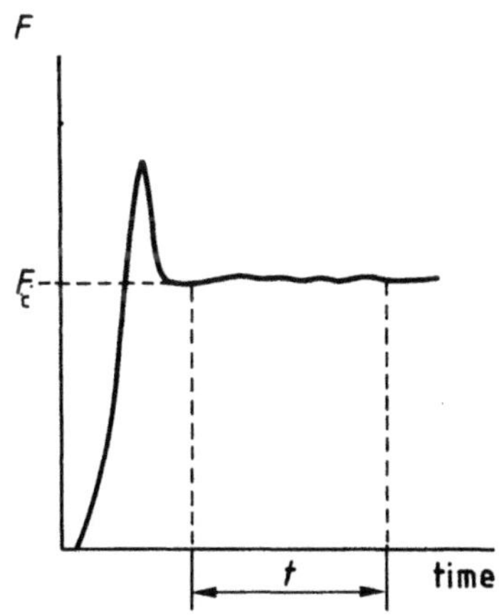

Fig. 7.14 Example of a force F *versus* time curve as observed in a wedge test or in a cutting experiment. F_c is constant value of F over time interval t.

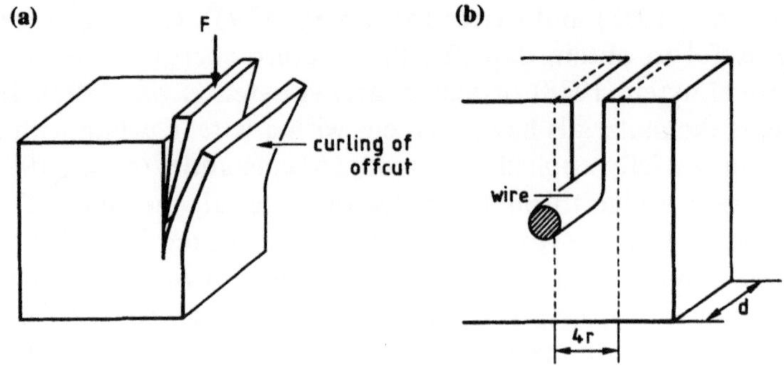

Fig. 7.15 Cutting experiments. (a) Cutting with a knife. (b) Cutting by a wire (after Luyten *et al.*, 1991). See text for explanation of symbols.

cracks may be formed, due to inhomogeneities in the structure of the material, especially if it is brittle. This means that the crack path is different from the imposed path of the wedge, and that no constant force is obtained. Then the fracture energy cannot be determined. (3) Friction between the wedge and the test piece may affect the measured force. Lubricating the wedge may help.

Cutting of food materials with an instrument, knife or wire (Fig. 7.15) is related to the wedge experiments described above. In this type of test, the specific fracture can be determined from the equation

$$W = W' + W_m''' + W_c'' + W_f + \text{curling energy.} \qquad (7.12)$$

The 'curling energy' stems from the curling of the piece of material that is cut of from the main block (Fig. 7.15); it can be determined by cutting slices of different thickness and extrapolating to zero slice thickness. When cutting a test piece of width d at speed v during a time t, the input energy W is $F_c vt$, where F_c is the required force for cutting. This energy is used to fracture an area dvt and to deform a volume of material around the wire of approximately $4rvt$ if the experiment is done by wire cutting. (Arbitrarily, it is assumed that the thickness of the disturbed layer around the wire is four times its radius r, but in reality this will depend on the rheological and fracture properties of the material; for cheese the assumption is roughly correct (Luyten *et al.*, 1991).) This gives

$$F_c vt = R_s vtd + (W' + W_m''' + W_c'')4rdvt. \qquad (7.13)$$

By performing experiments using wires of different thickness $F_c(r = 0)$ can be calculated by extrapolating to zero wire thickness so that $R_s = F_c(r = 0)/d$.

Advantages and drawbacks of cutting experiments are similar to those for wedge experiments.

7.3.8 *Less well-defined methods*

Certain fracture properties can also be studied by methods in which the deformation is less well defined. Examples are the determination of the difference between fracture initiation and propagation by 'biting' between wedges (Vincent *et al.*, 1991) and impact experiments which involve using an impacter (e.g. a wedge) falling in a controlled way (often via a circular trajectory) on to the test piece (Vincent, 1992).

Acknowledgement

The authors thank Professor P. Walstra for many valuable and stimulating discussions on fracture mechanics.

References

Atkins, A.G. and Mai, Y.-M. (1985) *Elastic and Plastic Fracture*, Ellis Horwood, Chichester.

Atkins, A.G. and Vincent, J.F.V. (1984) An instrumented microtome for improved histological sections and the measurement of fracture toughness. *J. Mater. Sci. Lett.*, **3**, 310–312.

Bremer, L.B.G., Bijsterbosch, B.B., Schrijvers, R., Vliet, T. van and Walstra, P. (1990) On the fractal nature of the structure of acid casein gels. *Colloids Surf.*, **61**, 159–170.

Casiraghi, E.M., Bagley, E.B. and Christianson, D.D. (1985) Behaviour of Mozzarella, Cheddar and processed cheese spread in lubricated and bonded uniaxial compression. *J. Texture Stud.*, **16**, 281–301.

Chu, C.F. and Peleg, M. (1985) The compressive behaviour of solid food specimens with small height to diameter ratios. *J. Texture Stud.*, **16**, 451–464.

Gent, A.M. and Lindley, P.B. (1959) The compression of bonded rubber blocks. *Proc. Inst. Mech. Eng.*, **173**, 111–122.

Gordon, J.E. (1978) *Structures, or Why Things Don't Fall Down*, Penguin Books, Middlesex, UK.

Khan, A.A. (1989) *Mechanical Properties of Fruit and Vegetables*, Ph.D. thesis, Reading University, UK.

Luyten, H. (1988) *Rheological and Fracture Properties of Gouda Cheese*, Ph.D. thesis, Wageningen Agricultural University, Wageningen, The Netherlands.

Luyten, H., Vliet, T. van and Walstra, P. (1991) Characterization of the consistency of Gouda cheese: fracture properties. *Neth. Milk Dairy J.*, **45**, 55–80.

Luyten, H., Vliet, T. van and Walstra, P. (1992) Comparison of various methods to evaluate fracture phenomena in food materials. *J. Texture Stud.*, **23**, 245–266.

McGarry, A. (1993) Influence of water status on carrot fracture properties. *J. Horticult. Sci.*, **68**, 431–437.

Montejano, J.G., Hamann, D.D. and Lanier, T.C. (1983) Final strength and rheological changes during processing of thermally induced fish muscle gels. *J. Rheol.*, **27**, 557–579.

Peleg, M. (1984) A note on the various strain measures at large compressive deformations. *J. Texture Stud.*, **15**, 317–326.

Vincent, J.F.V. (1992) Fracture, in *Biomechanics, Materials. A Practical Approach*, (ed. J.F.V. Vincent), IRL Press, Oxford, pp. 193–217.

Vincent, J.F.V., Jeronimidis, G., Khan, A.A. and Luyten, H. (1991) The wedge fracture test. A new method for measurement of food structure. *J. Texture Stud.*, **22**, 45–57.

Vliet, T. van (1991) Inventory of test methods, in *Rheological and Fracture Properties of Cheese*, IDF Bulletin 268, Brussels, pp. 16–25.

Vliet, T. van and Peleg, M. (1991) Effects of sample size and preparation, in *Rheological and Fracture Properties of Cheese*, IDF Bulletin 268, Brussels, pp. 26–29.

Vliet, T. van, Luyten H. and Walstra, P. (1991) Fracture and yielding of gels, in *Food Polymers, Gels and Colloids* (ed. E. Dickinson), Special Publication No. 82, Royal Society of Chemistry, Cambridge, pp. 392–403.

Vliet, T. van, Luyten, H. and Walstra, P. (1993) Time dependent fracture behaviour of food, in *Food Colloids and Polymers: Stability and Mechanical Properties* (eds. E. Dickinson and P. Walstra), Special Publication No. 113, Royal Society of Chemistry, Cambridge, pp. 175–190.

Young, W.C. (1989) *Roark's Formulas for Stress and Strain*, McGraw-Hill, New York.

Zoon, P., Vliet, T. van and Walstra, P. (1989) Rheological properties of rennet-induced skim milk gels. 5. Behaviour at large deformation. *Neth. Milk Dairy J.*, **43**, 35–52.

Chapter 8

Recent Developments in Infrared Spectroscopy and Microscopy

R.H. Wilson
Institute of Food Research,
Norwich, UK

8.1 Introduction

Infrared spectroscopy is an increasingly important and widely used technique for food analysis, with most applications involving the near-infrared (NIR). This region has been most exploited because the sample presentation is relatively easy, allowing complex heterogeneous systems to be readily analysed. The vast majority of NIR applications are, however, quantitative since the spectra are difficult to interpret qualitatively. In order to extract quantitative information from NIR data, multivariate statistical methods are often employed, and quite a large number of calibration samples are required.

The mid-infrared region (MIR) has only recently attracted the attention of food analysts. The development of Fourier transform instrumentation and new methods of sample presentation are the main reasons behind this increased interest, since, in combination, they allow spectra to be acquired rapidly and easily from most food samples. MIR spectra are easier to assign, and they can be interpreted to yield information on, for example, molecular structure, conformation and crystallinity. Furthermore, many more functional groups are seen in the MIR than in the NIR, which means that the former contains much more compositional information. However, MIR spectra can be difficult to interpret quantitatively because the infrared beam only penetrates a short distance into the sample (typically 20 μm) and so representative sampling of heterogeneous samples may not be possible. Nevertheless, the number of quantitative applications using MIR spectroscopy continues to increase, particularly as many of the multivariate methods originally developed for the NIR are being adapted for use in the MIR. This paper will deal exclusively with the newer MIR developments.

The development of multivariate statistical methods for use in the MIR has been one of the most important advances in the last few years.

Although quantitative analysis is still the main application for these methods, there is increasing interest in classification techniques which can be used to assign samples to defined groups. An example of this type of work is given in which infrared spectra are used to classify edible oils according to botanical origin. This example highlights the power of statistical methods, properly applied to spectroscopic data, for extracting information from spectra that are superficially very similar. A second example illustrates the potential for non-invasively probing molecular changes in the individual components of a complex, intact food system. The other recent development discussed is infrared microspectroscopy, a technique which allows infrared spectra to be obtained from very small samples. The use of microspectroscopy to provide insight into the composition and structure of the plant cell wall is described.

8.2 Theory

8.2.1 Origin of infrared spectra

Both infrared and Raman spectroscopies involve transitions between quantized vibrational energy levels and they are thus both forms of vibrational spectroscopy. The infrared region is divided into the near-, mid- (or middle) and far-infrared, a division based on instrumental factors as well as the types of vibration that occur in each region. The easiest to understand conceptually is the mid-infrared (MIR) which is usually considered to lie between 2.5 μm and 25 μm in wavelength. It is normal, however, for vibrational spectroscopists to use the term wave-number, with the units of cm^{-1}, which is actually a frequency term. The MIR then extends from 4000 cm^{-1} to 400 cm^{-1}.

A bond between two atoms can be considered to be like a spring with a specific elasticity or force constant k. The bond, or spring, can be stretched and caused to vibrate, and it will do so at some characteristic frequency f that depends upon k and the masses of the two atoms according to Hooke's law,

$$f = \frac{1}{2\pi}\sqrt{\frac{k}{\mu}}, \qquad (8.1)$$

where μ is the reduced mass defined as

$$\mu = \frac{m_1 m_2}{m_1 + m_2} \qquad (8.2)$$

and m_1 and m_2 are the masses of the individual atoms forming the bond. Equation (8.1) predicts that a particular bond will give rise to a characteristic frequency which depends upon the masses of the atoms and the strength of the bond. For example, a C=O bond, with a greater force constant than a C—O bond has a higher vibrational frequency despite both groups having identical reduced mass. In practice, different functional groups give rise to characteristic vibrational frequencies, and the exact frequency can be affected by groups adjacent to the vibrating bond due to vibrational coupling. This intramolecular coupling further increases the diagnostic power of the technique. The main use of vibrational spectroscopy is as a powerful technique for the identification of specific functional groups and for molecular structural determination. It has greater selectivity than, for example, UV/vis spectroscopy in this respect.

Equation (8.1) is derived from classical physics, and so it takes no account of quantum theory. In practice, the energy of molecular vibrations is quantized (Banwell, 1983) to satisfy

$$E_n = (n + \tfrac{1}{2})\, hf, \tag{8.3}$$

where n is the vibrational quantum number ($n = 0,1,2,3,\ldots$), h is Planck's constant, and f is the classical vibrational frequency in equation (8.1). At room temperature most molecules will be in the ground vibrational energy state E_0 given by

$$E_0 = \frac{1}{2}\frac{h}{2\pi}\sqrt{\frac{k}{\mu}}. \tag{8.4}$$

The energy E_1 of the first excited vibrational state is

$$E_1 = \frac{3}{2}\frac{h}{2\pi}\sqrt{\frac{k}{\mu}}. \tag{8.5}$$

A transition from the ground state to the first vibrational state may occur and the energy required for this is

$$\Delta E = E_1 - E_0 = \frac{h}{2\pi}\sqrt{\frac{k}{\mu}}. \tag{8.6}$$

The energy for this type of transition can be supplied by infrared radiation. However, for absorption to occur a selection rule must be satisfied which requires there to be a change in the dipole moment of the bond during the vibration.

The absorption bands in the MIR are known as fundamentals. However, various overtones and combinations of these fundamentals can arise. For example, a molecule with two fundamentals at frequencies v_1 and v_2 may give overtones at $2v_1$, $3v_1$, $4v_1$, or $2v_2$, etc., or combinations at say $v_1 + v_2$ or $2v_1 + 2v_2$. In practice, not all fundamentals give rise to overtones, usually only bonds in which a heavy atom such as nitrogen or oxygen is coupled to hydrogen. The overtone and combinations constitute the region of the near-infrared (NIR), 2.5–0.7 μm, which, despite the apparent complexity of the spectra, has found considerable application in food analysis. To complete the picture, at lower energy there is the far-infrared (400–10 cm^{-1}) which has major application in inorganic chemistry as bonds between metals and organic ligands appear here as well as skeletal vibrations of molecular backbones. This region has been of limited application in food and nutritional studies.

8.2.2 Absorption laws

In UV, NIR and MIR spectroscopy, the fundamental law governing absorption is the Beer–Lambert relationship. For a sample illuminated by radiation of intensity I_0 the amount transmitted, I, is given by

$$I = I_0\, 10^{-\varepsilon c l}, \tag{8.7}$$

where c is the concentration of absorbing species, l is the path length through which the light passes, and ε is the molar extinction coefficient.

For quantitative analysis, spectra are usually presented in terms of absorbance units, A, defined by

$$A = -\log \frac{I}{I_0} = \varepsilon c l, \tag{8.8}$$

so that the quantity A is directly proportional to concentration at constant path length.

8.2.3 Fourier transform infrared spectroscopy

The MIR region has largely been ignored in food studies, since in the past there have been a number of problems associated with the collection of good quality spectra suitable for quantitative work. However, using Fourier transform infrared (FTIR) spectroscopy, many of these problems can now be overcome (Wilson, 1990; Wilson and Goodfellow, 1994). FTIR spectrometers, using interferometers rather than conventional monochromators, have a number of advantages, including the ability to acquire information at all wavelengths simultaneously (the multiplex advantage), high wavelength stability, and the delivery of more power at the sample position (Griffiths and de Haseth, 1986). FTIR has allowed the use of a whole new range of sample presentation methods, and the most flexible of these is probably attenuated total reflectance (ATR). ATR overcomes the main sampling problems in the MIR which are the strong absorption of water, and the opacity of most samples. The short, reproducible optical path of an ATR device allows the spectrum of a wet sample to be collected without the complication of water bands totally absorbing at certain wavelengths. The spectral contribution of water can then be eliminated by dividing by, or subtraction of, the spectrum of pure water. This allows constituents to be analysed that would normally be obscured by the water absorption. ATR does not require the sample to be optically transparent in the conventional sense, and spectra can be obtained from materials as diverse as chocolate, oil, bread and aqueous solutions (Wilson and Goodfellow, 1994).

In comparison to the corresponding NIR spectrum, the MIR spectrum of a typical sample will tend to exhibit a higher degree of spectral resolution, and various peaks will be present that can be assigned to specific chemical entities or individual product constituents. Quantification can often be very easy, requiring only simple linear regression methods. A further advantage of the MIR region is that even quite similar molecules (e.g. sugars) give rise to characteristic MIR fingerprints as a result of vibrational coupling, so that quantitative analysis of complex mixtures using multivariate methods is also possible, and often quite straightforward. Unfortunately, because of the short optical paths used in the MIR region, sample homogeneity can be a severe problem. As a consequence, the most successful systems for quantitative work are usually homogeneous liquids. However, this is not always the case, as will become apparent.

8.3 Applications

8.3.1 Functional properties of food biopolymers: bread staling

Bread staling has been investigated using a range of physical methods, including thermal techniques such as differential scanning calorimetry (DSC) as well as simple rheological measurements of crumb firmness (Willhoft, 1976). During the gelatinization of starch, an order–disorder transition occurs. This process is reversed during the ageing of starch in gels or bread, as the amylopectin fraction crystallizes. When aged samples are measured by DSC, an endotherm (the staling endotherm, *s*) is observed, as the crystalline amylopectin becomes disordered once more. The time-scales for endotherm development and for the increase in crumb firmness are broadly similar in magnitude. However, techniques such as DSC are invasive, and in a complex system any order–disorder transition cannot be unambiguously ascribed to any one component. Spectroscopic techniques, on the other hand, can be non-invasive and allow direct observation of specific molecular functional groups in a system. The shape, and position of bands, obtained from these techniques can also give information about the physical state of the system. FTIR spectroscopy, in the MIR region, in combination with attenuated total reflectance (ATR), has been used to follow retrogradation in 20 wt%

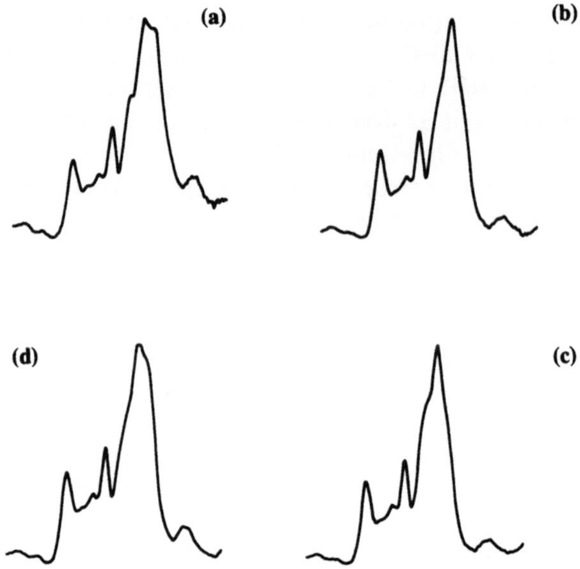

Fig. 8.1 Summary of changes occurring in MIR spectra of starches. Native starch in water (a) shows many peaks and shoulders. The shoulders are largely lost after gelatinization leading to a simplification of the spectrum (b). Retrogradation causes band-narrowing and a peak (shoulder) re-appears at the high frequency (left) side of the main peak (c). Dehydration of the sample leads to a shoulder forming on the low frequency (right) side of main peak (d).

waxy maize (Wilson *et al.*, 1987) and wheat starch gels (Wilson and Belton, 1988). During the storage of these materials, a line-narrowing process was observed in bands in the region 1200–1000 cm^{-1} (Fig. 8.1), assigned to coupled C—O and C—C vibrations. The time course of this process was found to be similar to that observed from changes in shear modulus. It was also shown that the FTIR/ATR methods could provide high-quality spectra of the starch fraction of bread (Wilson and Belton, 1988), from which processes could be observed that were similar to those seen in starch gels.

More recently, a direct comparison has been carried out of a range of potential physical techniques available for the study of the staling of bread. These techniques included DSC, NIR and MIR (Wilson *et al.*, 1991). The MIR results showed that the staling process in bread resulted in apparent band growth at 1047 cm^{-1} as a result of a general line-narrowing process. The cause of these changes was traced to processes occurring at the molecular level. In the disordered state amylopectin has a wide range of possible conformations. As retrogradation takes place, the system becomes more ordered and the range of conformations is reduced, resulting in a smaller distribution of bond energies, particularly

Table 8.1 Rate constant for staling process as obtained by various physical techniques

Technique	Initial value	Limiting value	Rate constant ($kh^{-1} \times 10^3$)
DSC (J g^{-1} sample)	0.00	1.71	16.06
FTIR[a]	1.04	1.12	12.90
NIR log(1/*R*)	1.52	1.33	16.31

[a]FTIR measurements were made, after Fourier self-deconvolution, by taking the ratio of an apperent peak at 1047 to one at 1150 cm^{-1}, the latter chosen to act as an internal reference.

hydrogen bond energies, compared to the initial state. This process is reflected in the MIR spectrum as a reduction in line width. With closely spaced peaks such line width reduction can give the impression of band growth. The rate constant obtained from the MIR data (band narrowing measurement) was found to be of the same order of magnitude as that from the other techniques, but the MIR value was found to be slightly lower. (These results are summarized in Table 8.1.) The reason for this was thought to arise from the different sensitivities of the techniques. NIR, for example, is strongly affected by changes in the scattering coefficient of the samples, reflecting changes in the physical state of the sample arising from the development of crystallinity. The crystallization of amylopectin may be a complex process in which the amylopectin chains undergo a coil-to-helix transition, followed by aggregation. Both DSC and NIR essentially measure the result of both of these processes, whereas MIR has been shown to be sensitive only to short range interactions and, consequently, measures only part of the retrogradation process. Nevertheless, the MIR results demonstrate that it is possible to follow retrogradation (staling) in bread using a spectroscopic technique in a relatively non-invasive way. The use of ATR allows the direct observation of changes in the MIR spectrum resulting from conformational transitions in the starch component alone.

8.3.2 Food authentication: edible oils

Adulteration of added value food products, usually involving the replacement of high-cost ingredients with lower grade, cheaper substitutes, is a potentially very lucrative practice for an unscrupulous manufacturer or raw material supplier. Composition, as declared on the label by the manufacturer, often legally determines product nomenclature and price. Chemical or sensory analysis of the product is often the only way of verifying the product composition. Substitution or adulteration of a product with a cheap ingredient is not only a major economic fraud, but can have severe health implications. Authentication of foods and raw materials is a major analytical challenge, and is often a time-consuming and laborious process. Very often, complex chemical treatment of the sample and the use of sophisticated instruments are required. Various physical and chemical tests have been proposed to establish the authenticity of olive oil, for example, and to detect the level of adulterants. UV spectroscopy based on the absorbance in the regions of 208–210 nm and 310–320 nm has been used to detect adulteration of virgin olive oil with refined olive oil. A second derivative spectrophotometric method was reported to be able to detect adulteration at a level of 6%. Analysis of the fatty acid profile by gas chromatography is a well established method for the quantitative analysis of mixtures of olive and other seed oils. HPLC analysis of the triglycerides and fatty acid composition is also widely used to detect adulteration in olive oil. Nuclear magnetic resonance and a spectrofluorometric method have also been reported.

Infrared spectroscopy has been used as an analytical tool in the oil and fat industries for many years. However, the MIR region has so far not been used widely for the detection of adulteration in olive oils, and the most likely reason is that most vegetable oils contain the same fatty acids (mainly with carbon number 16 or 18) and their triglyceride contents are also quite similar (carbon number 50, 52 or 54). As a result, the MIR spectra of most seed oils are superficially similar (Fig. 8.2). For oils as chemically alike as the various grades of olive oil, the spectral differences

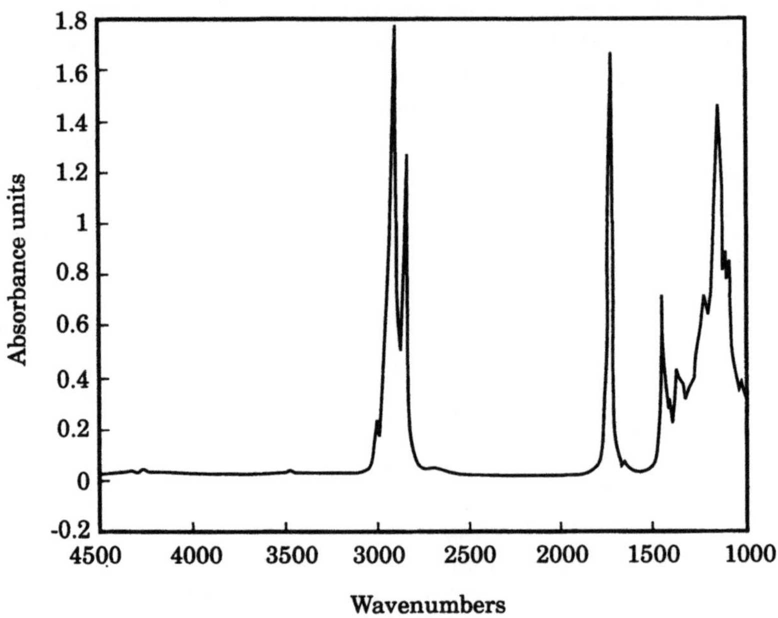

Fig. 8.2 Superimposed spectra of extra virgin olive oil and ordinary olive oil as acquired by FTIR with attenuated total reflectance sampling.

are very small indeed. Nevertheless, subtle differences between spectra do exist, in the structure of the polyethylene chains and the spectrum of an oil actually contains a great wealth and variety of information about the composition of the sample. The analytical problem is to extract the information from the spectra in such a way that it can be used to differentiate the oil types.

Principal components analysis (PCA) has been used as a first step since it is an efficient data reduction method. A typical spectrum contains several hundred data points (or variates); a typical data set comprises perhaps several tens of such spectra (or observations). Hence, the number of variates far exceeds the number of observations. This is not a propitious situation for the application of multivariate statistical procedures, which require that the number of observations exceeds the number of variates. PCA offers a means of overcoming this difficulty. The variates in the original data set will normally be correlated with one other, to a greater or lesser extent. PCA removes this redundancy by transforming the original data into a set of new, uncorrelated variates, termed the principal component (PC) scores. In doing so, a re-arrangement takes place, such that only the first few PC scores are required to

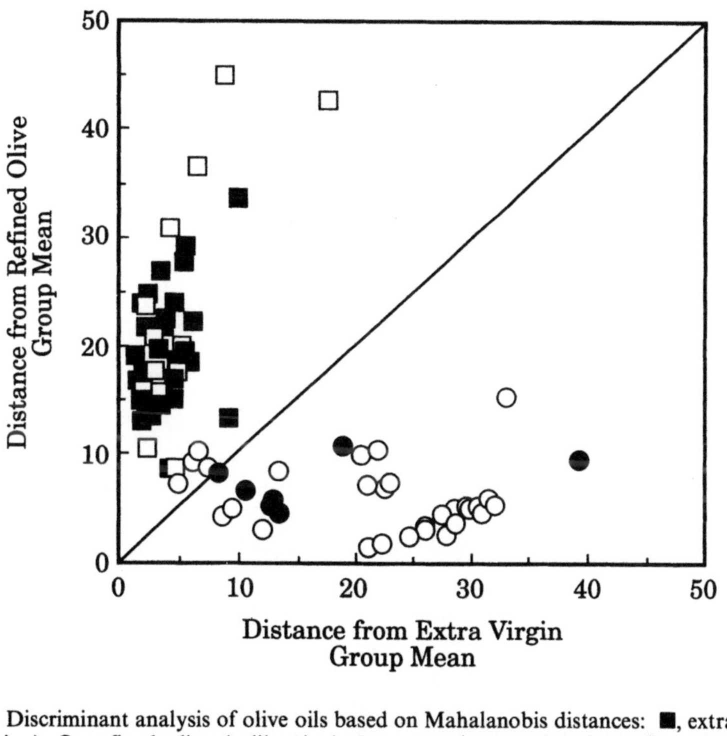

Fig. 8.3 Discriminant analysis of olive oils based on Mahalanobis distances: ■, extra virgin (calibration); ○, refined olive (calibration); □, extra virgin (validation); ● refined olive (validation). Reproduced from Lai *et al.* (1993) with permission.

describe the information contained in the many original variates. The number of significant PC scores is less than the number of observations, and multivariate methods (such as the calculation of Mahalanobis distances required in discriminant analysis) can proceed. A further advantage of the reduction in variates is the resultant simplification of the data set, enabling easier visualization of relationships within the data. Important relationships between observations may be revealed in bi-plots of PC scores against one another. (In contrast, pairs of original variates account for much smaller percentages of the total information present in the data set, and are usually correlated to some extent; bi-plots of these are unlikely to yield any interesting findings.)

Bi-plots of PC scores have been used to explore data clustering in a set of oils of differing botanical origin, as well as in a set of extra virgin olive oil and other olive oils (Lai *et al.*, 1994). In this study 93% of spectra in the extra virgin/other olive oils training set and 100% in the validation set were correctly classified according to type (Figs 8.3 and 8.4). It is interesting to note that these oils, which are chemically the most similar, and which were the most closely clustered in the exploratory PC analysis, can be almost completely discriminated using this technique. These two types represent the most challenging set for discriminant analysis. Discrimination between olive oils and other seed

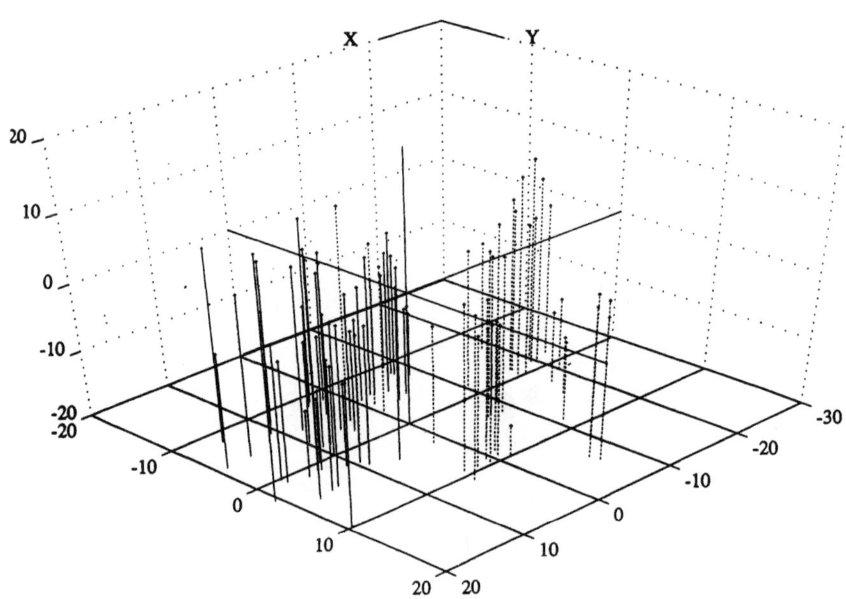

Fig. 8.4 Three-dimensional discriminant analysis using principal component scores showing clear separation of two oil groups (extra virgin and ordinary olive oils).

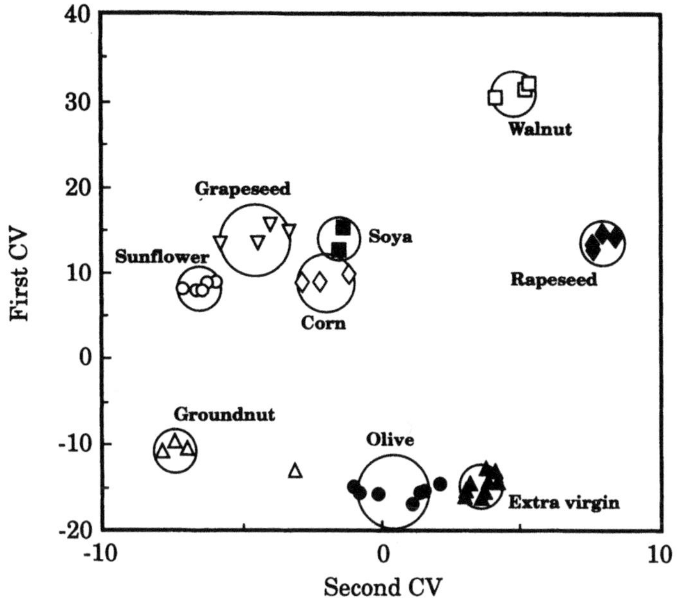

Fig. 8.5 Canonical variates (CV) analysis (based on eight principal components) showing clear clustering of oils from different botanical sources: m olive; ▽, grapeseed; ◆, rapeseed; ◇, corn; ▲, extra virgin; ▽, groundnut; ○, sunflower; ■, soya; □, walnut.

oils has proved to be even more conclusive (Fig. 8.5). Successful discriminant analysis is the first step towards producing a method which not only correctly classifies oils, but also reliably detects adulterated samples.

8.3.3 Determination of composition and structure of plant cell walls

Changes in cell wall composition and structure occur during the plant cell growth and differentiation cycle, and these changes may, in turn, feed back and influence events within the cell. To understand this dynamic relationship between the plant cell and its wall, a coherent model of wall architecture at the molecular level is required. At present such a conceptual framework does not exist, largely because appropriate structural methods and suitable probes are lacking. Accordingly, much molecular machinery underlying plant growth and development remains obscure. A minimal requirement is to understand how the three relatively invariant cell wall polymer classes, cellulose, hemicellulose and pectin, are put together in space, and for this a simple cell wall system and appropriate methods for uncovering its architectural principles are required. One approach to this problem has been to choose a chemically simple system, onion parenchyma cell wall, and to look at its architecture directly by way of fast-freeze, deep-etch rotary-shadowed replicas. This technique has enabled the definition of some basic parameters of wall construction. For example, cellulose microfibrils, diameter 5–12 nm are cross-linked by hemicellulose bridges about 20–40 nm long which prevent the lateral association of cellulose. Middle lamella pectins are Ca^{2+} cross-linked and ester-linked to primary cell wall pectins from a network

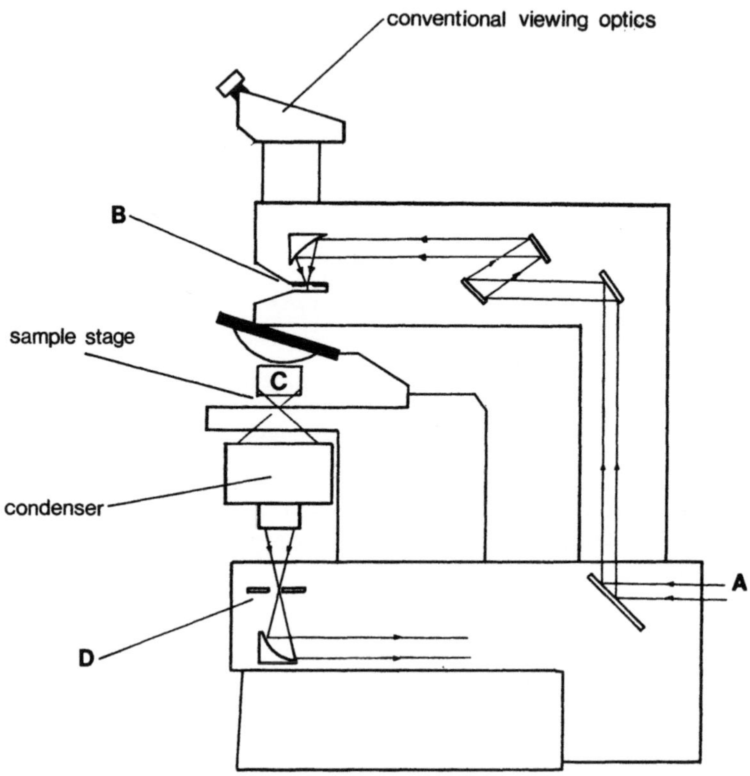

Fig. 8.6 The infrared microscope is externally very similar to an ordinary light microscope. Indeed, the internal structure is identical except that all-reflecting silver mirrors replace the conventional glass lenses to permit transmission of infrared light. The sample is mounted on an infrared transparent disc in the field of view using white light illumination. At an appropriate time, infrared light is brought into the microscope from the spectrometer at point A. After reflection from many mirrors, it is focused at point B. Here the part of the sample to be illuminated is controlled by adjusting an aperture. After passing through the aperture, the light is eventually focused by the objective C onto the sample. The light passing through the sample is then collected by the condenser and refocused through a second adjustable aperture D. The infrared light then arrives at the detector and a spectrum is generated. The smallest sample area from which a useful spectrum can be collected, limited by diffraction, is about 10×10 cm.

which is co-extensive with, but independent of, the cellulose/hemicellulose network. Pectins limit wall porosity and a neutral galactan is associated with microfibrils throughout extraction. There is no evidence for the existence of a cross-linked extensive network in this tissue.

FTIR spectroscopy has been a used as a probe of the various chemical and conformational changes that occur during sequential polymer extraction from the cell. The results indicate a wide potential for the use of FTIR microspectroscopy in cell wall research. For example, it can be used as a crude, but rapid chemical assay system for walls and polymers, to detect conformation changes in wall components, and to provide information about cross-links present in the wall. Infrared spectra can be used for both structural elucidation and compositional analysis. Coupling of vibrations can somewhat complicate spectra, but an important result of the coupling is to produce fingerprint patterns in the region 1200–900 cm^{-1} so that even complex polymers can be distinguished from one another.

Coupling the spectrometer to an appropriate microscope allows selection of a particular area (as small as 10 × 10 μm) in the field of view for microsampling (Fig. 8.6). Thus, spectra can be obtained from a defined region of a single cell wall as from bulk samples (Fig. 8.7). In scanning mode, the beam passing through the spectrometer sample compartment is diverted vertically downwards to pass through the sample on the stage and then it continues to a movable mirror assembly positioned to send it to the detector. Quality spectra from both single onion cell walls and their constituent polymers have been obtained under a variety of sampling conditions (McCann *et al.*, 1992). The specificity of the chemical extraction sequence used in the preparation of the material was confirmed; only pectins are extracted by cyclohexanediaminetetraacetic acid

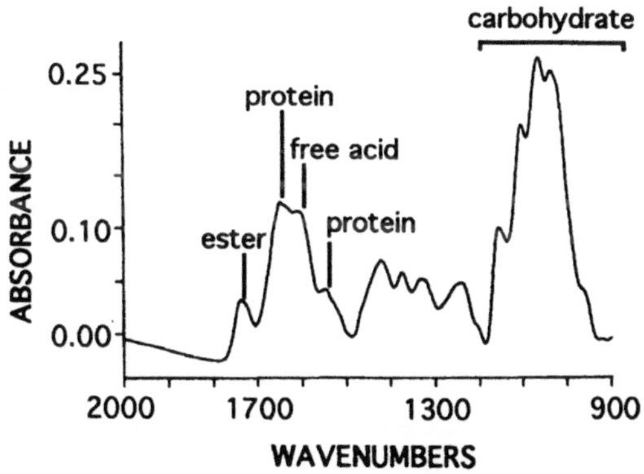

Fig. 8.7 Spectrum of small area (50 × 50 μm) of cell wall of tomato illustrating the main features of interest for analytical purposes.

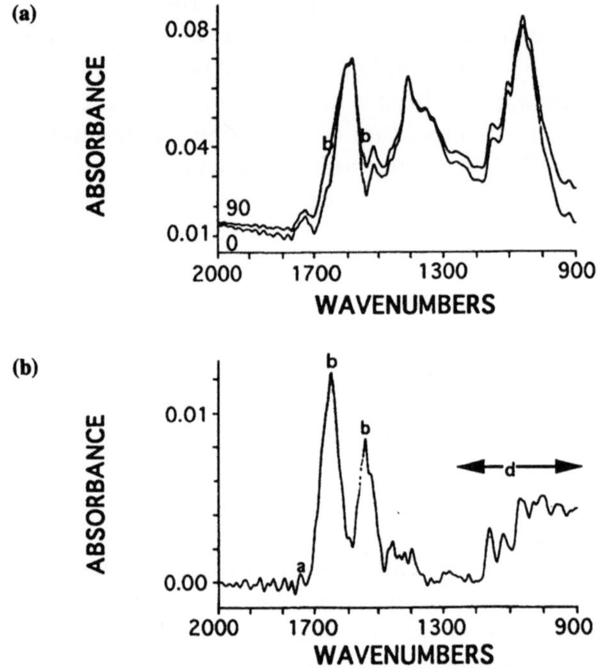

Fig. 8.8 Polarized infrared spectra of elongated carrot cell walls. The top spectra (a) were collected with the infrared light polarized at 0° and 90° to the direction of growth. Protein peaks (amide I and II) are indicated by symbol b. The difference between the two spectra can be obtained (b). This clearly shows that protein is preferentially aligned along the direction of growth. There is also some alignment of pectins, as shown by peak a and the peaks in the carbohydrate region (shown as d in spectrum (b)).

and sodium carbonate, whilst xyloglucans are extracted by increasing concentrations of potassium hydroxide. In the latter case, there was found to be very little contamination of the first potassium hydroxide extract with residual pectin. The low abundance of both phenolics and protein was also confirmed. The first sodium carbonate extraction almost completely de-esterifies the cell wall. MIR spectroscopy can detect large conformational changes in pectic polymers on removal from the cell wall and on drying. MIR spectroscopy provides a powerful and rapid assay for wall components and putative cross-links by identifying polymers and functional groups non-destructively *in muro*.

More recent work has shown that, if polarized infrared radiation is used, it is possible to show that biopolymers such as proteins, phenolics and pectins are orientated in cell walls (Fig. 8.8) and that changes in their orientation occur during growth (McCann *et al.*, 1993). It has also been possible to measure these changes in mixed cell systems such as stem

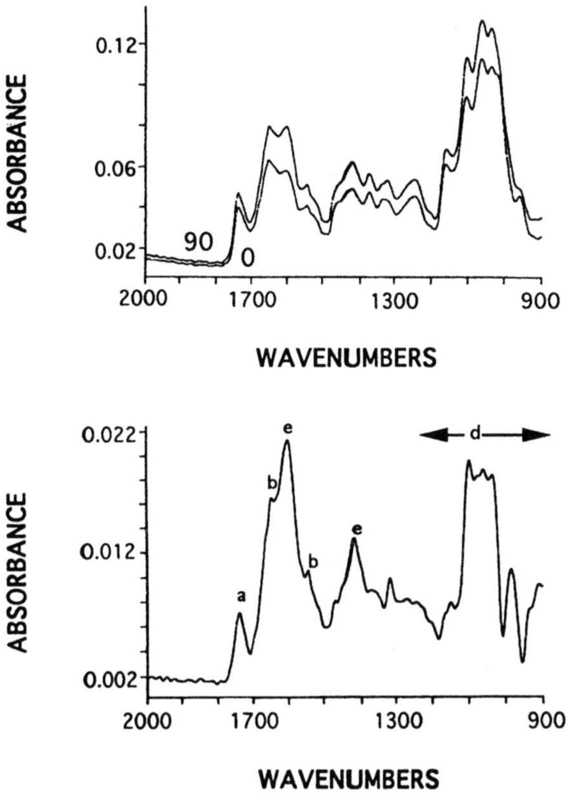

Fig. 8.9(a)

sections (Fig. 8.9). Polarized infrared spectra of cells from the epidermis, cortex and xylem of carrot root sections were acquired. The results showed that different molecules were aligned in the different cell types. In the epidermis, both acidic and esterified pectin was found to be aligned, as well as a small amount of protein. Inspection of the original spectra of the epidermis showed that a considerable amount of material was aligned. In contrast, the xylem contained less aligned material which appeared to be mainly protein, with some indication of oriented phenolic material. Evidence has been obtained that shows that, during growth, acidic pectins are converted to the methyl esterified pectins, which are then de-esterified following growth (Fig. 8.10). This is the first time that direct molecular evidence has been obtained for this process. The strength of the pectin network is an important factor in determining the ease of expansion of other molecules in the wall. During growth, the strength of the pectin network is reduced to facilitate expansion, with a rigid network re-formed after growth to give strength to the wall. This work has shed new light on cell wall structure and has led to a reinterpretation of growth mechanisms.

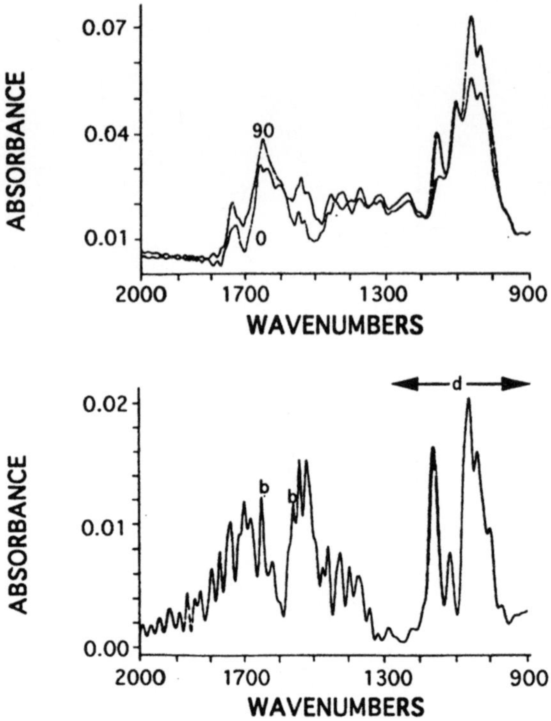

Fig. 8.9(b)

8.4 Conclusions

The use of the MIR region is now widespread in food analysis. Technical improvements continue to open new possibilities for applications to increasingly difficult problems. The application of chemometric methods, unlocking previously inaccessible information, has been perhaps one of the most important recent developments in MIR spectroscopy. Microspectroscopy is also an important new area that is ripe for exploitation and the potential of which is only just now being realized.

Acknowledgements

The author thanks E.K. Kemsley and M. McCann for their helpful contributions to the preparation of this work.

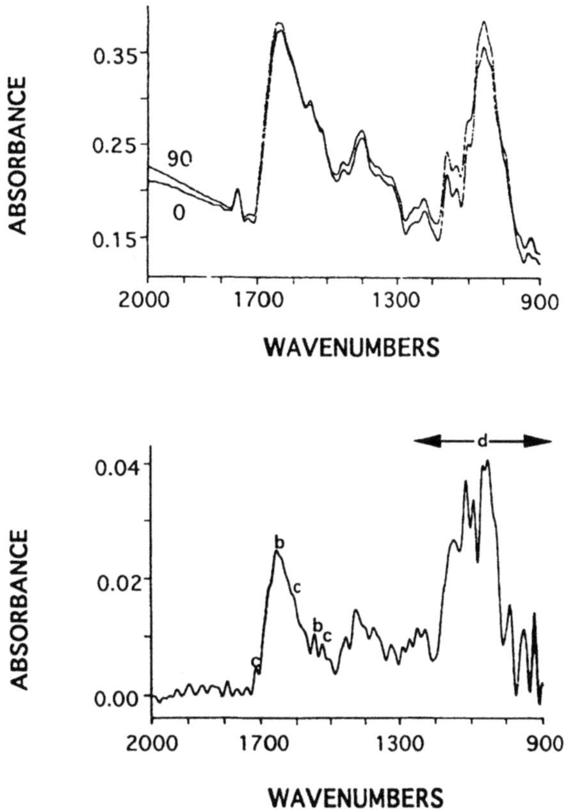

Fig. 8.9 Polarized infrared spectra of carrot stem section showing orientation of molecules in the (a) epidermis, (b) cortex and (c) xylem. In each case the upper spectra are the infrared spectra collected from 50 × 50 μm areas with the infrared radiation polarized both 0° and 90° to the direction of growth. The lower spectrum in each case is the difference spectrum (i.e. spectrum at 0° minus spectrum at 90°). The results show that different molecules are aligned in the different cell types. In the epidermis, pectin is aligned, as witnessed by the presence of peaks 'a' and 'e'. Peak 'a' arises from esterified pectin while the peaks labelled 'e' arise from acidic pectins. The peaks labelled 'b' show that some protein is aligned. Inspection of the original spectra of the epidermis shows that there is a considerable amount of material aligned. In contrast, the xylem contains less aligned material which appears to be mainly protein. The peaks labelled 'c' may be due to phenolic material. In the cortex there is also some aligned protein plus pectin. The pattern in the 'd' region is characteristic of polygalacturonic acid.

References

Banwell, C.N. (1983) *Fundamentals of Molecular Spectroscopy*, 3rd edn, Wiley, New York.

Griffiths, P.R. and de Haseth, J.A. (1986) *Fourier Transform Infrared Spectroscopy*, Wiley, New York.

Lai, Y.W., Kemsley, E.K. and Wilson, R.H. (1994) The potential of Fourier transform infrared spectroscopy for the authentication of edible oils. *J. Agric. Food Chem.*, **42**, 1154–1159.

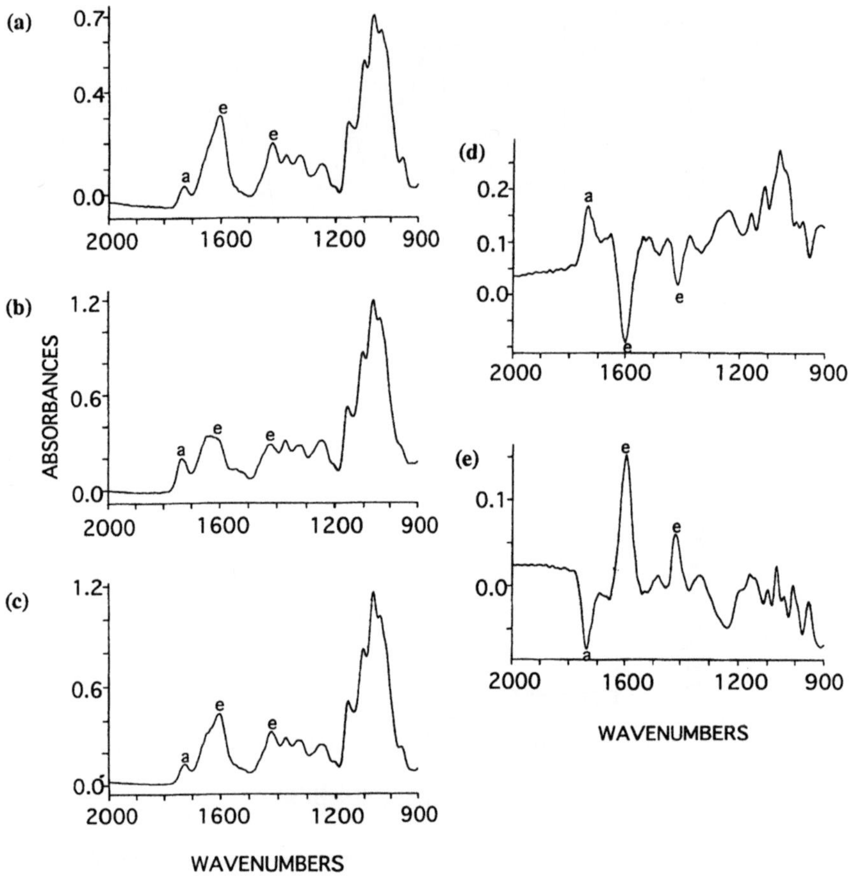

Fig. 8.10 Spectra of plant cell walls taken during growth. Spectra (a)–(c) were collected during growth and differences can clearly be seen in terms of changes in the ratio of peaks 'e' to peak 'a'. These differences are highlighted in spectra (d) and (e) which are the subtraction spectra corresponding to (b)–(a) and (c)–(b), respectively. This shows that during growth the pectin must be esterified. The pectin is de-esterified following growth. This result confirms that obtained using gold labelled antibodies, and it provides direct molecular evidence of the mechanism.

McCann, M., Hammouri, M.K., Wilson, R.H., Roberts, K. and Belton, P.S. (1992) Fourier transform infrared microspectroscopy: a new way to look at cell walls. *Plant Physiol.*, **100**, 1040–1047.

McCann, M., Stacey, N.J., Wilson, R.H. and Roberts, K. (1993) Orientation of macromolecules in the walls of elongating carrot cells. *J. Cell Sci.*, **106**, 1347–1356.

Willhoft, E.M.A. (1976) The role of starch in bread staling, In *Industrial Uses of Starch and its Derivatives*, (ed. J.A. Radley), Applied Science, London, pp. 117–147.

Wilson, R.H. (1990) Fourier transform mid-infrared spectroscopy for food analysis. *Trends Anal. Chem.*, **9**, 127–131.

Wilson, R.H. and Belton, P.S. (1988) A Fourier transform infrared study of wheat starch gels. *Carbohydr. Res.*, **180**, 339–344.

Wilson, R.H. and Goodfellow, B.J. (1994) Mid-infrared spectroscopic methods, in *Spectroscopic Techniques for Food Analysis* (ed. R.H. Wilson), VCH Publishers, New York, chap. 3.

Wilson, R.H., Kalichevsky, M.T., Ring, S.G. and Belton, P.S. (1987) A Fourier transform infrared study of the gelation and retrogradation of waxy maize starch. *Carbohydr. Res.*, **166**, 162–165.

Wilson, R.H., Goodfellow, B.J., Belton, P.S., Osbourne, B.G., Oliver, G. and Russell, P.L. (1991) A comparison of Fourier transform mid-infrared spectroscopy and near infrared spectroscopy with differential scanning calorimetry for the study of the staling of bread. *J. Sci. Food Agric.*, **54**, 471–483.

Chapter 9

Ultrasound Studies of Shelf-Life and Crystallization

M.J.W. Povey
Procter Department of Food Science,
University of Leeds, UK

9.1 Introduction

The standard ultrasound experiment is the acoustical time of flight measurement (ATOM), which is sometimes known as 'pitch and catch'. A pulse of ultrasound is generated by electrical stimulation of a piezoelectric crystal at frequencies between 30 kHz and 30 MHz. Normally the pulse contains only a few cycles.

An early example of this sort of measurement was reported in 1827 by Colladon and Sturm using a bell as the source transducer. The investigators used their ears as receivers and a stop-watch to time the sound of the clap of the bell as it crossed Lake Geneva. They measured the distance between the boats containing the bell and the stop watch and in this way determined the velocity of sound in the waters of the lake.

Today our equipment is a little more sophisticated but the principle is the same as for the experimenters of Lake Geneva. A block diagram of the acoustical time of flight measurement is given in Fig. 9.1. Here the transducers are piezoelectric ceramic disks made from lead zirconium titanate. Transducers such as these are widely available and can be excited with tone burst generators made by several manufacturers. The timing is made with a timer/counter which is started by a gate pulse from the signal generator and stopped by the output from the receiver/transducer. A block diagram of one of the experimental arrangements in our laboratory is presented in Fig. 9.2.

The advantage of this method is its simplicity and accuracy; a variation of just $1\ m\ s^{-1}$ in $1500\ m\ s^{-1}$ is easily achievable. In aqueous solutions with careful temperature control, accuracies ten times better than this can be achieved. Much higher accuracy values than this can even be obtained, to better than a few parts in a million, using interferometric techniques, but these experiments require more complex apparatus (del Grosso and Mader, 1972). Additionally, practical limitations on tem-

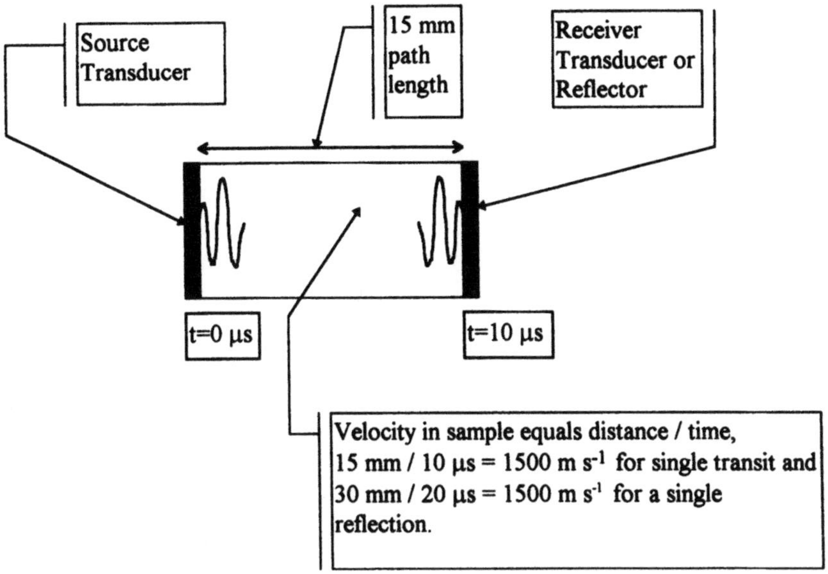

Fig. 9.1 Block diagram of an acoustic time of flight measurement.

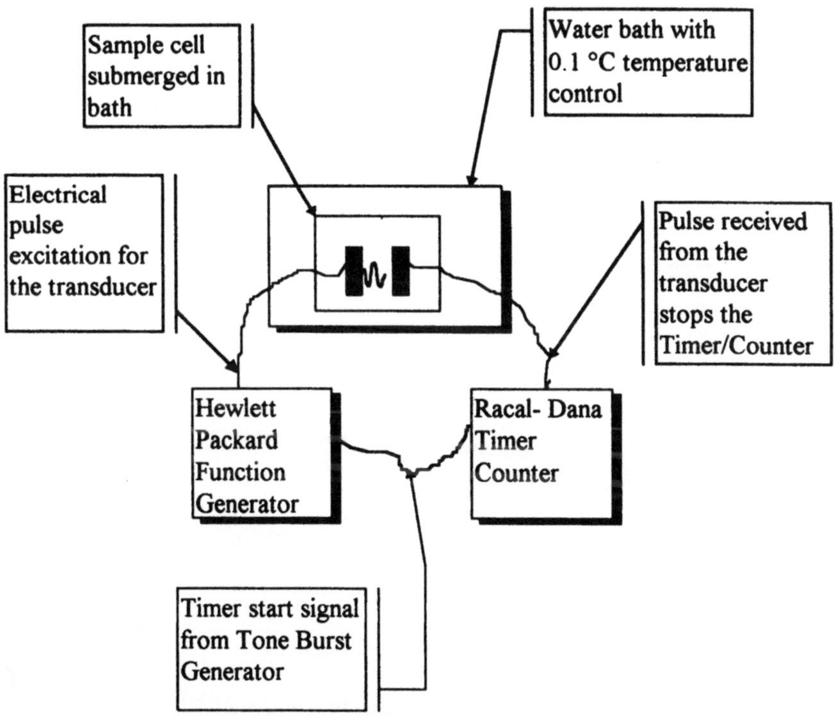

Fig. 9.2 Block diagram of two-transducer system as used in an ultrasonic scanner.

perature control and measurement normally limit the accuracy of sound velocity measurement to a few parts in 10^4. This is because at room temperature the temperature coefficient of the sound velocity in water is $\sim 3\ m\ s^{-1}$ (del Grosso and Mader, 1972). Since most food materials contain a large proportion of water, it is necessary to control or measure temperature to better than $0.3\ m\ s^{-1}$ to match the accuracy of most types of ultrasound apparatus. It is a relatively simple matter to improve the precision, as opposed to the accuracy, of ultrasound measurement to $0.1\ m\ s^{-1}$ or better, a level which is beyond all but the best temperature controllers. When interpreting acoustic velocity data from foods, it is important to know the temperature dependence of the velocity in water and in other common substances that make up the foods. In Fig. 9.3 the temperature dependence of the acoustic velocity at 1.25 MHz in water, fats and oil-in-water emulsions is shown. We note that the temperature coefficient of the velocity of sound in water around room temperature is positive. In this respect water is almost unique amongst liquids. In distilled water the velocity achieves a maximum value of $1554.8\ m\ s^{-1}$ at 74°C (del Grosso and Mader, 1972). Above this temperature the velocity starts falling again. The addition of solutes can have a relatively big effect on the velocity of sound and values for a typical aqueous phase found in margarine are given. The dominant modifiers of sound velocity in these instances are salt and surfactant. Although the shape of the curve changes little, the shift of the maximum in the curve to lower temperatures is significant. The next point to note is that the other liquids, fats in this case, have a temperature coefficient of velocity which

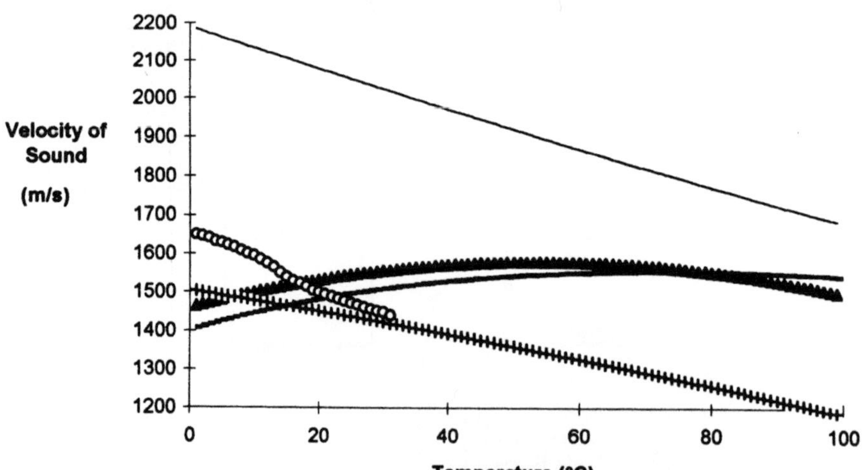

Fig. 9.3 Sound velocity plotted against temperature for some materials commonly found in food emulsions: ▲, aqueous phase of margarine; —, distilled water; +, pure liquid oil; –, pure solid fat; O, 80% fat/oil dispersed in the aqueous phase.

is of opposite sign to that of water. Moreover, there is a single temperature at which the velocity through the two liquids is equal.

9.2 Theoretical background

What can be learnt from data such as that presented in Fig. 9.3? First we can see that sound velocity in mixed liquids or dispersions does not depend on a mean value of the velocities in the components making up the material. The fundamental relationship between velocity, adiabatic compressibility and density is called the Urick equation (Urick, 1947),

$$v = \frac{1}{\sqrt{\kappa\rho}}, \qquad \kappa = \sum_j \phi_j \kappa_j, \qquad \rho = \sum_j \phi_j \rho_j, \tag{9.1}$$

which for a two phase system is

$$\kappa = \phi\kappa_2 + (1 - \phi)\kappa_1, \qquad \rho = \phi\rho_2 + (1 - \phi)\rho_1,$$

where v is the sound velocity in the dispersion, ϕ is the volume fraction of dispersed phase, and κ and ρ are the compressibility and density, respectively. The subscripts refer to the constituent phases; in particular 1 refers to the continuous phase, 2 to the dispersed phase in a two-phase system. (In a three phase system, 2 refers to the liquid dispersed phase and 3 to the solid dispersed phase.)

This equation has found wide use in the application of ultrasonic velocity measurement to the determination of dispersed phase volume fraction (McClements and Povey, 1988; Cao *et al.*, 1991; Fillery-Travis *et al.*, 1993). The Urick equation is an approximation, however, and there are cases where equation (9.1) is inadequate (McClements *et al.*, 1990; Holmes *et al.*, 1993). The description ignores the particulate nature of the system, and treats it as a continuous medium with modified density and compressibility. Pinfield *et al.* (1995) have shown, however, that many of the deviations from the Urick equation may be accommodated in a rewritten form of (9.1):

$$\frac{1}{v^2} = \frac{1}{v_1^2} + \alpha\phi_{\text{Urick}} + \beta\phi_{\text{Urick}}^2, \tag{9.2}$$

where v_1 is the velocity of sound in the pure continuous phase, and ϕ_{Urick} is the volume fraction (according to Urick) corresponding to a velocity v in the dispersion. The parameters α and β are related to the fractional differences between the density and compressibility of the two phases. Thus, the Urick equation tells us that the deviations from the velocity measured in the pure continuous phase when a dispersed phase is added are due to the fractional differences, $\Delta\kappa/\kappa_1$ and $\Delta\rho/\rho_1$, between the two phases, of the compressibility and density, respectively. Here we define $\Delta\kappa = \kappa_2 - \kappa_1$ and $\Delta\rho = \rho_2 - \rho_1$. The Urick equation can now be transformed into a more general form which accounts for the particulate nature of the dispersed phase:

$$\frac{1}{v^2} = \frac{1}{v_1^2} + \gamma\phi_{scatt} + \delta\phi_{scatt}^2 + O(\phi_{scatt}^3). \tag{9.3}$$

Here ϕ_{scatt} is the phase volume determined from a sound velocity measurement using a theory which takes account of scattering (McClements and Povey, 1989). The coefficients γ and δ may be viewed as effective Urick equation coefficients α and β. The reader is referred to Pinfield *et al.* (1995) for full details on the significance and application of the corrections; suffice it to say here that the modified Urick equation applies to a wide range of emulsions. It very satisfactorily allows for the effects of scattering where they occur, and it permits an account to be taken of these additional contributions without deploying 'full blown' scattering theory.

A further useful recasting of the Urick equation should be mentioned. Very often ultrasound velocity measurement is used to determine the volume fraction of a dispersed phase. In many stable emulsions and dispersions, the dispersed phase density is not very different from that of the continuum. This is not an accident, but is the result of the need to stabilize the dispersion against creaming or sedimentation: a simple way of doing this is to equalize the density of the phases. In the case where the density of the continuous and dispersed phases are exactly equal, equation (9.1) may be rewritten in the very convenient form

$$\frac{1}{v^2} = \sum_i \frac{\phi_i}{v_i^2}. \tag{9.4}$$

In the case of a three-phase system, say an oil-in-water emulsion in which solid fat is crystallizing in the oil phase, equation (9.4) becomes

$$\phi_S = \frac{\left[\left(\frac{1}{v^2} - \frac{1}{v_L^2}\right)\right]}{\left[\left(\frac{1}{v_S^2} - \frac{1}{v_L^2}\right)\right]}\Phi_S, \tag{9.5}$$

where ϕ_S is the volume fraction of dispersed phase which is solid, Φ_S is the total volume fraction of the dispersed phase if it were all solid, v is the measured ultrasonic velocity through the dispersion, and v_L and v_S are the velocities through the dispersion, according to the Urick equation, if the dispersed phase were all liquid or all solid, respectively. Note that the subscripts L and S refer here to the dispersion, whereas previously mentioned subscripts 2 and 3 refer to the pure phases. Thus, the solid volume fraction can be simply calculated from the ultrasound velocity. Pinfield *et al.* (1995) have shown that, for a number of reasons, equation (9.5) is likely to work very well, irrespective of the effects of scattering.

A further point, which relates not only to crystallizing systems, but also has a bearing on the effects of dissolved solutes, is that changes in compressibility are often much more dramatic than changes in density. Take, for example, the sunflower oil data in Fig. 9.3. At 20°C, the oil is almost totally liquid and its velocity is 1451.45 m s^{-1}. If it were possible for the oil to transform completely to solid at that temperature, then its velocity would become 2082.90 m s^{-1}, i.e. a 30.3% change. The corresponding densities are 920 kg m^{-3} and 1020 kg m^{-3}, corresponding to a 9.80% change. The compressibilities from these values are 5.16×10^{-10} Pa^{-1} and 2.27×10^{-10} Pa^{-1}, a 57% change. These figures are not untypical of the changes observed in many crystallizing systems. Indeed, it is possible to use compressibility data, obtained sometimes in combination with density measurements, to measure the interaction between a solute and the solvent in an aqueous solution. Professor Sarvazyan's group at Puschino has used this technique to obtain the thickness of the hydration layer around DNA (Sarvazyan, 1991), a method that can be readily extended to the study of proteins (Pavlovskaya *et al.*, 1992).

9.3 Crystallization in emulsions

From the preceding section, it is clear that ultrasound velocity measurement has much to offer in the study of crystallization processes. It is very sensitive to the liquid/solid phase transition, it is non-destructive, it can probe undiluted emulsions and it is easily automated.

9.3.1 Elimination of gas bubbles

Ultrasound velocity and attenuation are very sensitive to the presence of air bubbles in the sample. Consequently, it is of prime importance to remove the air from samples, particularly from aqueous materials. The quickest and best way of doing this is to sonicate the samples in an ultrasonic bath. Ultrasound baths employ continuous ultrasound at high power levels (~kW). This high power ultrasound causes cavitation which encourages dissolution of gas and the growth of larger bubbles which can then cream to the top of the sample. If the sample is so viscous that the bubbles cannot cream out, then some other way of removing them must be found, or the ultrasound experiment must be abandoned. If bubbles are present, it is likely that no signal will pass through the sample. Conversely, if a reasonable signal amplitude is obtained in a transmission experiment, then it is unlikely that bubbles are present.

9.3.2 Acoustic matching

The sample needs to be in good acoustical contact with the transducers and the sample should be continuous across the sample path. This might seem obvious, but it is not uncommon to see large voids appearing in crystallizing materials. The transducer should be well matched acoustically to the sample. This means choosing probes which are matched to water for the experiments referred to in this chapter. Care must be taken to ensure good thermal contact between the sample, the thermostatic bath, the probe, and the temperature measuring device. This is not so easy to achieve as it might appear. As systems crystallize, they tend to thicken and solidify. Heat can be evolved or absorbed during the crystallization process, thereby changing the sample temperature. For fats and other materials, the enthalpy of fusion can be very large indeed. Under these conditions, it may not be possible to change the temperature at the rate desired, and trial and error is needed to find suitable experimental conditions. These problems are not unknown to fats and oils technologists and the American Oil Chemists' Society and other bodies lay down complex rules for dilatometric determinations of solid content.

A further complication is due to the dynamics of crystallization. The velocity of sound can be affected by the process of crystallization itself, particularly when a significant proportion of the sample is participating in crystallization or melting. Under these circumstances the link between ultrasound velocity and solid content breaks down completely and none of the analysis presented earlier in this paper is applicable. (The situation has been analysed by McClements and Povey (1989).)

9.3.3 Equipment and some experimental results

An apparatus suitable for routine study of crystallization in emulsions, and indeed in any liquid sample in which sound velocity and temperature measurement are of interest, is shown in Fig. 9.4. Such an apparatus is available commercially from companies such as Cygnus Limited in Dorchester, UK, or Nusonics, USA. The equipment is capable of operating over a very wide temperature range, from −20°C to 150°C, and is very accurate. A high rate of data acquisition is implicit in the pulse echo technique, with repetition rates of between 300 Hz and 1 kHz. This means that the technique is well suited to rapidly changing conditions, e.g. in a pipeline. With apparatus such as this, new insights into crystallization may quickly be obtained. In Fig. 9.5 are shown the results of repeated heating and cooling of cocoa butter obtained with this equipment. The phase transition between 52°C and 60°C can clearly be seen. (The plot of data points was obtained directly from the instrument and the pure liquid line was added afterwards.)

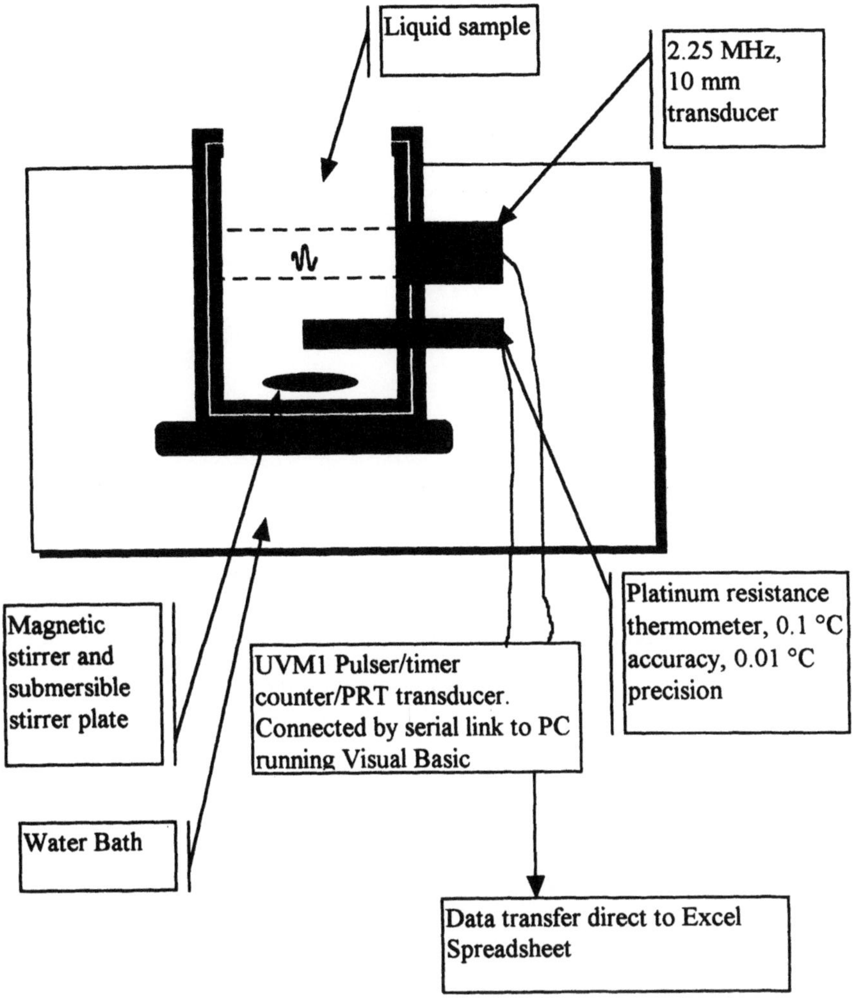

Fig. 9.4 Block diagram of an ultrasonic velocity meter.

Using equation (9.4), the solid content can be obtained from a plot such as that shown in Fig. 9.5. The velocity in the liquid is obtained by measuring the sound velocity at a temperature high enough to guarantee that the material is completely melted. The velocity in the pure solid normally has to be estimated because solid hard fats often expand on cooling and are capable of shattering the ultrasonic cell. The data in Fig. 9.5 are particularly interesting because they suggest that small quantities of solid are present in cocoa butter at temperatures well above 35°C, the quoted melting point. Moreover, a phase transition is indicated at 60°C. These data indicate that close study of the crystallization behaviour of

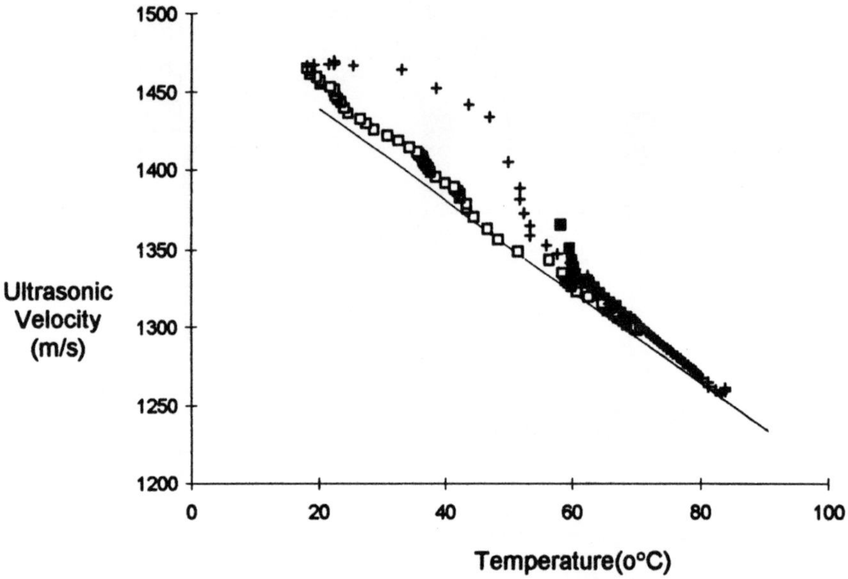

Fig. 9.5 Sound velocity measured against temperature in Malaysian cocoa butter. Heating and cooling rates were of the order of 0.5°C/min; ■, solid cocoa butter, melted and poured into the sample cell and then heated; □, cooling of the previously melted sample; +, reheating; —, fitted liquid data.

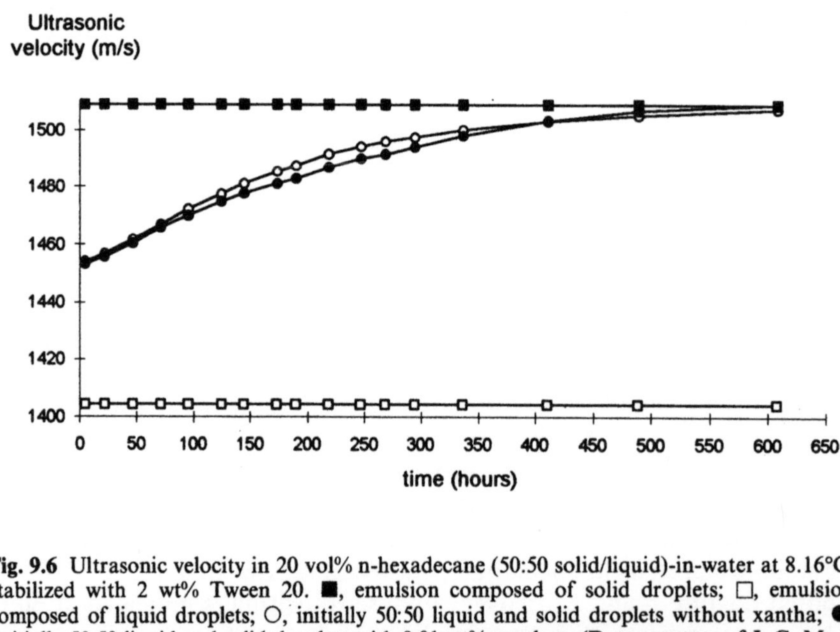

Fig. 9.6 Ultrasonic velocity in 20 vol% n-hexadecane (50:50 solid/liquid)-in-water at 8.16°C, stabilized with 2 wt% Tween 20. ■, emulsion composed of solid droplets; □, emulsion composed of liquid droplets; ○, initially 50:50 liquid and solid droplets without xantha; ●, initially 50:50 liquid and solid droplets with 0.01 wt% xanthan. (Data courtesy of J. G. Ma.)

chocolate at temperatures higher than the 35°C normally quoted as the melting point of chocolate may be worth while.

The results of a study of the interaction between solidified emulsion droplets and liquid droplets is shown in Fig. 9.6. Two emulsions were prepared from n-hexadecane and water, one of which contained droplets of oil which had completely solidified and the other containing wholly liquid droplets. The two previously prepared emulsions were monitored throughout the course of the experiment in order to demonstrate that, separately, they were stable at the temperature of the experiment. The third emulsion, prepared by mixing the original two, contained a 50:50 mixture of solid and liquid droplets. The sound velocity increases over a period of 600 h, until it reaches a value equal to the emulsion which contains completely solid droplets. This is consistent with the fact that the liquid drops have completely solidified. The effect of surfactants and stabilizers (e.g. xanthan) on this process can then be studied (see, for instance, Dickinson *et al.*, 1991; McClements *et al.*, 1993; Dickinson *et al.*, 1993)

9.4 Creaming in emulsions

9.4.1 Equipment and experimental procedure

Ultrasonic scanning is a powerful technique for the study of emulsion stability. This is because of its ability to produce a profile of the dispersed phase volume against height in a tube in which the emulsion is kept undisturbed. Additionally, it is not necessary to dilute or otherwise change the sample in order to make the measurement. Thus, the door is opened to studying emulsion stability under conditions similar to those in which the samples are normally stored. Moreover, the technique can detect changes in concentration long before they are visible by eye and so an early warning of emulsion instability may be obtained.

The apparatus is simple in concept and is illustrated in block diagram form in Fig. 9.7. The stepper motors for the transducer motion and the carousel are connected to a personal computer. The timer/counter and waveform generator (see Fig. 9.2) are also linked to the computer and all functions are controlled through an HP Instrument Bus. The interface card for the HP Instrument Bus system is a National Instruments card which acts as an IEEE controller. The whole instrument is controlled by a Visual Basic program which provides a simple user interface, controls the progress of the experiment, analyses the acoustic data, stores it on Excel spreadsheets, and presents the data in graphical form.

The ultrasonic scanner was first developed at the Institute of Food Research in Norwich (Gunning *et al.*, 1988). In its original manual and semi-automatic forms, however, it was difficult to realize the full

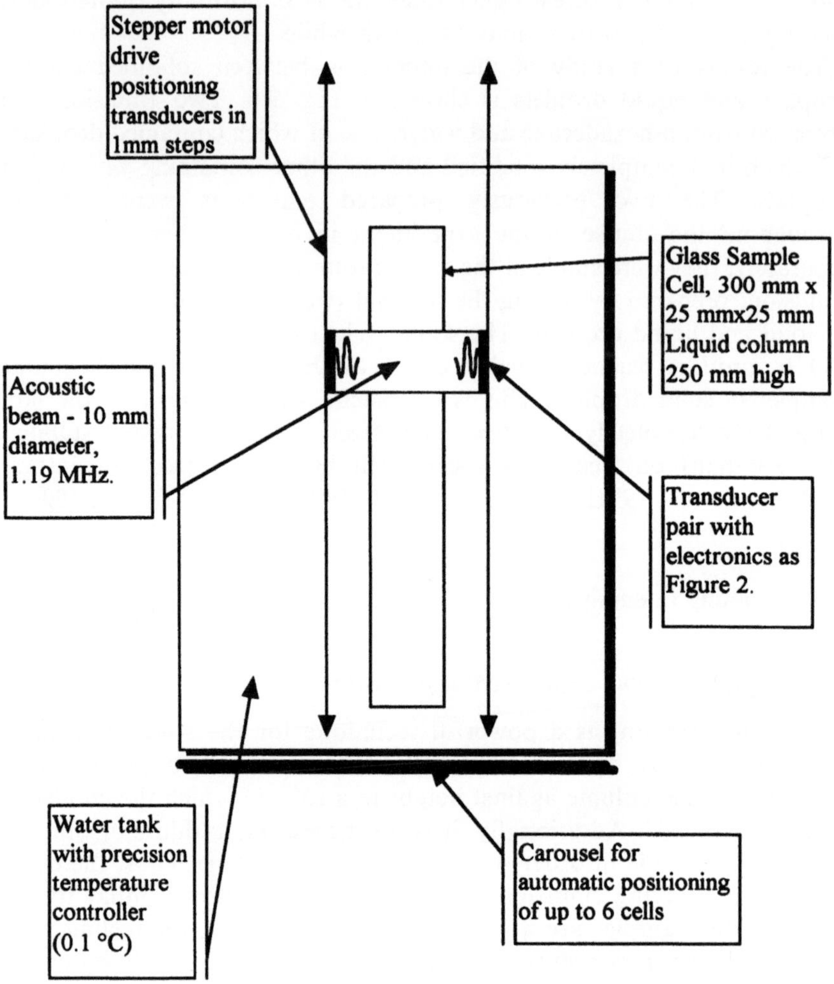

Fig. 9.7 Automated ultrasonic scanner for studying emulsion stability.

potential of the instrument. Our automated apparatus has advantages other than the obvious ones of less labour and higher productivity. Firstly, much greater quantities of data can be processed in far less time, and the statistical analysis of the raw data improves the accuracy and precision of the technique by a decade or so. Secondly, the much finer detail observable in data from a machine operator that does not get tired has transformed our view of a number of creaming processes. We have been able to detect events which we would likely to have missed if manual scanning had been performed. Finally, the quality of data presentation and the accuracy and veracity of the data are all greatly enhanced by full computer control.

To obtain good results from the ultrasound scanner instrument it is essential to be clear about some basic points. Referring to equation (9.2), it is important to avoid a situation where the compressibility (and density) of the dispersed and continuous phases are the same. In this situation it would be impossible to detect any changes in the concentration of the dispersed phase. In Fig. 9.3, for instance, the temperature at which the velocities through the two phases are equal is the temperature at which the measured velocity is equal to that through the continuous phase. This is clearly the region to be avoided in scanning for dispersed phase volume fraction changes. A practical conclusion is that it is desirable to obtain velocity/temperature profiles for the temperature range of interest from the pure continuous and dispersed phases. Apparatus such as that shown in Fig. 9.4 is suitable and convenient for this task. An additional benefit of preliminary measurements of velocity in the pure phases is that determination of the dispersed phase volume fraction can be carried out much more accurately. In addition, it may be convenient to improve the sensitivity of the experiment to changes in the dispersed phase concentration by altering the measurement temperature.

Each cell must be individually calibrated, and in the past this has been done with two separate calibrant fluids which are placed in the cells (Gunning *et al.*, 1988). Each cell must be scanned with both fluids. The drawback to this procedure is that, whilst the velocity of sound in distilled water is reliably and accurately known, the same cannot be said about other calibrant fluids. We have chosen therefore to employ distilled water only as the calibrant, and have used a double-echo technique (Fig. 9.8) to determine the acoustic path length within a cell. This is possible because the attenuation of the acoustic signal in water is much smaller than in other liquids. The double-echo technique greatly improves the accuracy of the velocity measurement, and we would prefer to use this technique in our emulsion sample measurements. However, the attenuation is generally too high in emulsion samples for us to measure a double transit time.

The calibration information for each cell is stored in a spreadsheet for each temperature at which experiments are to be conducted. When an experiment is being performed, the appropriate calibration sheet is invoked by the program and the data are automatically corrected to give a velocity reading for the sample at each height measured. The acoustic path in the scanner is complex and is subject to numerous sources of error. Whilst the calibration process takes care of the worst of the errors, it is dependent on a repeatable velocity in water in the thermostat tank, which occupies part of the acoustical path. We use double-distilled water for the tank fluid, and replace it as soon as there is any sign of contamination. The cells are made of glass, which is very convenient for cleaning and lack of interaction with samples. However, the glass walls

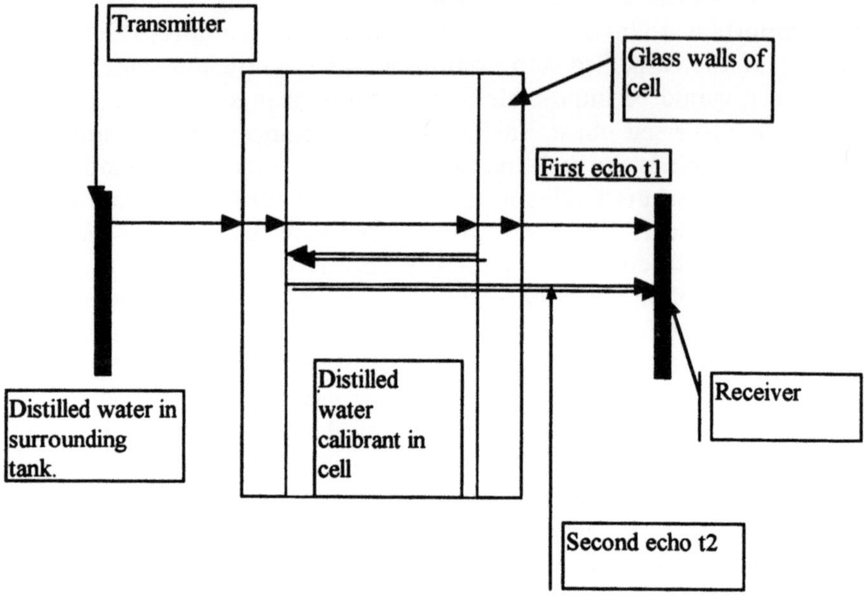

Fig. 9.8 Double-echo technique for cell calibration in an ultrasonic scanner.

are not particularly even, and this further degrades accuracy. Overall, our system is accurate to better than $3\ m\ s^{-1}$. This compares with an accuracy for the apparatus shown in Fig. 9.4 of better than $1\ m\ s^{-1}$, and a precision ten times better than that. The ultrasound scanner is capable of operating between approximately 4°C and 50°C.

9.4.2 Some experimental results

The conventional method of determining shelf-life involves visual inspection of a meniscus formed between a serum layer and the emulsion or between the emulsion and a cream. The rate of movement of this meniscus may then be plotted and conclusions drawn. The drawbacks of this technique are perhaps obvious. Firstly, this experiment is essentially a measurement of turbidity and can therefore only give quantitative results at very low concentrations. Secondly, processes which are precursors to the development of a visually detectable meniscus will be missed entirely. Thirdly, this is a very laborious experimental technique, prone to error, and often of dubious quantitative value. Figure 9.9 presents an ultrasound scan result for a gradually evolving cream layer. A boundary is not apparent visually until 45 hours, whereas the ultrasound scan has detected changes within the first few hours. The details of the evolution within the cream and serum layers are evident in this diagram. It is

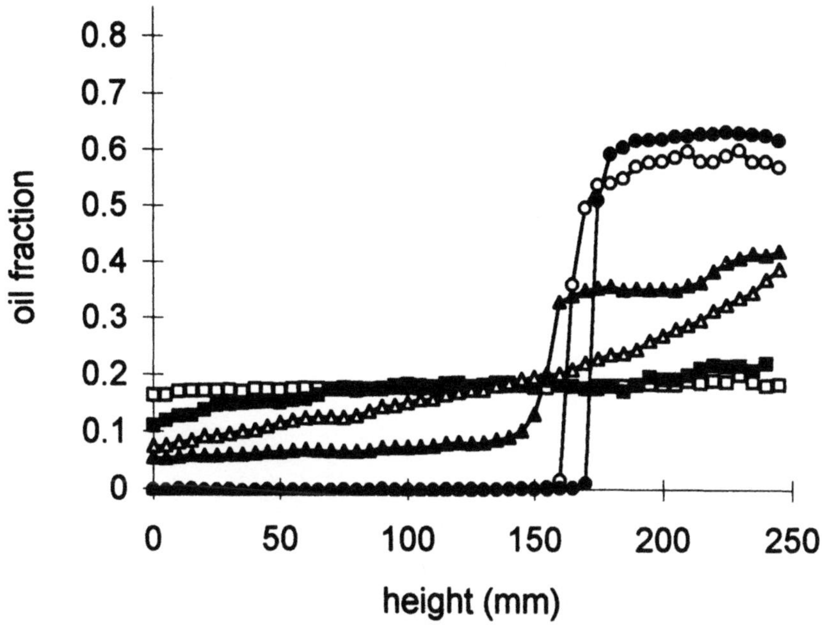

Fig. 9.9 Creaming profile of 18 vol% mineral oil/water emulsion containing 0.1733% (wt in aqueous phase) xanthan at 30°C: □, after 11 h; ■, after 20 h; Δ, after 30 h; ▲, after 45 h; ○, after 99 h; ●, after 188 h.

noticeable that in the early stages three regions exist: the beginnings of a cream, the beginnings of a serum layer, and the bulk of the emulsion.

In Fig. 9.10 a rather different situation is evident. Here the development of the cream layer was sudden and unexpected. The sharp peaks and troughs in the results are probably due to fractionation of droplet size in the cream layer and consequent scattering of the ultrasound (Pinfield *et al.*, 1994). Nevertheless, the parallel development of a serum layer and a cream layer are evident. We generally find that the effects of scattering can be particularly marked in the cream layer. One check of the methodology is to integrate the area under the oil volume fraction against height curve to see if it adds up to the original oil volume fraction. If it does not, then scattering can be presumed to be occurring. Techniques for adequately correcting for scattering were mentioned earlier in this chapter. Nevertheless, there are occasions when the scattering cannot be accounted for completely, and such a situation is shown in Fig. 9.10.

Following the rate of movement of a meniscus can be a 'hit and miss' way of determining shelf-life. The meniscus is observed by viewing light backscattered from the emulsions droplets. Concentrations above a few fractions of a percent will all look the same due to multiple scattering and obscuration if turbidity is measured. Additionally, the effects of

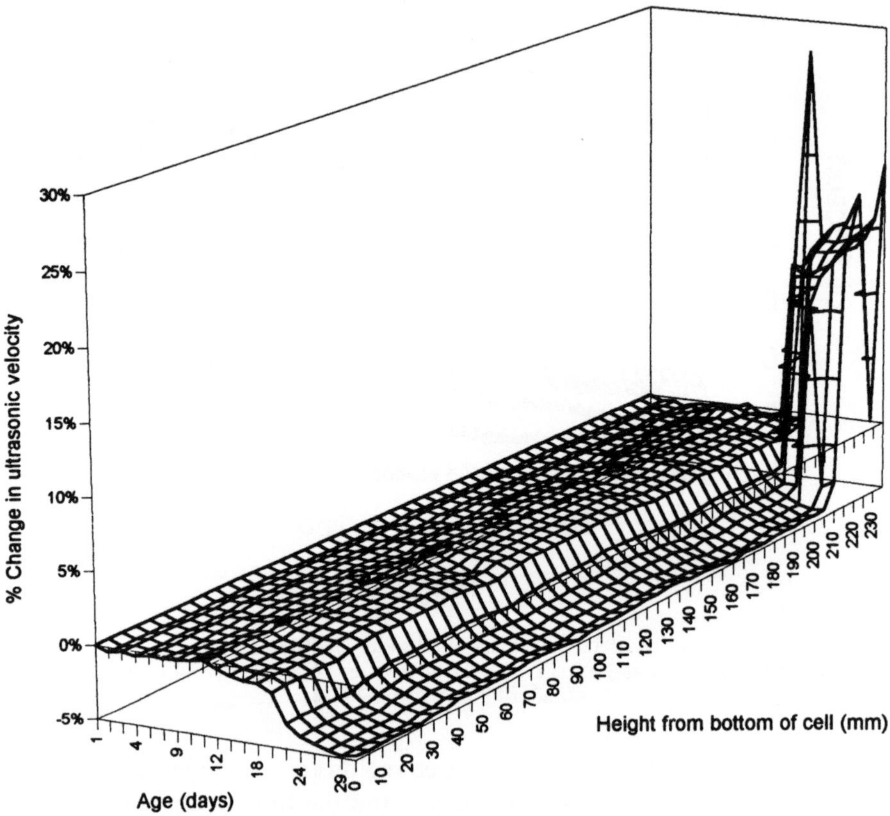

Fig. 9.10 Creaming in a soya bean oil-in-water emulsion, plotted as percentage ultrasound velocity deviation from zero-time values against height in a vertical sample tube against age of emulsion.

different particle sizes will be missed altogether. A possible alternative ultrasonic procedure could involve a modification of the statistical concept of skewness. If the emulsion remains stable, then each scan will be more or less the same as the previous one. Once creaming occurs, then a cream will start to appear at the top of the sample and a serum at the bottom and the profile will no longer be 'flat'. This is shown very well in the plots of Fig. 9.9. Rather than making an arbitrary decision about the position of the boundary between 'emulsion' and 'cream', we choose to use a measure of departure from 'flatness' as our measure of instability. Such a measure is illustrated in Fig. 9.11 which was derived from the data in Fig. 9.10. Here the absolute value of the percentage deviation of ultrasound velocity from its zero-time value was used to compute the 'skewness', a statistical parameter that quantifies departure from flatness.

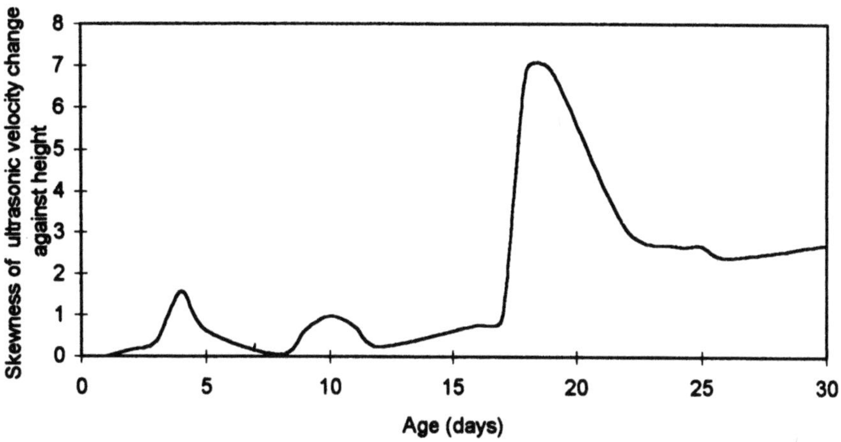

Fig. 9.11 Skewness of the percentage deviation in the ultrasound velocity, as an indicator of emulsion instability. (Skewness is a measure of the departure from 'flatness' of the curve.) The data for this plot were taken from Fig. 9.10.

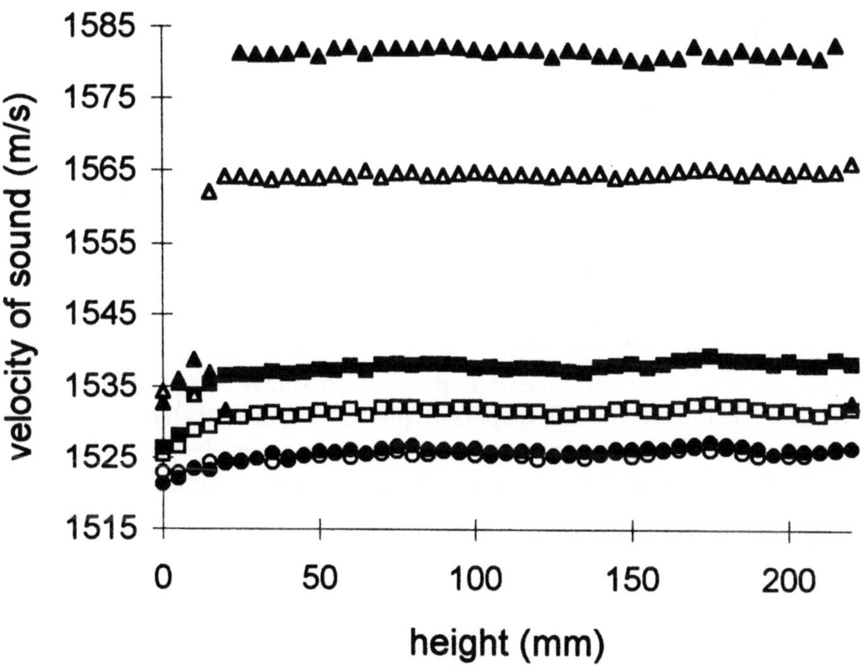

Fig. 9.12 Simultaneous creaming and crystallization in a 20 vol% hexadecane 50:50 solid/liquid emulsion with 2 wt% Tween 20 and 0.01 wt% xanthan in the aqueous phase. Measurements of velocity versus height were taken at 8.16°C: ○, after 34 h; ●, 43.5 h; □, 55 h; ■, 95 h; Δ, 261 h; ▲, 586 h.

The 'skewness' parameter has a value of zero or close to zero for a constant oil volume fraction across the height of the tube. As the serum and cream layers develop, they both contribute to the skewness, but it can be seen clearly that a very sudden breakdown occurs in the emulsion after about 16 days.

Finally, we mention how the investigation of crystallization and creaming can be brought together into a single experiment. In Fig. 9.12, a 50/50 mixture of solid emulsion droplets and liquid emulsion droplets has been mixed together in the presence of a small amount of xanthan gum. The xanthan polymer destabilizes the emulsion. We know from the earlier experiment (Fig. 9.6) that the liquid droplets have solidified completely after 600 h and that the presence of xanthan makes very little difference to the crystallization rate. The rise in the velocity of sound over the period of the experiment is consistent with solidification of the liquid droplets. Comparison of these results with measurements made on a wholly liquid sample under identical conditions indicates that, in this case, crystallization has very little effect on emulsion stability.

9.5 Conclusions

The ultrasound technique is a valuable addition to the armoury of instrumental techniques available to the food scientist and technologist. It is capable of giving quantitative measurements of changes in extents of crystallization and is particularly sensitive at low crystalline solid contents. This makes it well suited to studying the important nucleation and initial growth stages of crystals. The technique adapts easily to in-line use and it may be considered for process control applications.

When applied to emulsion stability and shelf-life determination, the ultrasound scanner is capable of providing information which is difficult to obtain otherwise. Because it is possible to follow detailed changes in the distribution of the oil dispersed phase in concentrated oil-in-water emulsions, new insights into the process of emulsion destabilization can be obtained. In particular, quantitative data are now becoming available which can be matched to numerical models of concentration profiles in creaming emulsions (Pinfield *et al.*, 1994).

References

Cao, Y., Dickinson, E. and Wedlock, D.J. (1991) Influence of polysaccharides on the creaming of casein-stabilized emulsions. *Food Hydrocolloids*, **5**, 443–454.

Colladon, J.D. and Sturm, J.C.F. (1827) Memoir on the compression of liquids. *Ann. Chim. Phys.*, **36**, 225–227.

del Grosso, V.A. and Mader, C.W. (1972) The speed of sound in pure water. *J. Acoust. Soc. Am.*, **52**, 1442–1445.

Dickinson, E., Goller, M., McClements, D.J. and Povey, M.J.W. (1991) Monitoring crystallization in simple and mixed oil-in-water emulsions using ultrasonic velocity measurement, in *Food Polymers, Gels and Colloids* (ed. E. Dickinson), Royal Society of Chemistry, London, pp. 171–179.

Dickinson, E., Kruizenga, F.-J., Povey, M.J.W. and van der Molen, M. (1993) Crystallization in oil-in-water emulsions containing liquid and solid droplets. *Colloids Surf. A*, **81**, 273–279.

Fillery-Travis, A.J., Gunning, P.A., Hibberd, D.J. and Robins, M.M. (1993) Coexistent phases in concentrated polydisperse emulsions floculated by non-adsorbing polymer. *J. Colloid Interface Sci.*, **159**, 189–197.

Gunning, P.A., Hibberd, D.J., Howe, A.M. and Robins, M.M. (1988) Gravitational destabilization of emulsions floculated by non-adsorbed xanthan. *Food Hydrocolloids*, **2**, 119–129.

Holmes, A.K., Challis, R.E. and Wedlock, D.J. (1993) A wide bandwidth study of ultrasound velocity and attenuation in suspensions: comparison of theory with experimental measurements. *J. Colloid Interface Sci.*, **156**, 261–268.

McClements, D.J. and Povey, M.J.W. (1988) Comparison between pulsed NMR and ultrasonic velocity techniques for determining solid fat contents. *Int. J. Food Sci. Technol.* **23**, 159–170.

McClements, D.J. and Povey, M.J.W. (1989) Scattering of ultrasound by emulsions. *J. Phys. D., Appl. Phys.*, **22**, 38–47.

McClements, D.J., Povey, M.J.W., Jury, M. and Betsanis, E. (1990) Ultrasonic characterization of a food emulsion. *Ultrasonics*, **28**, 266–272.

McClements, D.J., Dickinson, E., Dungan, S.R., Kinsella, J.E., Ma, J.G. and Povey, M.J. W. (1993) Effect of emulsifier type on the crystallization kinetics of oil-in-water emulsions containing a mixture of solid and liquid droplets. *J. Colloid Interface Sci.*, **160**, 293–297.

Pavlovskaya, G.E., McClements, D.J. and Povey, M.J.W. (1992) A preliminary study of the influence of dextran on the precipitation of legumin from aqueous salt solutions. *Int. J. Food Sci. Technol.*, **27**, 629–635.

Pinfield, V.J., Dickinson, E., and Povey, M.J.W. (1994) Modelling of concentration profiles and ultrasound velocity profiles in a creaming emulsion: importance of scattering effects. *J. Colloid Interface Sci.*, **166**, 363–374.

Pinfield, V.J., Povey, M.J.W. and Dickinson, E. (1995) The application of modified forms of the Urick equation to the interpretation of ultrasound velocity in scattering. *Ultrasonics*, in press.

Sarvazyan, A. (1991) Ultrasonic velocimetry of biological compounds. *Annu. Rev. Biophys. Chem.*, **20**, 321–342.

Urick, R.J. (1947) A sound velocity method for determining the compressibility of finely divided substances. *J. Appl. Phys.*, **18**, 983–987.

Chapter 10

Dynamic Surface Tension and Dilational Interfacial Properties

A. Prins
Department of Food Science,
Wageningen Agricultural University, The Netherlands

10.1 Introduction

Dispersions such as foams and emulsions obtain their properties to a large part from the properties of their surfaces and interfaces. Not only during making, but also during storage and manipulation, these dispersions are not in complete equilibrium. Mass transport by diffusion and convection and mechanical and thermal disturbances result in non-equilibrium situations. The stability of these dispersions is governed largely by the stability of the thin liquid films between the dispersed particles. Properties of the film surfaces play a crucial role in the stability of the thin films.

In order to understand the physical and physico-chemical behaviour of surfaces, interfaces and thin liquid films under non-equilibrium conditions, a number of experimental measuring techniques have been developed which are in everyday use in our laboratory. What these techniques have in common is that the mechanical behaviour of the surfaces and their liquid films is measured. Deformation and rate of deformation, as well as the nature of forces acting on surfaces and thin films, are important parameters in these techniques. The mechanical couplings between surface deformation and the deformation of the contacting liquid layers are also important considerations, knowledge about which is necessary for a complete understanding of the surface layers.

The following techniques are described: the overflowing cylinder; the free falling film; the ring trough; the caterpillar trough; and the Mannheimer trough. These techniques enable us to measure dynamic surface and interfacial behaviour close to and far away from equilibrium. It will be indicated which of these techniques can be used to explain specific aspects of foaming and emulsifying behaviour. In order to obtain relevant physical data, these techniques are used in combination with other experimental methods such as ellipsometry, laser Doppler anemometry, and the Wilhelmy plate technique.

10.2 Overflowing cylinder technique

10.2.1 Principles

In this apparatus, the liquid under investigation is pumped from below to the top of a vertical cylinder, where it is allowed to flow over the top rim and downwards along the outside wall of the cylinder, thereby returning to the pump again (Fig. 10.1). In this way a radial expanding liquid surface is made at the top of the cylinder; after the start of liquid flow, this reaches a steady state (Bergink-Martens, 1993; Bergink-Martens *et al.*, 1990, 1994).

The radial movement of the surface ensures that the surface expansion is purely dilational, which means that no shearing deformations are present. Experimentally it is found that the radial surface velocity, which is zero in the centre of the surface, increases linearly with the distance from the centre. Near the rim the velocity increases faster than proportional to the distance to the centre. The relative expansion rate of the surface is

$$\frac{1}{A}\frac{\mathrm{d}A}{\mathrm{d}t} = \frac{\mathrm{d}\ln A}{\mathrm{d}t}, \tag{10.1}$$

where A is the surface area and t is the time. It can be calculated from the radial velocity, V_r according to

$$\frac{\mathrm{d}\ln A}{\mathrm{d}t} = \frac{\mathrm{d}V_r}{\mathrm{d}r} + \frac{V_r}{r}, \tag{10.2}$$

where r is the distance to the centre. In the 'linear' part of the surface, this equation simplifies to

$$\frac{\mathrm{d}\ln A}{\mathrm{d}t} = \frac{2V_r}{r}. \tag{10.3}$$

The surface velocity can be measured by means of a laser Doppler technique. Two laser beams cross each other at a small angle at the liquid surface, from which they are reflected to a photo cell (Fig. 10.2). Due to interference at the point of intersection, a pattern of light and dark fringes is formed in such a way that a particle floating on the surface and moving through that pattern causes a flickering light whose frequency is a measure of the velocity of that particle. The frequency can be measured by using a spectrum analyser (Bergink-Martens *et al.*, 1994).

(a)

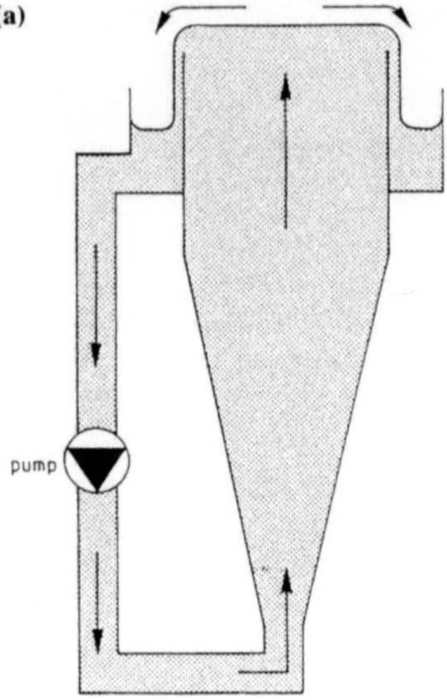

(b)

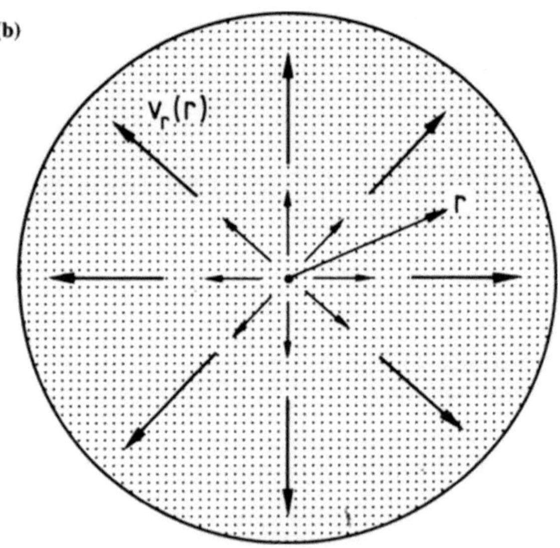

Fig. 10.1 Schematic representation of overflowing cylinder technique: (a) side view, (b) radial velocity $V_r(r)$.

The surface tension of the continuous expanding surface – the dynamic surface tension γ_{dyn} – can be measured using the Wilhelmy plate technique. The force exerted on a vertical rectangular sand-blasted glass plate just contacting the liquid surface, by which it is completely wetted, is a

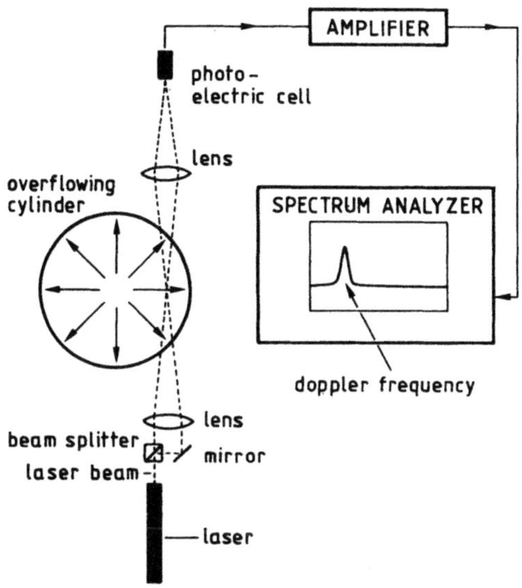

Fig. 10.2 Schematic representation of the laser Doppler anemometer technique applied to the overflowing cylinder technique.

measure of the surface tension: the horizontal circumference times the surface tension equals the force.

Investigating the overflowing cylinder technique, it is found that, by using an aqueous surfactant solution instead of water, the surface velocity and consequently the expansion rate increases considerably: for instance, by using a surfactant (Teepol) solution, it increases by up to a factor of 10 (Fig. 10.3). The explanation of this phenomenon also elucidates the basic principles on which this technique is founded. At the outside wall of the cylinder close to the rim, the falling film causes an expansion of the surface and consequently a high surface tension. This high surface tension causes a surface tension gradient along the horizontal top surface pointing from the centre to the rim. This surface tension gradient causes a surface movement in the same direction and, due to the viscosity of the liquid, a thin liquid layer close to the surface is also drawn in the same direction (Bergink-Martens *et al.*, 1990, 1994). That is, the weight of the falling film causes the surface to expand, and the resulting surface tension gradient pulls the horizontal surface and adhering liquid over the rim of the cylinder.

From the above, it follows that the higher the value of the generated surface tension gradient, the higher the value the expansion rate of the surface will be. From a hydrodynamic analysis the following relationship between these two parameters of the overflowing cylinder is found:

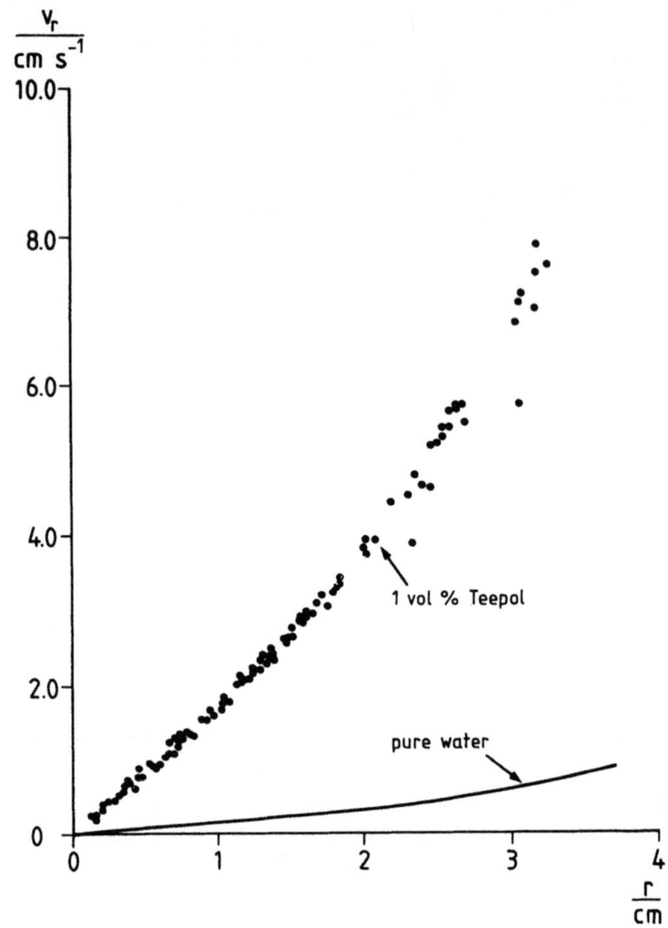

Fig. 10.3 Radial surface velocity V_r as a function of the radial distance r for water and an aqueous surfactant (Teepol) solution measured at the same flow rate with the overflowing cylinder.

$$\frac{\mathrm{d}\ln A}{\mathrm{d}t} = (0.415r\sqrt{\eta_b\rho_b})^{-2/3}\left(\frac{\mathrm{d}\gamma}{\mathrm{d}r}\right)^{2/3} + \left(\frac{\mathrm{d}\ln A}{\mathrm{d}t}\right)_0. \tag{10.4}$$

Here, η_b and ρ_b are the viscosity and density of the bulk liquid, and $(\mathrm{d}\ln A/\mathrm{d}t)_0$ is the expansion rate of the surface in the absence of surfactant, which for water is of the order of 1 s^{-1} (Bergink-Martens *et al.*, 1994).

The surface expansion rate can be changed in three ways: (1) by the flow rate of the liquid; (2) by the height of the falling film; and (3) by the surfactant concentration. By increasing the flow rate, the surface expansion rate increases until at a certain flow rate a plateau value of the expansion rate is reached. In general, the plateau value is reached when the mean

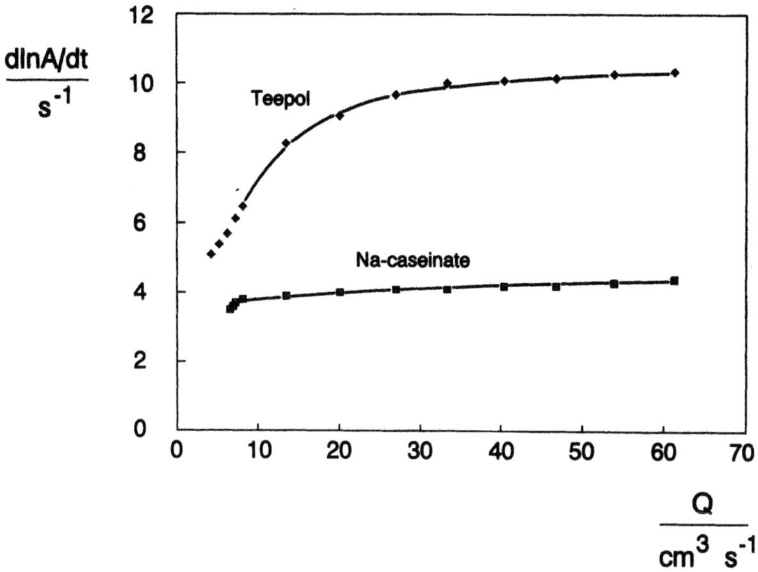

Fig. 10.4 The relative surface expansion rate d ln $A/\mathrm{d}t$ as a function of the flow rate Q for aqueous Teepol and sodium caseinate solutions.

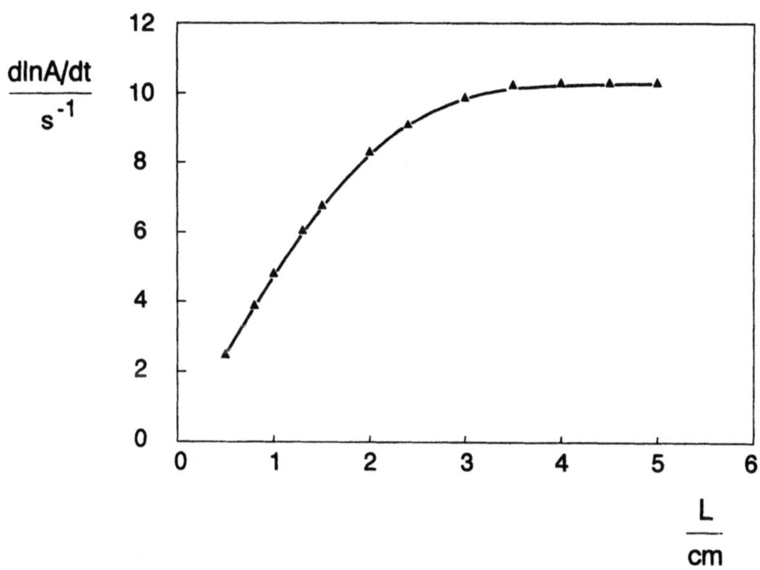

Fig. 10.5 The relative surface expansion rate d ln $A/\mathrm{d}t$ as a function of the height L of the falling film for an aqueous Teepol solution.

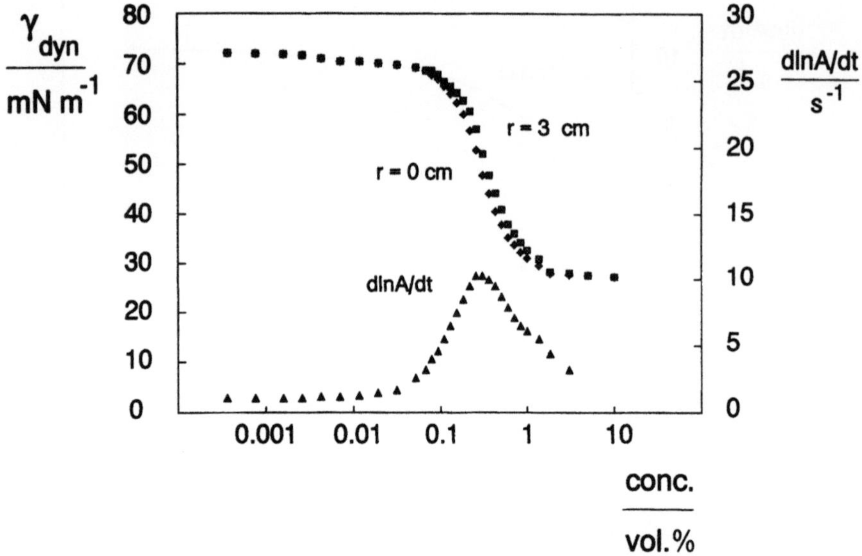

Fig. 10.6 The relative surface expansion rate $d \ln A/dt$ and the dynamic surface tension γ_{dyn} as a function of the Teepol concentration.

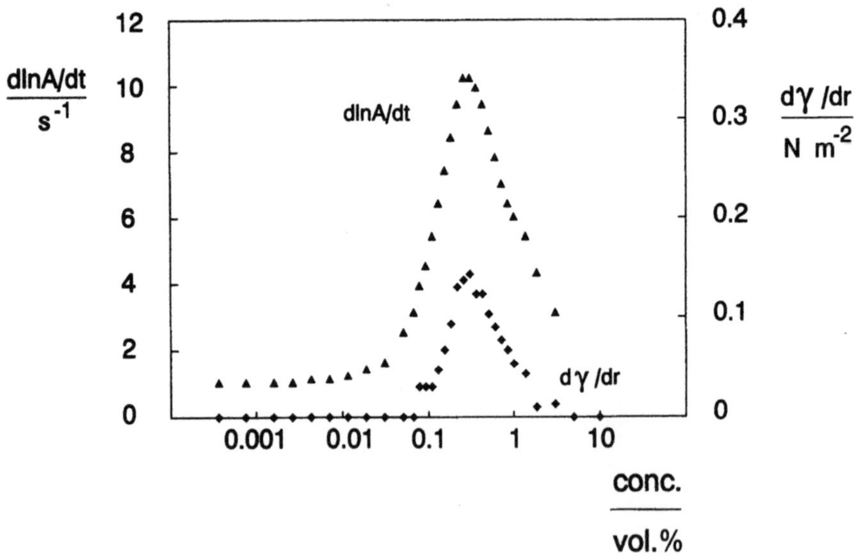

Fig. 10.7 The relative surface expansion rate $d \ln A/dt$ and the surface tension gradient $d\gamma/dr$ along the expanding surface as a function of the Teepol concentration.

velocity in the cylinder is approximately 1 cm s^{-1} (Fig. 10.4). By increasing the height of the falling film, the surface expansion rate increases and it also reaches a plateau value at a certain height of the falling film. It is found, in agreement with a hydrodynamic analysis of the liquid motion, that for falling films longer than 4 cm there is no longer an effect on the surface expansion rate (Fig. 10.5). By increasing the surfactant concentration, the surface expansion rate passes over a pronounced maximum. This maximum coincides with the maximum slope in the γ_{dyn} *versus* log (concentration) curve and according to equation (10.4) runs parallel with the value of $d\gamma/dr$ over the expanding surface (Figs 10.6 and 10.7). In practice, the best way to change the surface expansion rate is by changing the falling film height. This is because, by changing the flow rate, the complete wetting of the outside wall of the cylinder cannot be ensured when the flow rate becomes small.

The overflowing cylinder technique has been successfully applied also to oil–water interfaces (Fig. 10.8). To this end the outer cylinder is

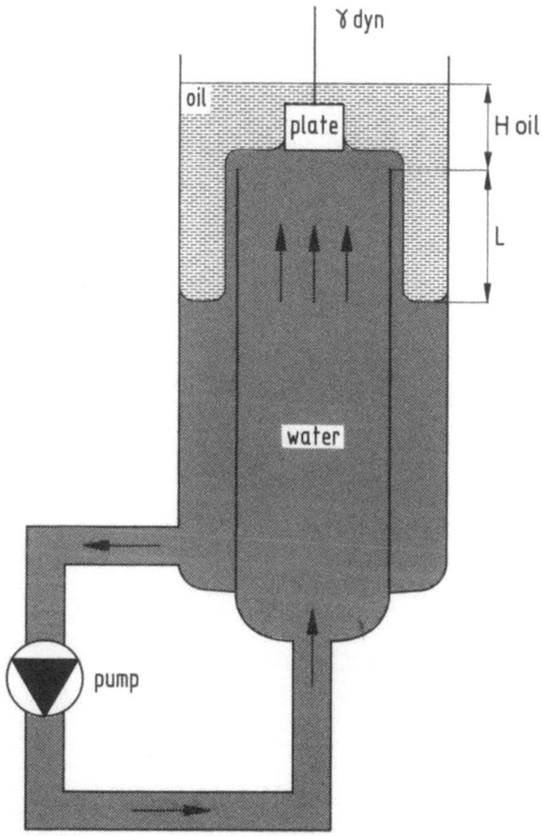

Fig. 10.8 Schematic representation of the water–oil overflowing cylinder.

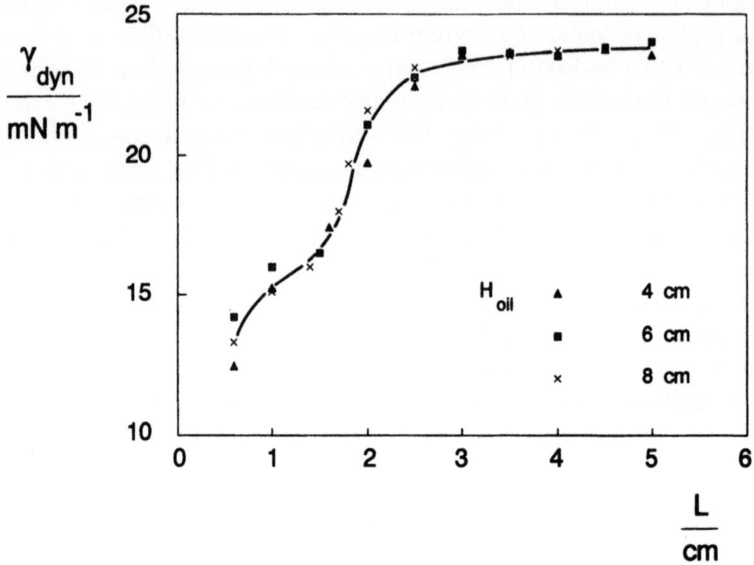

Fig. 10.9 The dynamic interfacial tension γ_{dyn} of water against soya oil as a function of the length L of the falling water film.

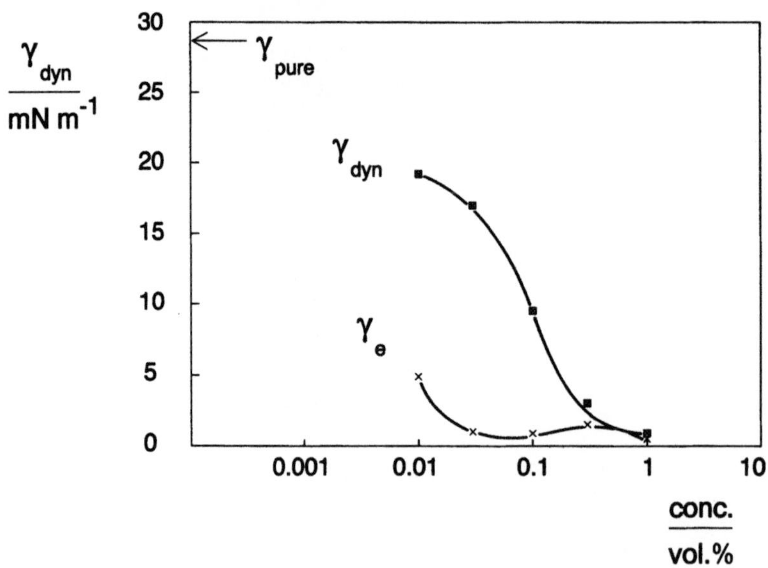

Fig. 10.10 The dynamic interfacial tension γ_{dyn} and equilibrium interfacial tension γ_e as a function of the Teepol concentration for the soya oil–water interface.

extended about 10 cm above the inner cylinder. An oil layer of about 6 cm height is put on top of the overflowing water layer. The interfacial tension can be measured by means of the Wilhelmy plate technique using a roughened glass plate that is completely wetted by the aqueous phase (Bergink-Martens *et al.*, 1993).

Surprisingly, the oil–water model operates in the same way as the water–air model; in the oil–water situation a height of more than 5 cm of the falling water film was also found to be enough to be in the plateau region (Fig. 10.9) and the dynamic interfacial tension between household soya oil and water was also $\sim 10\ \mathrm{mN\ m^{-1}}$ higher than the equilibrium interfacial tension (Fig. 10.10). On the other hand, some aspects are quite different: in the oil–water situation the aqueous film at the outside of the cylinder is much thicker than in the water–air situation and the rate of expansion of the oil–water interface is much lower than that of the water–air surface.

10.2.2 Applications

Foaming and emulsification of liquids take place under conditions where the surfaces and thin liquid films are subjected to mechanical agitation and consequently are in a state far from equilibrium. Especially during production of these systems, new surfaces of bubbles, emulsion droplets and thin films are made. It is well known that the connected increase in surface area affects the stability of thin films between bubbles or droplets in a negative way (Prins, 1976). Film stability can only be ensured during surface expansion when enough surfactant can be supplied to the expanding surface. The ability of a surfactant solution to supply surfactant to an expanding surface can be measured by means of the overflowing cylinder technique: the lower the dynamic surface tension, the better is the supply of surfactant to the expanding surface.

Foamability of a surfactant solution can be measured by shaking the surfactant solution under standardized conditions in a graduated cylinder and measuring the foam height formed in the steady state. It appears that this foamability increases considerably with the surfactant concentration and that this is consistent with the lowering of the dynamic surface tension measured with the overflowing cylinder (Bergink-Martens *et al.*, 1992, 1993). Other examples of a close relationship between the lowering of the dynamic surface tension and the increased stability of a thin liquid film are: the manufacturing of stickers, the formulation of aqueous liquids for the controlled wetting of offset printing plates, and the development of plant spray liquids that spread over leaves (de Ruiter *et al.*, 1990). In all these cases, a thin liquid film has to spread over a solid surface in such a way that it is stable during the expansion of the film surface.

Recently it has been found that the overflowing cylinder technique can give worthwhile information about the physical background of the transient foaming behaviour of water + alcohol mixtures. These short-lived foams – foam life-times are measured in seconds – can be measured in the steady state by means of a bubble column. In a closed circuit arrangement, saturated vapour is pumped around through the solution. By standardizing the vapour flow rate, the foam height reached in the steady state is a measure for the foam life-time and it can be used to compare foam life-times of different samples.

Alcohol + water mixtures are not able to form equilibrium films due to lack of disjoining pressure. Therefore it is assumed that the foam life-time is governed by the drainage rate of the thin liquid films between the bubbles. It is well known that drainage can be counteracted by surface tension gradients opposing the liquid flow, not only in the thin films, but also in the Plateau borders (Prins, 1990). This means that the ability to generate a surface tension gradient will give rise to a longer foam life-time.

According to equation (10.4), the measurement of $d \ln A/dt$ in the overflowing cylinder gives direct information about $d\gamma/dr$, the ability of the liquid to generate a surface tension gradient. For butanol + water systems in Figs 10.11 and 10.12, $d \ln A/dt$ and foam height values as a function of the butanol activity (mol/L) are given as an example. Similar (unpublished) results have been obtained in our laboratory for ethanol,

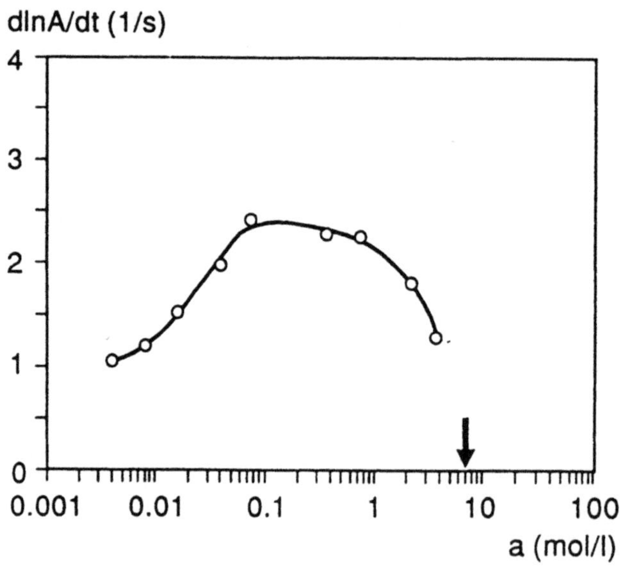

Fig. 10.11 The relative expansion rate $d \ln A/dt$ of the surface of aqueous butanol solutions as a function of the butanol activity a.

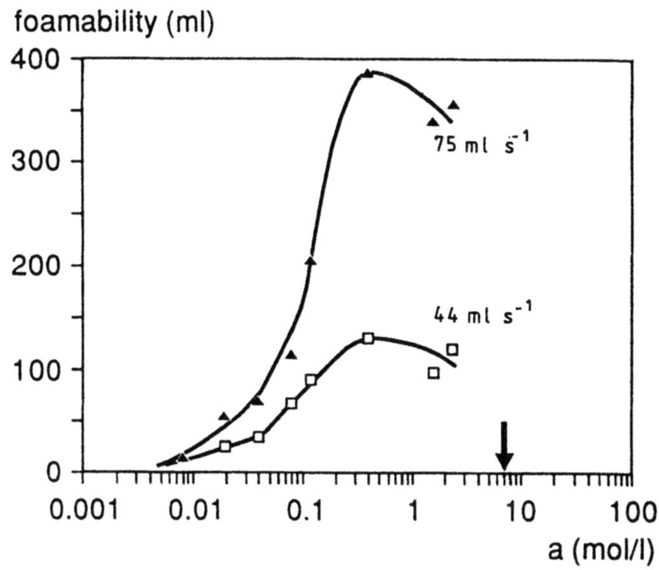

Fig. 10.12 Foamability of butanol + water solutions at two vapour flow rates as a function of the butanol activity *a*. The arrow indicates the limit of solubility of butanol. The maximum lifetime of the foam generated at 75 ml s^{-1} is 5 s.

hexanol and octanol. The arrows in Figs 10.11 and 10.12 indicate the limit of solubility of butanol in water above which no foam stability is found whatsoever. This is due to the presence of alcohol droplets, which, due to their low surface tension, spread over the thin aqueous film and cause the rupture of these films. The close relationship between the ability to generate a surface tension gradient, as indicated by $d \ln A/dt$ values, and the life-times of these transient foams measured by their stationary foam heights, are taken as an argument in favour of the idea that the foam life-time for aqueous alcohol solutions is governed by the foam draining behaviour.

Following the same line of reasoning as applied to oil–water interfaces, it is to be expected that the emulsifying capacity is related to the dynamic interfacial tension between oil and water (Bergink-Martens, 1993). The emulsifying capacity can be measured as follows: 10 cm^3 of soya oil is emulsified in 90 cm^3 Teepol solution in water by shaking it in a 110 cm^3 graduated cylinder by hand. The amount of emulsified oil is measured by measuring the amount of creamed oil that is not emulsified after 1 min. The results show that here also the emulsifying capacity continues to increase when the equilibrium interfacial tension in already constant (Fig. 10.13). The dynamic interfacial tension, however, decreases when the emulsifying capacity increases, which is analogous to

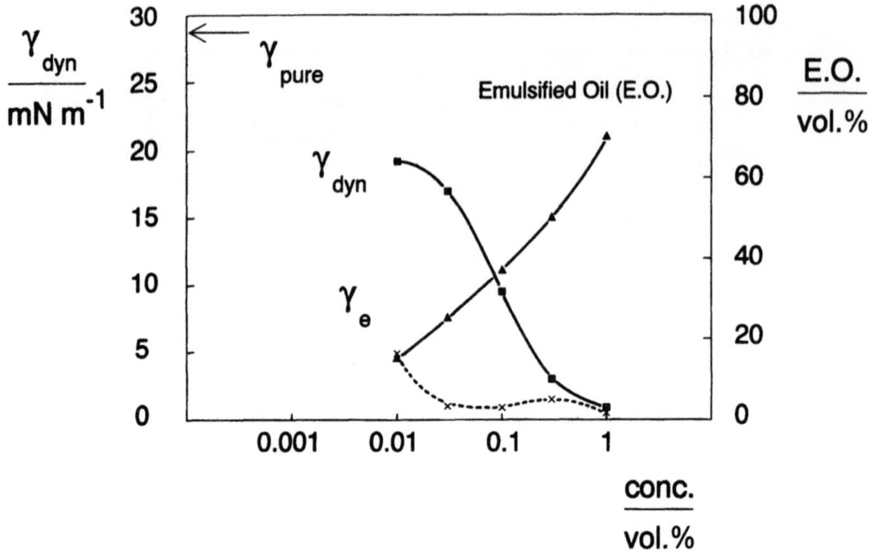

Fig. 10.13 The equilibrium interfacial tension γ_e and dynamic interfacial tension γ_{dyn} and the fraction of emulsified soya oil (E.O.) as a function of the Teepol concentration.

what is found in the relation between foamability and the dynamic surface tension.

For future applications of the overflowing cylinder technique, it is worthwhile noticing that, in addition to the dynamic surface tension and the expansion rate, it is now possible to measure the dynamic adsorption of the surfactant directly at the expanding surface. This measurement can be carried out with ellipsometry. Especially for high-molecular-weight surfactants such as proteins, this offers the possibility to study the relationship between the dynamic surface tension and the dynamic adsorption under physically well-defined circumstances and to compare these results with their equilibrium values. From this it is hoped that information can be obtained about the parameter – surface tension or adsorption – that is mainly responsible for the observed foamability of macromolecular systems. A preliminary result obtained with BSA (Fig. 10.14) demonstrates that the same surface tension measured under equilibrium conditions takes place at an adsorption value that is about 50% higher than under dynamic conditions. It must be noted, however, that the concentration needed in the dynamic experiment is a factor of 100 higher than in the equilibrium experiment. One may speculate that the conformation of this protein at the surface is different under dynamic conditions compared to equilibrium conditions.

The overflowing cylinder technique can be used to measure the surface dilational viscosity η_s^d defined by

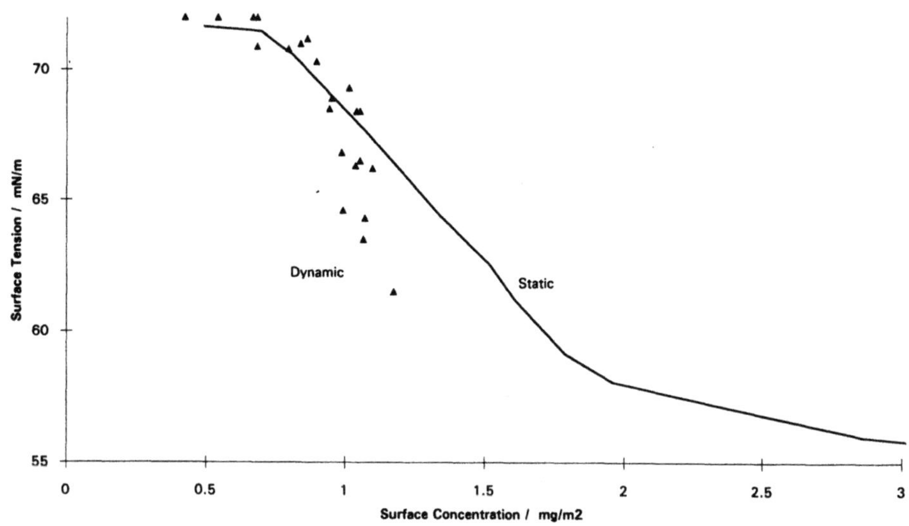

Fig. 10.14 Equilibrium and dynamic surface tension of aqueous BSA solutions measured as a function of the adsorbed amount of BSA. Static data are from Graham and Phillips (1979); dynamic data are obtained by combining ellipsometer and overflowing cylinder results.

$$\eta_s^d = \frac{\gamma_{dyn} - \gamma_{eq}}{d \ln A/dt} = \frac{\Delta\gamma}{d \ln A/dt}, \quad (11.5)$$

where γ_{eq} is the equilibrium surface tension of the system. This physically well-defined dynamic surface property is a measure of the stiffness of the surface when it is subjected to a continuous rate of expansion or compression. For the measurement in compression, the overflowing cylinder has to be replaced by the caterpillar trough technique described below.

Use of this surface stiffness parameter to get a better understanding of, for instance, the foaming behaviour of practical systems must be made with some care for the following two reasons:

(1) The value of η_s^d depends strongly on the value of $d \ln A/dt$. For all practical systems investigated up to now (such as beer, milk, manure and Teepol), we found a power-law relationship between η_s^d and $d \ln A/dt$ which is valid in compression as well as in expansion:

$$\log \eta_s^d = m \log \frac{d \ln A}{dt} + n. \quad (10.6)$$

Here, m and n are constants appropriate to a particular system. The quantity m mostly has values close to -1, which indicates that by increasing $d \ln A/dt$ the value of η_s^d decreases.

(2) In the definition of η_s^d given above, the equilibrium surface tension plays an important role. This surface tension depends strongly on the presence of so-called 'minor components' mostly present in practical systems. Minor components are highly surface-active, poorly soluble impurities, mostly present at low concentration, which lower the surface tension considerably when they have the opportunity (time) to reach the surface. When, as in foam-making, the surface is expanded, these minor components have no time to reach the surface and so no correlation is found between the value of η_s^d and phenomena like foamability. It may be that this can explain the observation that for beer foam the power-law model can be successfully applied to explain the slowing down of the disproportionation process (see below), because in that case minor components will become accumulated at the surface of the shrinking bubble.

10.3 Free falling film technique

In the free falling film technique, the liquid under investigation is allowed to flow through a narrow slit in the bottom of a container (Fig. 10.15). Nylon wires at both sides of the film stabilize it during its downward movement. Due to the increased velocity of the film when it falls down, the film becomes thinner the longer it gets. At the bottom the film liquid is collected and it is pumped up to the slit again. In this way, in a steady state, a free falling liquid film can be made. Films 1 m high and 15 cm wide can easily be made, especially when surfactants are present.

Due to the increased velocity of the film, the film surface is expanded according to

$$\frac{\mathrm{d} \ln A}{\mathrm{d}t} = \frac{\mathrm{d}V_z}{\mathrm{d}z}, \tag{10.7}$$

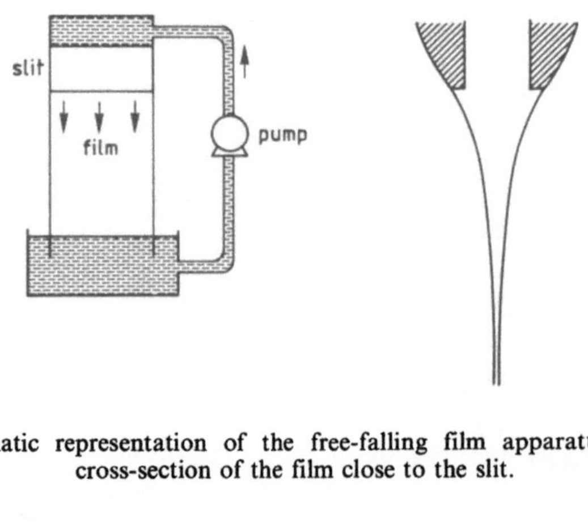

Fig. 10.15 Schematic representation of the free-falling film apparatus and a vertical cross-section of the film close to the slit.

where V_z is the surface velocity in the vertical z-direction. When the film falls free – that means that there is no reduction of the velocity due to the presence of a surface tension gradient along the height of the film – the velocity V_z at a distance z from the slit is given by:

$$V_z = V_0\left(1 + \frac{2gz}{V_0^2}\right)^{1/2}. \tag{10.8}$$

In equation (10.8), V_0 is the mean velocity of the liquid at the outlet of the slit. Differentiation with respect to z gives the relative expansion rate of the surface,

$$\frac{\mathrm{d}\ln A}{\mathrm{d}t} = g(V_0^2 + 2gz)^{-1/2}. \tag{10.9}$$

Just after the slit at $z = 0$, we have $\mathrm{d}\ln A/\mathrm{d}t = g/V_0$. For a typical value of $V_0 = 0.1\ \mathrm{m\ s^{-1}}$, this gives a value of $\mathrm{d}\ln A/\mathrm{d}t = 100\ \mathrm{s^{-1}}$, which is much higher than can be obtained at the horizontal surface of the overflowing cylinder. At a distance of about 5 cm from the slit, the $\mathrm{d}\ln A/\mathrm{d}t$ value is reduced to $10\ \mathrm{s^{-1}}$.

The dynamic surface tension of the free falling film can be measured by means of an adapted Wilhelmy plate technique (Bergink-Martens *et al.*, 1992). It appears that, for the same surfactant solution, the γ_{dyn} log *versus* log (concentration) curve follows about the same line as is found in the overflowing cylinder technqiue, indicating that both techniques are based on the same physical principle (Fig. 10.16). Further, it appears that

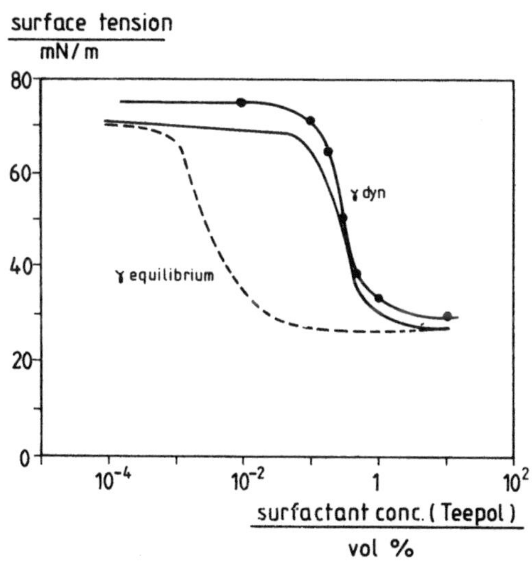

Fig. 10.16 Equilibrium and dynamic surface tension as a function of the Teepol concentration for the overflowing cylinder (—) and the free falling film (●).

the dynamic surface tension is rather constant over the height of the film from 5 cm to 20 cm from the slit.

The free falling film technique is especially useful for studying the rupture of thin liquid films as a result of hole formation. This can be realized by introducing fatty particles in the film, which, by hydrophobic action or through spreading phenomena, can cause the rupture of the film.

Most of our research has been devoted so far to the spreading of particles over the film surface. Spreading is only possible when the particle surface contacts the film surface, and when the film surface tension is larger than the sum of the surface tension of the particle and the interfacial tension between the particle and the film liquid. When the particle is coming from the interior of the film to the film surface, contact between particle surface and a film surface can only take place when the thin liquid film covering the particle is broken. It is found that the drainage of the particle covering film determines the time needed for the de-wetting of the particle. When the particle spreads over the film surface, two phenomena can be distinguished: hole formation, and the appearance of 'sparkling light spots' on the film surface.

10.3.1 Hole formation

Hole formation is the result of a local thinning of the film caused by the radial movement of the spreading layer away from the spreading particle (Fig. 10.17). Due to this radial movement of the film surface, the film

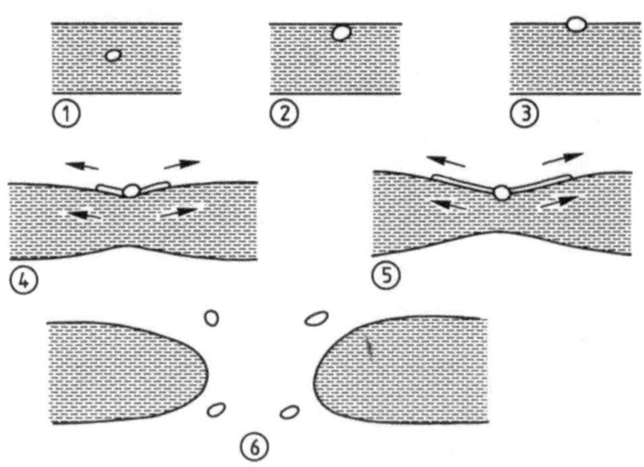

Fig. 10.17 Schematic representation of film rupture due to spreading from a particle. Not to scale. (1) Particle in film; (2) particle close to surface; (3) particle becomes dry and contacts surface; (4), (5) spreading causes local thinning of film; (6) when process proceeds far enough film ruptures.

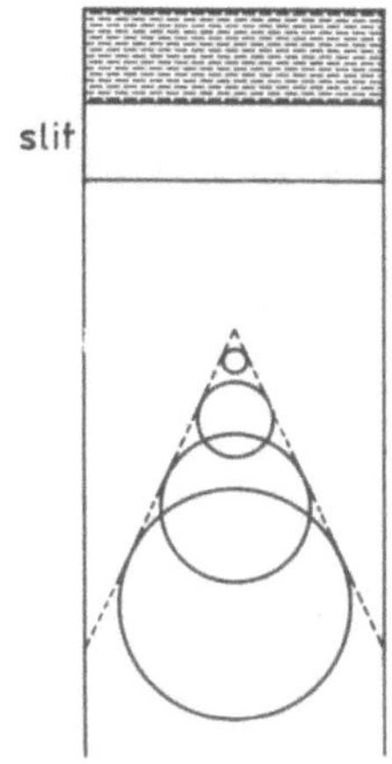

Fig. 10.18 Hole formation in a free falling film due to the presence of a spreading particle as seen in stroboscopic light. The envelope is indicated by a dotted line.

liquid is also squeezed away causing the local thinning of the film. When this process proceeds far enough, the film becomes so thin that due to thermal motion the film ruptures spontaneously. Once the hole is formed, it expands circularly in the falling film. The combined action of the expansion and the downward motion, observed in reflected stroboscopic light, gives a picture which is reproduced in Fig. 10.18. The envelope of the falling holes forms two straight lines which cross each other at the point where hole formation was initiated. The angle between these lines (also called the bow wave or tear angle) depends on the film thickness and its surface tension. So, by knowing the film thickness, the surface tension of the film can be deduced from this angle.

The advantage of this method is that under suitable conditions the film restores itself after hole formation has taken place. By introducing particles of different sizes and compositions, the conditions for appearance (or otherwise) of hole formation can be investigated by simple visual observation. The film thickness can be changed simply by adjusting the flow rate through the apparatus because the two are directly proportional to one another.

10.3.2 Sparkling light spots

Spreading of particles over a film surface does not necessarily result in hole formation in the film. This can be due to the fact that the particles are too small or the film is too thick. Still, it should be convenient to have an analytical tool that can be used to indicate whether spreading takes place or not. We have observed that spreading is always accompanied by an optical phenomenon which is best described as 'sparkling light spots'. The likely explanation is that, due to the difference in refractive index between the spreading layer (oil) and the film liquid (water), some extra light is reflected to the observer. Anyhow, it is an easily observable phenomenon, and we have experience that it can be used to indicate unambiguously the occurrence of the spreading process. This is an interesting piece of information because spreading seems to be a much more effective way to cause coalescence of bubbles in a foam than the presence of hydrophobic particles. The latter are able to rupture films which have the same thickness as the hydrophobic particles. In contrast, spreading particles can cause rupture of films which are ten times thicker than the particle size.

In addition, when sparkling light spots are observed, coalescence of bubbles in a foam made out of that liquid can be one of the destabilizing mechanisms. When sparkling light spots are absent, any coalescence observed must be the result of another destabilizing mechanism, such as lack of disjoining pressure, the presence of hydrophobic particles, or as a result of disproportionation (see below).

10.4 Ring trough technique

The ring trough technique is an improved Langmuir trough technique; instead of a rectangular trough with barriers moving parallel to the surface, a ring, completely wetted by the contacting liquid, is moved perpendicular to the surface (Kokelaar *et al.*, 1991). In this way a pure dilational deformation of the liquid surface inside the ring is ensured, and any leakage of the barrier, which is one of the problems encountered when using the Langmuir trough technique, is completely ruled out.

By moving the ring up and down in a sinusoidal way, the surface area of the liquid inside the ring is also deformed in a sinusoidal way. The resulting sinusoidal change in surface tension is a measure of the surface dilational modulus E defined by

$$E = \frac{d\gamma}{d \ln A}. \tag{11.10}$$

Most practical systems show visco-elastic surface behaviour. Therefore this modulus can be considered to be composed out of an elastic part (the storage modulus) and a viscous part (the loss modulus). Visco-elastic surface behaviour means that the values of these three moduli depend on the frequency of the applied sinusoidal deformation. Moduli measurements carried out over a wide frequency range are a fingerprint of the surface rheological properties of a particular system close to equilibrium. Therefore, much more discriminating than the surface tension, moduli measurements are a sensitive tool for indicating changes in the surface properties that can arise as a result of changes in, for instance, temperature, pH, salt concentration, surfactant concentration and ageing.

Moduli measurements are especially relevant in relation to one of the main foam destabilizing mechanisms: disproportionation (Ronteltap and Prins, 1990; Ronteltap *et al.*, 1990). This process (Ostwald ripening) is transport of gas from smaller bubbles to bigger ones. The driving force for the process is the Laplace pressure difference between the gas in the smaller bubble and that in the bigger one. Laplace's law indicates that, in a bubble of radius R having a surface tension γ, the pressure of the gas is larger by an amount Δp than that outside the bubble:

$$\Delta p = \frac{2\gamma}{R}. \qquad (10.11)$$

According to Henry's law the maximum solubility of a gas in a liquid is proportional of the gas pressure. So, in the neighbourhood of the smaller bubble, the gas concentration in the liquid is higher than around the bigger one, and this causes gas transport through the liquid – for instance by diffusion. This process is self-accelerating because the radius of the smaller bubble becomes smaller and smaller, and according to equation (10.11) the driving force correspondingly increases. At the end of the process, the disappearing bubble even implodes. When this process takes place in the honeycomb structure of a polyhedric foam an unstable configuration is formed which rather rapidly jumps over into a new stable one (Fig. 10.19). This transition, however, is accompanied by the rapid formation of a new film and this film may easily be ruptured under these conditions due to one of the mechanisms discussed earlier.

The disproportionation process can be stopped when, due to the compression of the shrinking bubble surface, the surface tension drops to such a low value that the decrease in radius R is compensated for. Lucassen has calculated from a fundamental point of view that this will

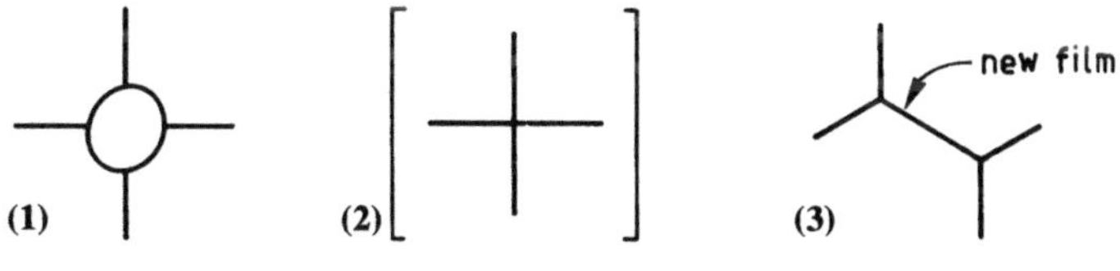

Fig. 10.19 When a small bubble in a polyhedric foam (1) disappears due to gas diffusion, an unstable situation is formed (2) where four films meet at one point. The configuration moves over rapidly to the equilibrium situation (3). During this transition a new film is formed rather rapidly, which may cause the rupture of this film.

be the case when the surface behaves completely elastically and the surface dilational modulus fulfils the following relationship:

$$E > \frac{\gamma}{2}. \tag{10.12}$$

Measurements of the surface dilational modulus can therefore give worthwhile information about whether the liquid surface is able to stop disproportionation. In aerated systems such as instant whipped cream and beer foam, the presence of highly soluble gases such as CO_2 and N_2O, in combination with the fact that the surfaces of these liquids do not fulfil the requirement described above, explains the observation that disproportionation is to a large extent responsible for the instability of these foams.

10.5 Caterpillar trough technique

The caterpillar trough technique uses a Langmuir trough provided with a number of barriers (4–6) which move over the liquid surface one after the other by means of an endless band (Fig. 10.20). In this way the liquid surface can continuously be expanded or compressed depending on the direction in which the barriers move. The resulting change in surface tension of the liquid in the steady state can easily be measured by means of the Wilhelmy plate technique. The method is rather rough because the rate of compression or expansion during one stroke of a barrier is not constant and leakage along the barriers cannot be ruled out completely. In addition, every time that a barrier contacts or leaves the surface it gives an extra disturbance of the surface. However, presumably due to damping phenomena, the method works reasonably well in practice.

This method contributes to the understanding obtained from the dilational facilities of the overflowing cylinder technique because it can apply compression to the surface as well. However, it can only be used in the range of relatively small deformation rates, up to $0.2\,s^{-1}$, because otherwise the moving barriers disturb the bulk liquid too much. In this

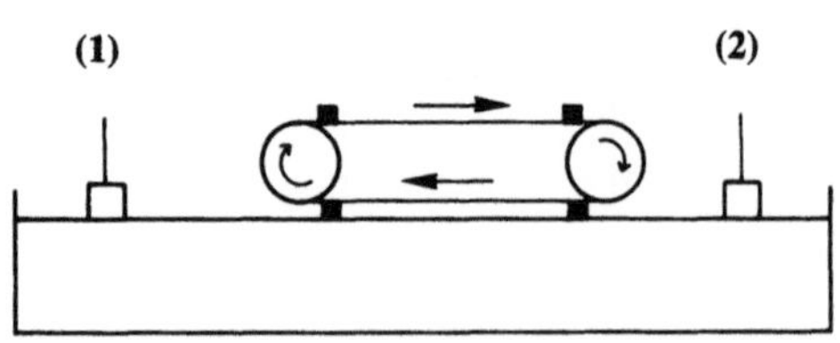

Fig. 10.20 Schematic representation of the caterpillar trough showing two Wilhelmy plates (1) and (2) and four barriers mounted on an endless band moved in the direction indicated. In this way the surface is compressed at (1) and expanded at (2).

way it fills up the range of small expansion rates not easily accessible by the overflowing cylinder technique.

In an alternative approach to disproportionation, it is realized that during shrinkage of the smaller bubble the bubble surface is continuously compressed and in a state far from equilibrium. It can be proved that, in the disproportionation process, the shrinkage of the smaller bubble is more important than the expansion of the bigger bubble. The behaviour of this shrinking surface can be described by means of the power-law relationship between the surface dilational viscosity η_s^d and the relative rate of compression of the surface, d ln A/dt, as described above. In this way, it is possible to predict the shrinking behaviour of a gas bubble in beer which is confirmed by measurements carried out on the shrinking bubble.

In still another approach to disproportionation, it is realized that, in order to stabilize one bubble (1) which is 10 times smaller than another bubble (2), the surface tension of the smaller bubble must be so small that the Laplace pressure differences over both bubble surfaces is the same, i.e.

$$\Delta p_1 = \Delta p_2 = \frac{2\gamma_1}{R_1} = \frac{2\gamma_2}{R_2}. \qquad (10.13)$$

When $R_1 = (1/10)R_2$ it follows that $\gamma_1 = 1/10\,\gamma_2$. Assuming a value for γ_2 of 40–50 mN m^{-1} this means that we must have γ_1 = 4–5 mN m^{-1} in order to stabilize the system.

The caterpillar trough provided with one barrier can be used to carry out a one-stroke compression test of the surface to measure the value of the lowest surface tension obtained in the compressed surface. After the compression is stopped, the relaxation of the surface tension to higher values can be observed. This kind of measurement gives additional information about the ability of surfaces to stabilize against disproportionation.

10.6 Mannheimer trough technique

The Mannheimer trough technique (Mannheimer and Schechter, 1970) is one of the techniques developed to measure shear properties of liquid surfaces. Shear properties arise in a surface when a kind of network structure is present in that surface. So interacting macromolecular compounds such as proteins and polysaccharides are able to generate substantial shear properties in a surface.

Just as in bulk rheology, with surface shear properties one can distinguish between properties close and far from equilibrium, and between viscous and elastic behaviour. The Mannheimer trough is

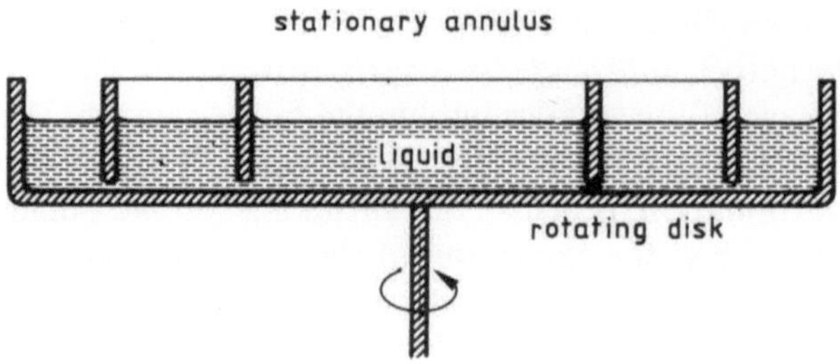

Fig. 10.21 Schematic front view of Mannheimer trough showing the stationary circular canal (the annulus) and the rotating dish. The movement of the liquid surface can be observed from above.

especially developed to measure the surface shear viscosity in a steady state. Further it can also be used to measure elastic surface properties and yield phenomena as explained below.

The Mannheimer apparatus consists of a ring-shaped canal, called the annulus, with a rectangular cross-section that is provided with the liquid under investigation (Figs 10.21 and 10.22). The vertical side walls of the canal are stationary, but the bottom floor of the annulus can be rotated at a known speed. By means of a hydrodynamic analysis, the shear stress exerted by the moving bottom via the viscous liquid on the surface of the liquid can be calculated. As a result, a measurement of the surface velocity in the centre of the canal is a measure of the surface shear viscosity of that liquid. So far the method is restricted to Newtonian liquids.

In order to visualize the surface shear viscosity, consider a surface element bounded by two parallel barriers. One barrier is motionless and the other moves with a constant velocity v_x in the direction of the barrier

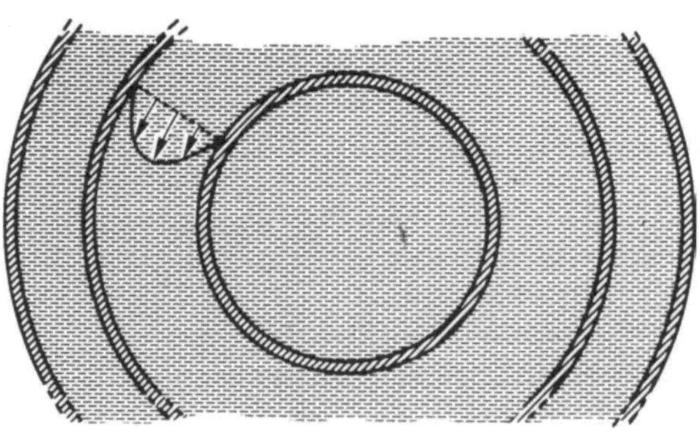

Fig. 10.22 Schematic top view of Mannheimer trough showing the 'parabolic' velocity profile of the liquid surface in the steady state.

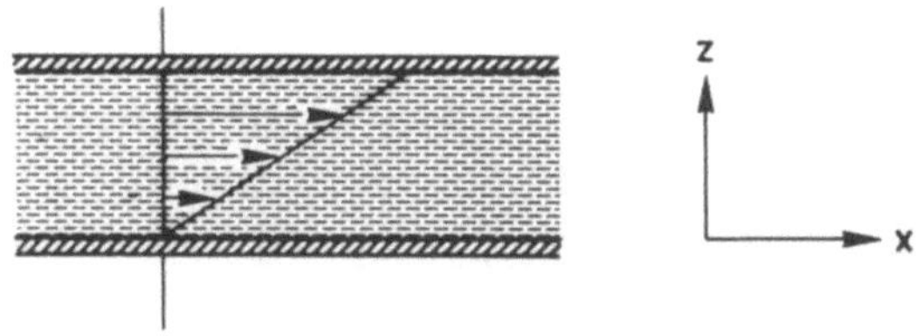

Fig. 10.23 A liquid surface element bounded by two straight parallel barriers subjected to a shearing deformation. For the surface velocity profile indicated, the rate of deformation is $\mathrm{d}V_x/\mathrm{d}z$.

(see Fig. 10.23). The shear rate exerted on the surface element is $\mathrm{d}V_x/\mathrm{d}z$, where z is the coordinate in the plane of the surface perpendicular to x. Under these conditions a force per unit length, the line tension ($\mathrm{N\,m^{-1}}$), has to be exerted on the motionless barrier in order to keep it motionless. The surface shear viscosity η_s^s is defined now as

$$\eta_s^s = \frac{T_{zx}}{\mathrm{d}V_x/\mathrm{d}z}, \tag{10.14}$$

where T_{zx} is the line tension acting in the x-direction on a line in the surface having a normal pointing into the z-direction.

One of the problems encountered in performing surface shear measurements is that, due to the viscosity of the liquid, not only the surface is set in motion but also some liquid close to that surface is set in motion. For a correct measurement, this effect has to be accounted for, which, in the case of the Mannheimer method, is carried out in the correct way. When the network structure in the surface gives rise to yield phenomena, the Mannheimer method enables us to measure the yield stress in a simple way. As long as the stress exerted by the rotating bottom on the surface remains below the surface yield stress, the liquid surface remains motionless. By increasing the speed of rotation step by step, the stress at which the movement of the surface sets in equals the yield value of the surface. After the surface has been yielded, the measurement can be continued at a lower speed of rotation to investigate how the surface structure reestablishes itself under these conditions.

Arguments to study shear phenomena in liquid surfaces are that they can serve as an analytical tool to indicate the presence of network structures in that surface, and that the mechanical properties of a liquid surface subjected to shearing deformations can be measured and expressed in physically well defined parameters. However, as can be expected, network structures also contribute to dilational surface properties, which can also give information about surface structures. In addition, as we are discussing the stability of thin liquid films, it is well known that a mechanical expansion of the films can rupture the film, whereas a film broken by a shearing deformation has never been reported.

A phenomenon often related to surface shear properties is the drainage of thin films. Marginal regeneration is one of the mechanisms that contribute considerably to film drainage. Marginal regeneration is a process in which thicker and thinner film parts are exchanged with the film surrounding Plateau border: the thin parts are drawn out of the border and thicker parts are sucked into it. This process can easily be observed in macroscopic films (say, 1 × 1 cm) and it gives rise to a vivid colour spectacle close to the Plateau border. Since thicker and thinner film parts move along past each other, shear properties become involved, and they can affect the rate of the process. In the ultimate case, the process of marginal regeneration stops when the surface shear resistance becomes too large; a so-called 'rigid' film is then formed. The film parts involved in this process have dimensions of the order of millimetres. So this process can develop itself when the film has dimensions of at least centimetres. The consequence is that in fine foams, where the films are smaller than 1 mm, the process of marginal regeneration is absent.

References

Bergink-Martens, D.J.M. (1993) *Interface Dilation: the Overflowing Cylinder Technique*, Ph.D. thesis, Wageningen Agricultural University.

Bergink-Martens, D.J.M., Bos, H.J., Prins, A. and Schulte, B.C. (1990) Surface dilation and fluid dynamical behaviour of Newtonian liquids in an overflowing cylinder. 1. Pure liquids. *J. Colloid Interface Sci.*, **138,** 1–9.

Bergink-Martens, D.J.M., Bisperink, C.G.J., Bos, H.J., Prins, A. and Zuidberg, A.F. (1992) Surface dilational behaviour of surfactant solutions: a comparison between the overflowing cylinder and the free falling film technique. *Colloids Surf.*, **65,** 191–199.

Bergink-Martens, D.J.M., Bos, H.J., van Kalsbeek, H.K.A.I. and Prins, A. (1993) Dilation of fluid interfaces in a vertical overflowing cylinder, in *Food Colloids and Polymers: Stability and Mechanical Properties* (eds. E. Dickinson and P. Walstra), Royal Society of Chemistry, Cambridge, pp. 376–386.

Bergink-Martens, D.J.M., Bos, H.J. and Prins, A. (1994) Surface dilation and fluid dynamical behaviour of Newtonian liquids in an overflowing cylinder. 2. Surfactant solutions. *J. Colloid Interface Sci.*, **165,** 221–228.

de Ruiter, H., Uffing, A.J.M., Meinen, E. and Prins, A. (1990) Influence of surfactants and plant species on leaf retention of spray solutions. *Weed Sci.*, **38,** 567–572.

Graham, D.E. and Phillips, M.C. (1979) Proteins at liquid interfaces. 2. Adsorption isotherms. *J. Colloid Interface Sci.*, **70,** 415–425.

Kokelaar, J.J., Prins, A. and de Gee, M. (1991) A new method for measuring the surface dilational modulus of a liquid. *J. Colloid Interface Sci.*, **146,** 507–511.

Mannheimer, R.J. and Schechter, R.S. (1970) An improved apparatus and analysis for surface rheological measurements. *J. Colloid Interface Sci.*, **32,** 195–211.

Prins, A. (1976) Dynamic surface properties and foaming behaviour of aqueous surfactant solutions, in *Foams* (ed. R.J. Akers), Academic Press, London, pp. 51–60.

Prins, A. (1990) Liquid flow in foams as affected by surface rheological properties: a contribution to a better understanding of the foaming behaviour of liquids, in *Hydrodynamics of Dispersed Media* (eds. J.P. Hulin, A.M. Cazabat, E. Guyon and F. Carmona), Proceedings of the 4th E.P.S. Liquid State Conference, Elsevier Science Publishers PV, Amsterdam, pp. 5–15.

Ronteltap, A.D. and Prins, A. (1990) The role of surface viscosity in gas diffusion in aqueous foams. 2. Experimental. *Colloids Surf.*, **47**, 285–298.
Ronteltap, A.D., Damsté, B.R., de Gee, M. and Prins, A. (1990) The role of surface viscosity in gas diffusion in aqueous foams. 1. Theoretical. *Colloids Surf.*, **47**, 269–283.

Chapter 11

Light Scattering Studies of Colloid Stability and Gelation

D.S. Horne
Hannah Research Institute,
Ayr, UK

11.1 Introduction

Historically food products were originally prepared on an empirical basis, frequently in the home or nearby, for immediate consumption. There was therefore no pressing need for the food manufacturer to gain a detailed picture of the molecular structure of the product. With the growth of the food industry, however, has come an increasing demand for consistency in quality. This can only be achieved through a full understanding of the structure/function relationship of all ingredients in the product so that their behaviour and performance can be controlled through all the stages of processing, storage and eventual consumption.

At the simplest level, most food systems can be thought of as a collection of colloidal particles dispersed in a continuous medium. The dispersion can be fluid in which case the particles are often stabilized and protected against aggregation by adsorbed layers of biopolymers, as in emulsion systems. On occasion this biopolymer layer may constitute an integral part of the dispersed particle, as in the casein micelles of milk, which will provide us with many examples in this chapter. In other food products the dispersion may be structured and the properties of these structures may determine the characteristics of the final food product such as texture, fat- or water-holding capacity, stability and shelf-life. In many cases the food processor will create such structure by inducing gel formation in a fluid system, as in the manufacture of yoghurt or cheese. Control of this reaction is therefore important in achieving the desired properties in the final product. Techniques are therefore required to obtain detailed quantitative information on the colloidal particles themselves, on their interactions, and on the changes occurring in these and their structural relationship as the gelation reactions progress. Light scattering techniques provide such a molecular probe. For electromagnetic radiation, the ruler or distance scale is set in reciprocal space by the wave vector, q, which is inversely proportional to the wavelength of the incident radiation in the scattering medium. For visible light, we have $q \approx 0.001 \rightarrow 0.05\,\text{nm}^{-1}$. Light scattering is thus especially suited for studies of particle size or of structure with long range order in the range 20–1000 nm.

In this chapter we discuss the general basis of light scattering and the equipment required for its observation, and especially the origins and techniques of dynamic light scattering (DLS). Intially confining ourselves to dilute suspensions, we demonstrate how information on particle size and the details of particle size distribution in a colloidal suspension can be extracted from observations of dynamic light scattering behaviour. Concentration of these fluid suspensions introduces the problems of multiple scattering and interactions between the particles. We describe new extensions of DLS employing fibre optics which allow us to circumvent these problems, outlining the principles under which they operate, together with their shortcomings. We present specific applications of these techniques to particle sizing in emulsions and aggregating micelle systems producing renneted and acid milk gels. The further complications in interpreting DLS behaviour introduced by the transition to the gel state will be covered in this section.

11.2 Light scattering background

11.2.1 Basic principles

The interaction of light with matter is ubiquitous. We are able to see and register objects near and far because essentially all materials scatter and/or absorb light. The interaction takes place with the electrons bound in the material. These may re-radiate the light as scattered light: if the light experiences no energy loss then the scattering is termed 'elastic'. If the light changes the energy state of the electron, that portion of light is lost from the system, effectively absorbed, although part of that energy may re-appear as light re-emitted at a longer wavelength or perhaps part or all of it may appear as heat.

The essential main elements of the apparatus required to observe the scattering process are shown in Fig. 11.1. First, we have a well-focused or collimated stable source of light. A laser fulfils these requirements as well as providing coherence, an essential for dynamic light scattering. The well-collimated beam enables the direction of scattering to be established with precision; stability is always important when comparing scattering from different samples or under different conditions. Emission over a narrow line width or wavelength range is another important attribute of laser light sources allowing the wave vector to be defined precisely.

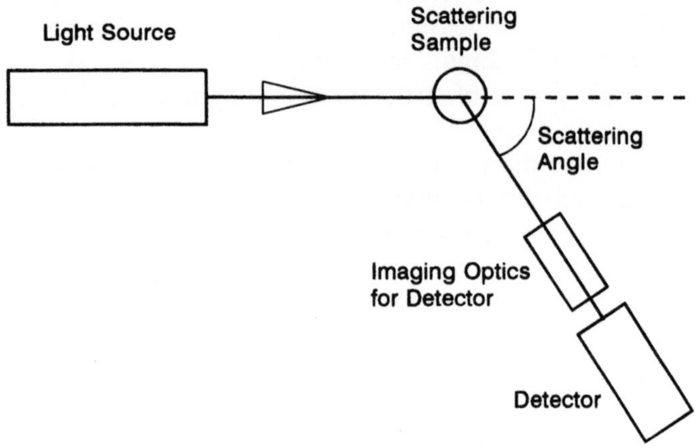

Fig. 11.1 Block diagram of apparatus required for light scattering studies.

Light scattering will generally take place from any medium through which the light passes, and so careful design of the instrument is required to ensure that the scattering observed originates only from the 'sample' of interest. This is frequently achieved by suspending the cuvette in a bath filled with a 'refractive-index matched' fluid which minimizes the change in index at the cuvette boundary. This fluid, often water, also serves to thermostatically couple the cuvette to the rest of the measurement cell, which is usually temperature stabilized.

The detector, usually a photon-counting photomultiplier, is mounted on a goniometer arm. In front of it are mounted some optics, a lens and pair of apertures, which define an optical axis at the scattering or collection angle. Careful alignment is necessary to ensure that the detector is viewing the centre of the probe volume defined by the laser source, that the scattering angle is well-defined, and that the maximum power from the light source is being utilized.

11.2.2 Dynamic light scattering

The experimental configuration in Fig. 11.1 can be employed to make the simplest type of light scattering measurement, that of the (time-averaged) intensity of the light scattered as a function of scattering angle and solute concentration. In this chapter, however, we are interested primarily in the dynamic nature of the scattered light, the rapid fluctuations in the intensity which arise as a result of the random diffusional motion of the colloidal scattering particles. By analysing these fluctuations, information can be extracted on the mobility of the scatterers, in particular the translational diffusion coefficient. Thomas (1991) has given the following phenomenological explanation of the origin of these fluctuations. The

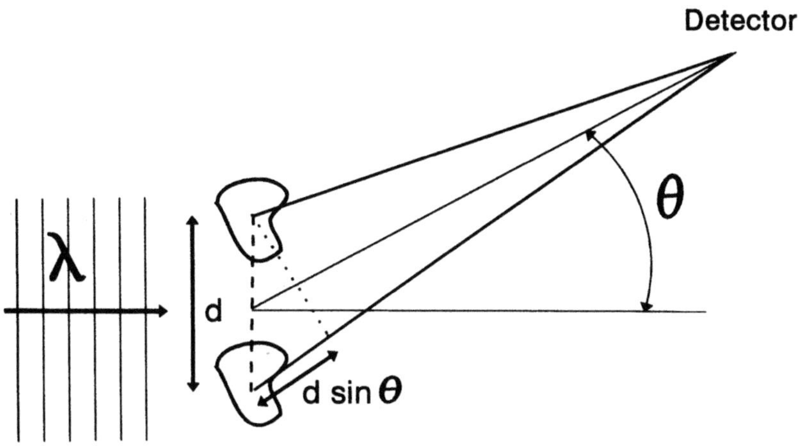

Fig. 11.2 The scattering of light waves from two particles illustrating the origin of phase differences and interference effects.

total intensity from a collection of scatterers results from the interference of the light scattered from each of the scatterers in the scattering volume. The phase of the light scattered by each particle is determined by its position and orientation. Brownian motion of the colloidal scatterers causes the phase of the light scattered from each of them to change with time so that the total scattering intensity will fluctuate with time. The time-scale of the particle movements clearly defines the time-scale of the intensity fluctuations. The origin of the intensity fluctuations can be understood by reference to the scattering diagram of Fig. 11.2, which shows a laser beam of wavelength λ incident on two identical particles. Light scattered by the particles is collected by the detector. If the scattering angle is θ and the distance between the two particles is d, then the path length difference between the wavelets scattered by the two particles is $d \sin \theta$. When this path length difference is equal to an integer number of wavelengths ($d \sin \theta = m\lambda$), the wavelets arrive at the detector in phase, giving rise to constructive interference, i.e. the total intensity will be twice that from a single particle. When the path length difference is a half-integral number of wavelengths ($d \sin \theta = \lambda/2$, $3\lambda/2$, $5\lambda/2$, etc.), then the two wavelets will be precisely out of phase, giving rise to total destructive interference, i.e. the total intensity at that moment will be zero. For other path length differences, the intensity will lie between these two extremes. If the particles are undergoing Brownian motion so that their relative position and orientation change randomly with time, the path length difference between the two scattered waves also changes randomly, leading to a randomly varying phase relationship. This results in a total intensity which fluctuates with time from zero to double the

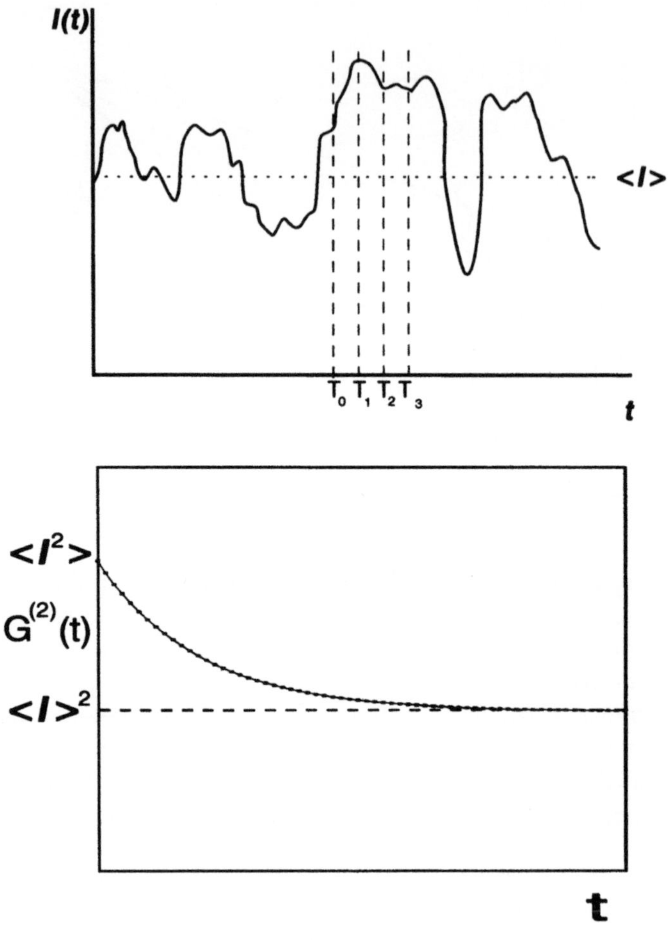

Fig. 11.3 Derivation of the intensity autocorrelation function. The top diagram illustrates a portion of the rapidly fluctuating signal (scattered intensity) seen by the detector as a function of time. The lower plot is a typical correlation function derived from this intensity behaviour as outlined in the text.

single particle scattered intensity. In a real experiment with perhaps as many as 10^8 particles in the scattering volume, the total intensity results from interference between the scattered fields of all of these. In this case the intensity will be a randomly fluctuating signal as exemplified in Fig. 11.3, with a well-defined mean value $\langle I \rangle$, the time-averaged signal measured in a static light scattering experiment. This description emphasizes two essential requirements in the apparatus, namely the coherent nature of the laser beam which ensures that all incident light arrives in phase, and the photon-counting ability of the detector which must be capable of following the rapidly fluctuating signal.

The intensity signal is analysed digitally, by creating from it a correlation function, as follows. The time axis (Fig. 11.3) is divided into a series of very narrow time slices of length equal to the correlator 'sample' time. An arbitrary slice is taken, labelled T_0 and its intensity multiplied by the intensity in the next slice, T_1. The product is placed in the first bin or correlator store channel. Note that if the sample time is short, intensities will hardly change and the product will be effectively intensity squared, I^2. Into the next store channel is placed the product of slice T_0 and slice T_2, and so on through the channels until the last available which receives the product of slices T_0 and T_N. We then shift up one slice, effectively T_1 becomes the new T_0 and the whole cycle is repeated, perhaps several million times building up the correlation function also shown in Fig. 11.3. This effectively decays from a time-average of the light scattering intensity squared to the square of the average intensity. The channel time is selected so that this decay is essentially completed over the time span of the correlation function, i.e. so that the correlation function decays into its own noise level. Mathematically the intensity autocorrelation function is written as

$$G^{(2)}(\boldsymbol{t}) = \langle I(t)^* I(t + \boldsymbol{t}) \rangle$$

$$= \frac{1}{N} \sum_{i=1}^{\infty} I(T_i) \times I(T_i + \boldsymbol{t}). \tag{11.1}$$

Here the angle brackets indicate a time average, N is the number of correlator channels, and $\boldsymbol{t}$ is the delay time between channels. At very short delay times, it is clear that

$$\lim_{\boldsymbol{t} \to 0} \langle I(t) \times I(t + \boldsymbol{t}) \rangle = \langle I(t)^2 \rangle. \tag{11.2}$$

At the other extreme, when $\boldsymbol{t}$ becomes very large, the signal becomes uncorrelated and

$$\lim_{\boldsymbol{t} \to \infty} \langle I(t) \times I(t + \boldsymbol{t}) \rangle = \langle I(t) \rangle^2, \tag{11.3}$$

which is the average intensity squared. This intensity correlation function (equation (11.1)) is generally normalized as

$$g^{(2)}(\boldsymbol{t}) = \frac{G^{(2)}(\boldsymbol{t})}{\langle I(t) \rangle^2}. \tag{11.4}$$

The correlation function which carries the information regarding particle mobility is the electric field correlation function, describing the fluctuations which give rise to the light scattering behaviour. Intensity and field are related and, in the case of Gaussian statistics, their correlation functions are related by the Siegert equation (Siegert, 1943),

$$g^{(2)}(\boldsymbol{t}) = 1 + b\,|g^{(1)}(\boldsymbol{t})|^2, \tag{11.5}$$

where b is an instrumental constant, of order 1, whose value depends on scattering volume, detector efficiency, apparatus geometry, and so on.

If we have a dilute suspension of non-interacting particles all of the same size (monodisperse), the electric field correlation function decays as a single exponential:

$$|g^{(1)}(t)| = \exp(-\Gamma t). \tag{11.6}$$

A plot of the logarithm of this correlation function against delay time will therefore be a straight line of gradient Γ. The result of the experiment will therefore not be a size but a mobility, which is a measure of the particle's ability to move through the suspension. It is important to stress this from the outset. In a dynamic light scattering experiment, the quantity measured is particle mobility, sometimes expressed as its reciprocal, a relaxation time. For this dilute suspension, where single scattering prevails, this mobility is related to the translational diffusion coefficient of the particle, D_T, by

$$\Gamma = D_T q^2, \tag{11.7}$$

q being the scattering wave vector given by

$$q = 4\pi n \sin(\theta/2)/\lambda_0, \tag{11.8}$$

which emphasizes that the actual relaxation time ($1/\Gamma$) observed in the correlation function will depend on both the angle of observation, θ, and the wavelength of illumination in the scattering medium, λ_0/n, n being the index of refraction of the liquid.

For spherical particles, the translational diffusion coefficient is directly related to their radius, r, by the Stokes–Einstein relationship,

$$D_T = k_B T/6\pi\eta r. \tag{11.9}$$

Here k_B is the Boltzmann constant, T is the absolute temperature, and η is the viscosity of the suspending liquid.

Rod-like and plate-like particles can also undergo rotational movement which introduces another term into the correlation function. Flexible particles (like polymer chains) can also possess internal motions which give rise to correlation terms. In both instances specialized data treatments are required to extract useful information on such systems. These are beyond the scope of this chapter but further information may be found in texts on dynamic light scattering (Chu, 1974; Berne and Pecora, 1976; Brown, 1993).

11.2.3 Particle size measurement

As can be seen by the substantial literature associated with the technique, dynamic light scattering has a very wide field of application. However, the single most important application is measurement of colloidal particle size. For sub-micrometre particles accurate measurements of average size with an indication of the width of the size distribution can be obtained rapidly with the minimum of sample preparation and data analysis.

Equation (11.6) indicates that for a monodisperse sample, the correlation function exhibits a single exponential decay. Most systems of interest are not monodisperse. The particles have a range of sizes and their correlation functions consequently exhibit a range of decay times, with each size contributing according to its intensity-weighted mass fraction in the system. This manifests itself as a non-linear decay of the correlation function in the semi-log plot, as illustrated in Fig. 11.4. The smaller particles introduce the faster decay, and the larger ones the slower decay.

Mathematically the correlation function can be expressed as

$$g^{(1)}(t) = \int G(\Gamma)\exp(-\Gamma t)\,d\Gamma, \qquad (11.10)$$

where $G(\Gamma)$ is the intensity-weighted distribution of decay constants which must be determined to obtain the underlying particle size distribution function. Over the years, as dynamic light scattering has developed,

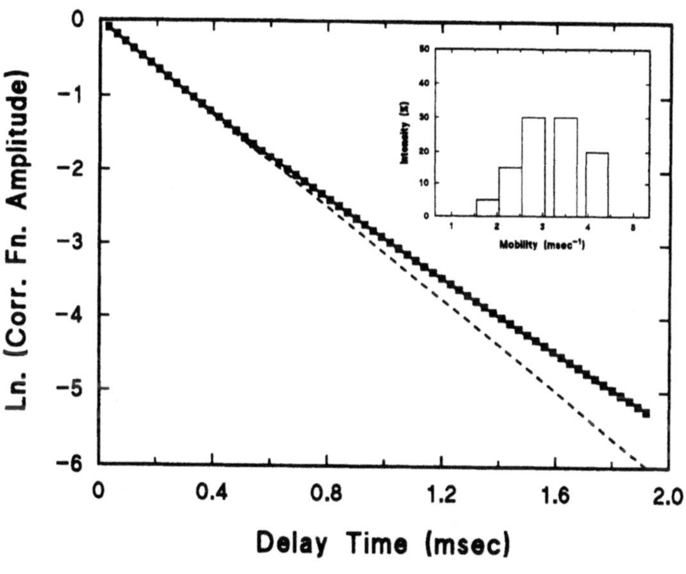

Fig. 11.4 A typical semi-log plot of the normalized correlation function decay obtained from a polydisperse suspension. The inset shows the histogram of the contributions to the overall amplitude as a function of their respective decay times. The dashed line illustrates the linear behaviour from a monodisperse suspension, reinforcing the positive deviations from linearity brought about by the widening of the particle-size distribution function.

considerable effort has been expended in attempting to deconvolute or invert this integral. Most instrument manufacturers now provide software packages to perform this exercise. There is not space here to describe each of these fully, nor to discuss their relative merits or failings. Instead, the so-called cumulants analysis will be considered in detail. Performing this is perhaps the best starting point for analysis of an unknown sample. It gives a good idea of the mean size and spread of the size distribution. Moreover, since the cumulant expansion defines a smooth function, the residual plot gives a feel for the amount of noise present in the measurement. This exposé of the cumulants analysis is followed by a few words on the exponential sampling analysis package of Ostrowsky *et al.* (1981), together with some quite general comments which can be applied to any of the inversion techniques.

11.2.4 Cumulant expansion

The cumulant method assumes nothing about the size distribution function. First described by Koppel (1972), the essence of this analysis is captured in the following outline, due to Brown and Pusey (1975). The mean value of the mobility, $\bar{\Gamma}$, is defined by

$$\bar{\Gamma} = \int_0^\infty \Gamma G(\Gamma)\, \mathrm{d}\Gamma \tag{11.11}$$

and $\exp(-\Gamma \boldsymbol{t})$ can be written as the product of exponentials

$$\exp(-\Gamma \boldsymbol{t}) = \exp(-\bar{\Gamma}\boldsymbol{t}) \exp(-(\Gamma - \bar{\Gamma})\boldsymbol{t}), \tag{11.12}$$

where the second exponential can be expanded in a Taylor series, giving

$$\exp(-\Gamma \boldsymbol{t}) = \exp(-\bar{\Gamma}\boldsymbol{t})\left[1 - (\Gamma - \bar{\Gamma})\boldsymbol{t} + \frac{(\Gamma - \bar{\Gamma})^2 \boldsymbol{t}^2}{2!} - \frac{(\Gamma - \bar{\Gamma})^3 \boldsymbol{t}^3}{3!} + \cdots\right]. \tag{11.13}$$

Substitution of this equation into equation (11.10) gives

$$|g^{(1)}(\boldsymbol{t})| = \exp(-\bar{\Gamma}\boldsymbol{t})\left[1 + \frac{\mu_2 \boldsymbol{t}^2}{2!} - \frac{\mu_3 \boldsymbol{t}^3}{3!} + \cdots\right], \tag{11.14}$$

where the μ_n are the normalized moments of the distribution $G(\Gamma)$ about the mean value $\bar{\Gamma}$. The term $\ln|g^{(1)}(\boldsymbol{t})|$ is thus reduced to a polynomial expansion in **t**, the delay time, as

$$\ln|g^{(1)}(\boldsymbol{t})| = -\bar{\Gamma}\boldsymbol{t} + \frac{\mu_2 \boldsymbol{t}^2}{2!} - \frac{\mu_3 \boldsymbol{t}^3}{3!} + \cdots. \tag{11.15}$$

Generally this series is truncated at the quadratic or cubic term; and, on fitting the data, we obtain an average size from the linear term and a measure of the polydispersity or width of the size distribution from the variance defined by $\mu_2/\bar{\Gamma}^2$. The cubic term gives a third moment which can be related to the skewness of the size distribution.

In principle, cumulants analysis as an infinite series would define a monomodal size distribution exactly. In this respect the truncated expansion is valid only for small polydispersity, typically $\leqslant 0.1$. For broader distributions, the choice of polynomial order appropriate to a given correlation function is not obvious. Too low an order will introduce a systematic error, but too high an order will give statistical errors perhaps larger than the values of the coefficients themselves. Brown and Pusey (1975) have shown that a third-order polynomial provides a reasonable compromise in the trade-off between systematic and statistical error if the correlation function does not probe too far out in delay time, where any truncated expansion must fail to fit the data.

These arguments apply to a cumulants analysis of data generated by a monomodal distribution. The same analysis can be applied to any correlation function. A multimodal distribution could easily have the same first and second moments and hence the same average size and distribution width. Indications that a multimodal distribution should be suspected in an unknown sample can be gleaned from the goodness-of-fit parameter obtained from the polynomial fitting procedure. Extending the measurement period should reduce the statistical error here. Should this parameter fail to decline, it is a strong indication that the assumptions inherent in the cumulants analysis are failing and that a more sophisticated analysis should be undertaken.

11.2.5 Polydispersity analysis – a general approach

The development of analytical methodology to deconvolute equation (11.10) and to derive the underlying size distribution has remained an active area of research for many years. Basically all of these inversion techniques are attempting to solve a mathematically ill-conditioned problem where no unique solution exists when there is noise present in the data. The number of equally correct solutions, which may differ substantially in form, increases with the magnitude of the noise term. The methods of tackling the problem fall into two main categories. These are (a) truncated expansion of the function $G(\Gamma)$ in the eigenfunctions of the Laplace transform operator, the so-called exponential sampling technique (Ostrowsky *et al.*, 1981), and (b) the CONTIN package of Provencher (1979, 1982) wherein the derived curvature in $G(\Gamma)$ is explicitly restricted to yield a smooth function. CONTIN therefore imposes a bias against peaks and shoulders and could discriminate against the selection of a multimodal solution. Without doubt in all cases, selection of the 'true' solution is aided by the introduction of prior knowledge. Knowing what the solution might look like is a great help in finding it! A less subjective approach (Horne and Dalgleish, 1985) which has been successfully applied to analyses based on the exponential sampling package is to compare solutions from correlation functions generated from the same sample at a series of different scattering angles. The rationale behind this procedure is outlined as follows. The result of the inversion procedure is an 'intensity' distribution, the total amount of light scattered as a function of particle size, whereas what is sought is a number, $N(r)$, or weight fraction distribution. For spherical rigid particles, these are related by the simple transform function

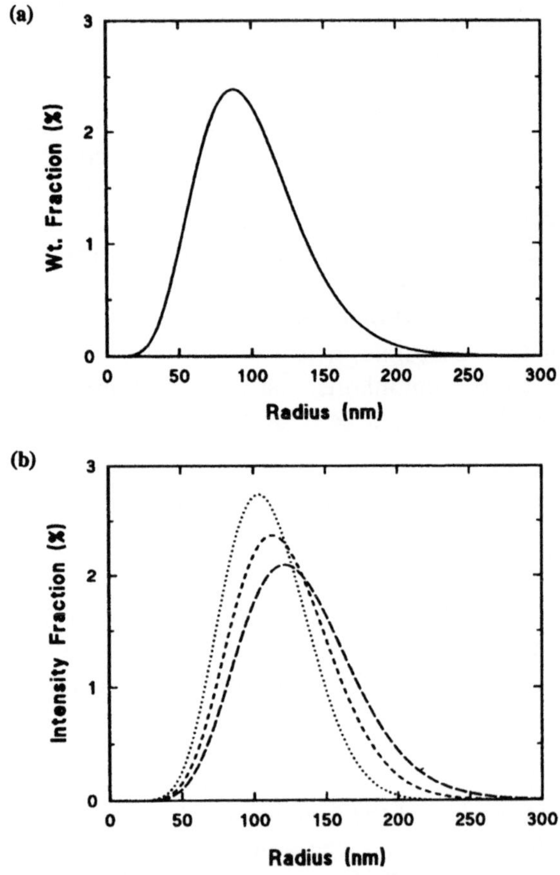

Fig. 11.5 (a) An illustrative weight-fraction distribution as a function of particle radius for a colloidal suspension, such as may be encountered with the casein micelles of milk. (b) Again as a function of particle radius, the scattered intensity distributions that would be observed for this weight-fraction distribution at scattering angles of 30° (– – – –), 60° (- - - - - -) and 90° (··········). Without further correction for the scattering power of the individual particle, this intensity distribution would be the output of the deconvolution packages.

$$G(\Gamma)\,\mathrm{d}\Gamma = N(r)I(r, q)(\mathrm{d}\Gamma/\mathrm{d}r)\,\mathrm{d}r, \tag{11.16}$$

where $I(r,q)$ is the intensity of light scattered by a single particle of radius r at scattering wave vector q. For small particles, much smaller than the wavelength of light, $I(r,q)$ can be calculated using the simple Rayleigh formula (see Kerker, 1969), or more generally for particles of arbitrary size and known refractive index by Mie theory (Kerker, 1969). The amount of light scattered by a single particle is a strong function of size and scattering angle, and the influence of this on the intensity distribution is illustrated in Fig. 11.5. We have taken the weight fraction distribution plotted in Fig. 11.5(A) and have calculated from it the expected intensity distributions at three scattering angles, 30°, 60° and 90°. Note how the peak shifts to lower size values and narrows markedly as the angle of scattering is increased from 30° to 90°. Note also that when this procedure is operated in reverse on experimental data, because of intensity weighting, the 'correct' solution from the inversion procedures will vary with scattering angle for the same sample. The requirement on these calculations of producing the same weight fraction distribution after correction for intensity weighting is a powerful constraint on the range of possible solutions, provided that the scattering particles are themselves large enough to exhibit angular variation in their scattering. Cummins and Staples (1987) have advocated similar usage of angular dependent changes in the correlation functions but employ a forward calculation of intensity weighted correlation functions to evaluate the distribution of sizes. Such a procedure works well with multimodal mixtures in predicting both sizes and relative scattering distributions.

11.3 Dynamic light scattering from concentrated opaque suspensions

The extension of dynamic light scattering to concentrated suspensions introduces a series of problems. The first of these is multiple scattering. Conventional dynamic light scattering demands operation under single scattering conditions. The detected photon can be scattered only once by the scattering particle on its route to the detector in Fig. 11.1. This is because the angle of scattering defines the wave vector, the ruler against which particle size can be determined. In a multiple scattering situation this information is lost, i.e. it becomes inaccessible. Several approaches, which unfortunately are not always generally applicable, have been devised to circumvent this problem.

In our discussion of the underlying theory of light scattering, it was noted that scattering only occurs when the refractive index of the scatterer is different from that of the suspending medium. One approach to overcoming the problem of multiple scattering is to match the refractive index of the dispersed particles and dispersant and then to observe the scattering from tracer particles added in small amounts to this suspension (van Megen *et al.*, 1987; van Veluwen *et al.*, 1987). Because of the restrictions on available solvents, this technique is not of general applicability. Another approach, first described by Drewel *et al.* (1989), employs two lasers of different wavelengths and cross-correlates the scattered light at angles chosen in such a way that the scattering vectors for both beams coincide. As well as requiring a complex light scattering apparatus with fairly prodigious alignment problems, this technique is still limited to fairly low concentrations as the light still has to traverse a scattering volume. This latter problem has been circumvented in other approaches by employing optical fibres as the light carriers and operating in a back-scattering configuration, directly immersing the fibre optic probe in the sample. Again two techniques have been developed, one using a single fibre, the other a multi-fibre bundle, the latter technique being known as diffusing wave spectroscopy (DWS).

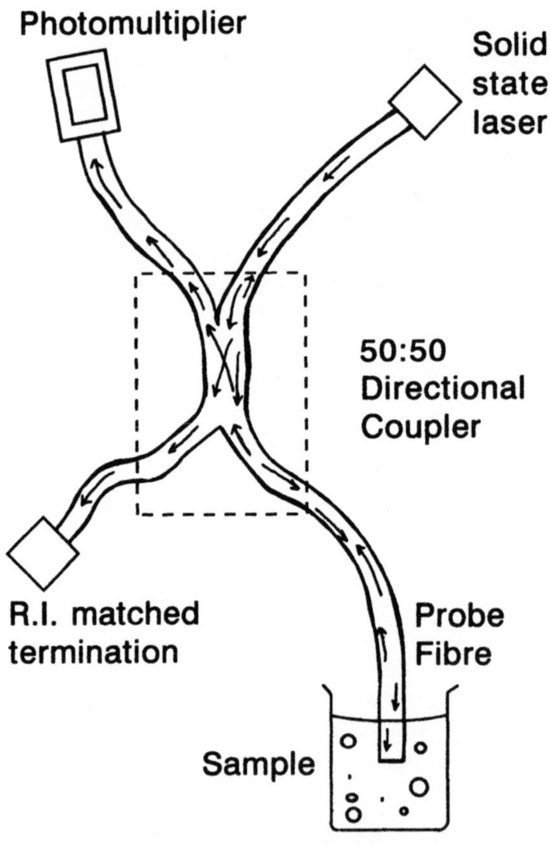

Fig. 11.6 Schematic diagram of Hi-C fibre optics arrangement.

Since the original introduction of fibre optics to dynamic light scattering by Tanaka and Benedek (1975), several groups have refined the single fibre technique and instruments using it are now commercially available from several manufacturers. Amongst these is the Malvern Hi-C (Malvern Instruments, Malvern, Worcs.) which uses a single monomode fibre as both the transmitter and receiver of scattered light, as shown schematically in Fig. 11.6. The acceptance optics of the narrow fibre are such that only photons singly scattered straight back into the fibre at 180° reach the detector (Thomas and Tjin, 1989; Wiese and Horn, 1991, 1992). Any multiple scattering is almost entirely discriminated against. The correlation functions developed can then be analysed as for dilute suspensions, with modifications as explained below.

As we saw earlier, the fundamental quantity measured in a dynamic light scattering experiment is the normalized intensity autocorrelation function of the scattered light,

$$g^{(2)}(t) = \frac{\langle I(0) \times I(t)\rangle}{\langle I(0)\rangle^2}, \tag{11.17}$$

$I(t)$ being the scattered light intensity measured at delay time t. Our earlier analysis considered only the situation where scattered light is detected. This is homodyne detection and here $g^{(2)}(t)$ is related to the normalized electric field correlation function $g^{(1)}(t)$ by the Siegert relation (equation (11.5)). An alternate form of detection, known as heterodyne detection, illuminates the detector simultaneously both by scattered light and by a coherent local oscillator signal, usually a portion of the incident beam which has not undergone scattering. In this case $g^{(2)}(t)$ may be written as (Oliver, 1974)

$$g^{(2)}(t) = 1 + b\left[2\frac{\langle I_S\rangle I_{LO}}{\langle I\rangle^2} g^{(1)}(t) + \frac{\langle I_S\rangle^2}{\langle I\rangle^2} |g^{(1)}(t)|^2\right]. \tag{11.18}$$

Here I_{LO} is the (constant) local oscillator intensity, $\langle I_S\rangle$ is the averaged scattered intensity, and

$$\langle I\rangle = I_{LO} + \langle I_S\rangle \tag{11.19}$$

is the total intensity. When I_{LO} is very much greater than $\langle I_S\rangle$, the last term on the right-hand side of equation (11.18) becomes vanishingly small and equation (11.18) reduces to

$$g^{(2)}(t) = 1 + b\frac{2\langle I_S\rangle}{I_{LO}} g^{(1)}(t). \tag{11.20}$$

Conversely, when I_{LO} is vanishingly small, the middle term disappears and equation (11.18) reduces to the Siegert relation (equation (11.5)) above.

As previously described, the dynamic information on the scattering system is contained in the electric field correlation function, $g^{(1)}(t)$. Note that because of the squaring inherent in the Siegert relation, the homodyne function decays at twice the rate of the heterodyne function and would be perceived as arising from a particle half the size, were a heterodyne-behaving system to be incorrectly analysed using the homodyne equation.

Both heterodyne and homodyne situations have been encountered with single mode optical fibres. The problem which arises from the local oscillator signals generated by reflection of the incoming light from the front face of the fibre has been tackled in two ways. In the first, the instrument has been biased towards homodyne mode by slanting the front face. Thus, Wiese and Horn (1991, 1992) showed that, with an angle of 10°, the reflected wave is not propagated in the fibre since its angle of incidence at the core/cladding interface is now greater than the critical angle. However, problems have been noted when the scattered signal contribution is weak, as with low refractive index material or small particles at relatively low concentrations, since 'background' light is likely to be present and any 'in-phase' contribution will constitute a reference, i.e. local oscillator signal (Bremer *et al.*, 1993).

The alternative is then to operate in a situation where the local oscillator signal is maximized, with the front face polished and normal to the incident light. This implies that the ratio $\langle I_S \rangle / I_{LO}$ in equation (11.18) is low. Generally a figure of 0.05 is acceptable to permit analysis of the signal as solely heterodyne. This situation is favoured at a low volume fraction (ϕ) of scatterers. As ϕ increases, however, the scattered light intensity also increases whereas the local oscillator signal remains constant. Thus, with increasing ϕ, a transition is observed from a predominantly heterodyne mode at low volume fractions to almost pure homodyne at very high volume fractions (Thomas, 1989). Since scattered intensity is also a strong function of particle size, the transition region is defined by the particle and cannot be pre-programmed. The most sensible approach would seem to be to recognize the existence of the heterodyne/homodyne mixing phenomenon and attempt to take this into account in the data analysis package. One such route (Bremer *et al.*, 1993) has been successfully demonstrated for polystyrene latex suspensions and their mixtures but its extension to a naturally occurring polydisperse colloidal suspension has yet to be attempted.

Practical applications of the single mode fibre techniques have therefore been limited to situations where either heterodyne (Horne and Davidson, 1993a; van der Meeren, 1993) or homodyne (Lilge and Horn, 1991) behaviour predominates. The outcome from such analyses is again a diffusional mobility, but in concentrated suspensions this mobility depends not only on the particle size but also on the particle's interactions with its neighbours. This is seen in Fig. 11.7, with casein micelles in milk as an example, as a concentration dependence in the apparent particle size, if the Stokes–Einstein equation (equation (11.9) is blindly

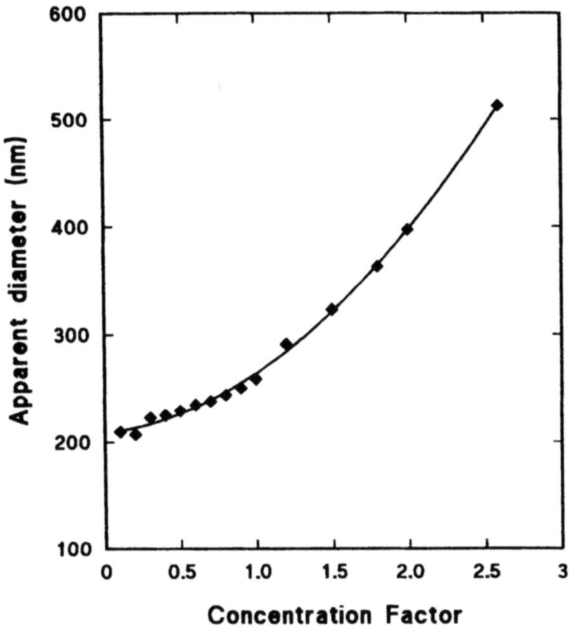

Fig. 11.7 Concentration dependence of apparent casein micelle diameter in skim milk measured using Hi-C system at the plotted concentration factor. Concentrations are relative to the original native skim milk and were achieved by ultrafiltration of the milk (> 1.0) or by dilution with UF permeate (< 1.0).

applied to the measured diffusional mobility. To obtain a true size from this instrument, one should measure size as a function of concentration and extrapolate the value to infinite dilution. In practice, the measured sizes generally become independent of concentration at volume fractions less than 1%, i.e. for still considerably turbid suspensions. However, since one reason for attempting dynamic light scattering in concentrated suspensions is to avoid the dilution step where the stability or integrity of the system may be challenged, another method of removing the effects of particle–particle interactions would seem preferable. Described in the next section, this was first authenticated (Horne and Davidson, 1993b) for mobilities using diffusing wave spectroscopy, but the truth of the underlying physical principles means the approach is equally applicable to the correction of measurements obtained using single fibre apparatus.

11.3.1 Diffusing wave spectroscopy

Diffusing wave spectroscopy (DWS) (Maret and Wolf, 1987; Pine *et al.*, 1988, 1990a, b; Weitz *et al.*, 1993) approaches the problem of analysing the dynamic light scattering behaviour of an opaque suspension from an entirely different viewpoint, turning multiple scattering into a necessity rather than something to be avoided or eliminated. In strongly scattering media, the propagation of light can be described very well by the diffusion approximation. Each detected photon is assumed to execute a random walk through the sample, the detected light being the incoherent sum of contributions scattered through all possible paths (MacKintosh and John, 1989). The non-exponential character of the correlation function measured for such systems betrays the existence of multiple time constants, indicative of the differing decay rates of the paths of various lengths. Physically, the decay of the correlation function reflects the time it takes for the total path length to change by one wavelength. This change results from the cumulative movement of a large number of particles. Paths involving large numbers of scattering events decay the most rapidly, because each individual particle in the chain only needs to move a short distance and takes a short time to do so. By contrast paths made up of few scattering events decay more slowly as the lower number of particles must move relatively further to reach an accumulated path length change of one wavelength. The contribution from single scattering therefore decays slowest of all.

To obtain the full autocorrelation function, the contributions from all paths, weighted by the probability $P(s)$ that the light follows that path, are summed. This integration over all possible light diffusion paths is highly influenced by the geometry of the experiment. Size, shape and positioning of sample, source and detector all play a part. The overall problem of integration then reduces to a solution of the diffusion equation for the light subject to the boundary conditions set by these geometrical constraints.

Two experimental geometries are particularly favourable and result in reasonably simple analytical expressions. These are forward transmission and backscattering. Our apparatus, which has been described previously (Horne, 1989), operates in backscattering mode. Shown schematically in Fig. 11.8, the instrument incorporates a bifurcated bundle of optical fibres in a Y-shaped configuration. Half of the fibres, distributed randomly over the front face of the common leg, bring light from a randomly polarized He–Ne laser down one arm of the Y. When this common leg, the base of the Y, is dipped into the scattering suspension, the other half of the fibres carry backscattered light up the second arm of the Y to the photomultiplier. There, masking by slit and pinhole ensures that light from only a small area impinges on the detector. The launching and receiver fibres on the buried face of the common leg are physically separated by a finite distance. Light can reach the receiver fibre only when the bundle is dipped in the suspension and only after having followed a tortuous multiple scattering path through the suspen-

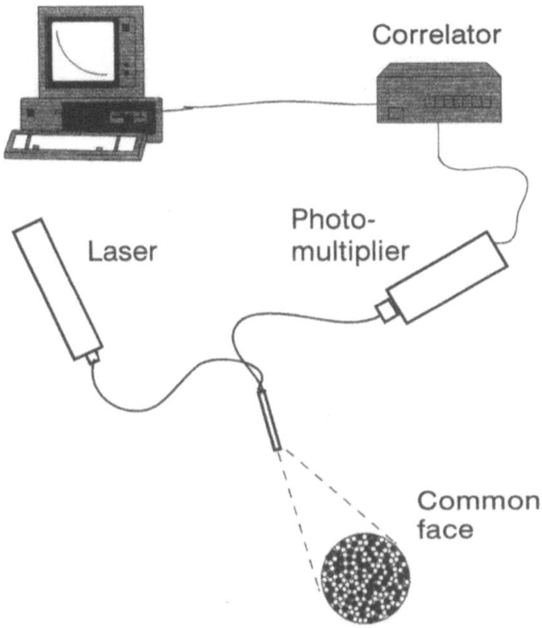

Fig. 11.8 Schematic diagram of the multi-fibre apparatus for diffusing wave spectroscopy in backscattering mode. Enlargement shows the common face of the fibre-optic bundle and the random distribution of launching (black) and receiving (white) fibres.

sion. Multiple scattering is therefore essential to the operation of this DWS technique. The signal is processed by a Malvern 7032 digital autocorrelator, as for conventional dynamic light scattering, though the analysis of the resulting correlation function is modified to accommodate the multiple scattering diffusion behaviour of the light.

Our experimental configuration approximates to an extended plane-wave light source incident on the front face of a slab of thickness L (the depth of the solution) and effectively infinite lateral extent, with the scattered light collected at a point on the same front face. Pine *et al.* (1988) have shown that under such experimental geometry constraints, the electric field correlation function for the multiply scattered light can be written as

$$g^{(1)}(t) = \frac{1}{(1-\gamma l^*/L)} \frac{\sinh[(L/l^*)(6t/\tau)^{1/2}(1-\gamma l^*/L)]}{\sinh[(L/l^*)(6t/\tau)^{1/2}]}$$

$$\approx \exp[-\gamma(6t/\tau)^{1/2}], \qquad \text{for } L >> l^*, \tag{11.21}$$

where l^* is the diffusive mean free path for light in the system, the length scale over which the direction of light propagation is randomized. The parameter γ is introduced as a multiplicative factor to locate the conversion plane, parallel to the bundle face, within the sample at which the passage of light can be considered diffusive, providing a physically meaningful boundary condition to permit analytical solution of the diffusion equation. In experiments using monodisperse polystyrene lattices of known diameter, a value of approximately 2.0 was obtained for this parameter (Pine *et al.*, 1988).

When an infinite depth (effectively several centimetres) of solution is employed, the derived equation predicts linear behaviour when the logarithm of the correlation function is plotted against the square root of the delay time, the gradient of the line being $2\gamma(\tau/6)^{-1/2}$. This behaviour has been verified for a selection of polystyrene lattices ranging in diameter from 0.1 to 5.0 μm at concentrations sufficiently high to satisfy the multiple scattering requirements, but low enough to avoid major interaction effects between the particles (Horne, 1989, 1991).

When the concentration of scatterers is raised, the mobility observed by DWS becomes concentration dependent, a function of both particle size and interparticle interactions, decreasing as ϕ is increased. The similarity between this behaviour and that of the particle self-diffusion coefficient has been noted (Horne, 1990; Wiese and Horn, 1991). Typical examples are shown in Fig. 11.9 where the inverse of mobility, the relaxation time τ, as measured by DWS, is plotted for a series of protein-stabilized oil-in-water emulsions as a function of emulsion concentration.

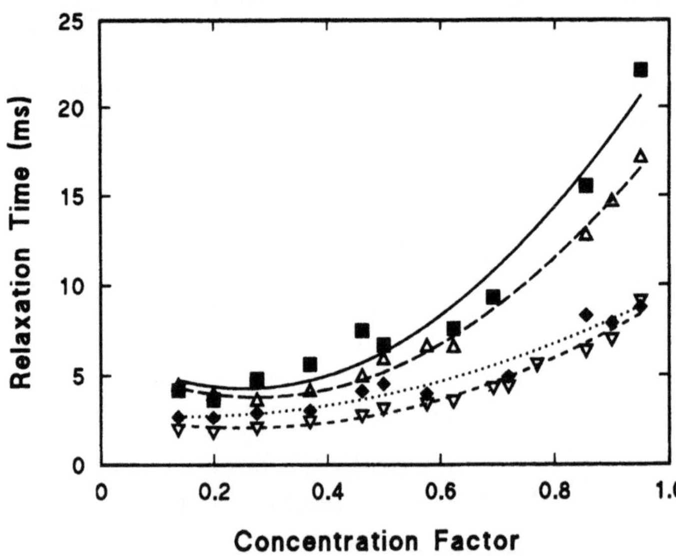

Fig. 11.9 Relaxation times measured by DWS for a series of opaque emulsions as a function of emulsion dilution (1.0 equals concentration of original emulsion). All four emulsions are milk protein-stabilized oil-in-water emulsions prepared under slightly differing processing regimes to achieve different droplet-size distributions. Line are drawn to guide the eye.

If the DWS technique is to find use as a means of measuring size in opaque suspensions, some method of estimating or correcting for particle interactions must be devised. The following suggests an empirical route to achieving this objective. In the limit of infinite dilution, as we noted earlier, the self-diffusion coefficient D_0 of a spherical particle is related to the particle radius, r, and the pure solvent viscosity, η_0, by the Stokes–Einstein equation, repeated here for convenience,

$$D_0 = \frac{k_B T}{6\pi\eta_0 r}, \tag{11.22}$$

where k_B is the Boltzmann constant and T is the absolute temperature. In concentrated systems this diffusion coefficient can be expressed as a function of the scattering wave vector according to the relation (Pusey and Tough, 1981):

$$D(q) = D_0 \frac{H(q)}{S(q)} \tag{11.23}$$

where $S(q)$ is the static structure factor, related to the radial distribution function of the material distributed around the particle, and $H(q)$ is a function describing the effect of hydrodynamic interactions between the colloidal particles. For sufficiently large particles, greater than approximately 0.5 μm and for volume fractions up to $\phi = 0.4$, $S(q)$ does not deviate significantly from unity. Variations in the diffusion coefficient with concentration in such systems would therefore be dominated by hydrodynamic effects. As a simple first order approximation, Horne and Davidson (1993b) suggested that these hydrodynamic effects can be taken into account by replacing solvent viscosity with the actual viscosity of the suspension in the Stokes–Einstein equation. Measured DWS relaxation time, τ, is thus predicted to be directly proportional to the product of particle size and emulsion viscosity, i.e.

$$\tau = a\eta r, \tag{11.24}$$

where a is a constant.

Independent measurement of the viscosity of the emulsions at the concentrations shown in Fig. 11.9 confirmed a positive linear relationship between relaxation time and viscosity (Fig. 11.10) with intercepts close to zero. Since these dairy-based emulsions were reasonably stable to dilution, conventional low-angle laser light scattering techniques (Malvern Mastersizer) gave good estimates of the particle diameters. These were found to correlate well with the gradients of the τ *versus* η plots (Fig. 11.11), as further predicted by equation (11.24). Since this equation predicts direct proportionality between relaxation time and viscosity, it is not essential to perform measurements of both these parameters at a series of suspension concentrations. A measurement of

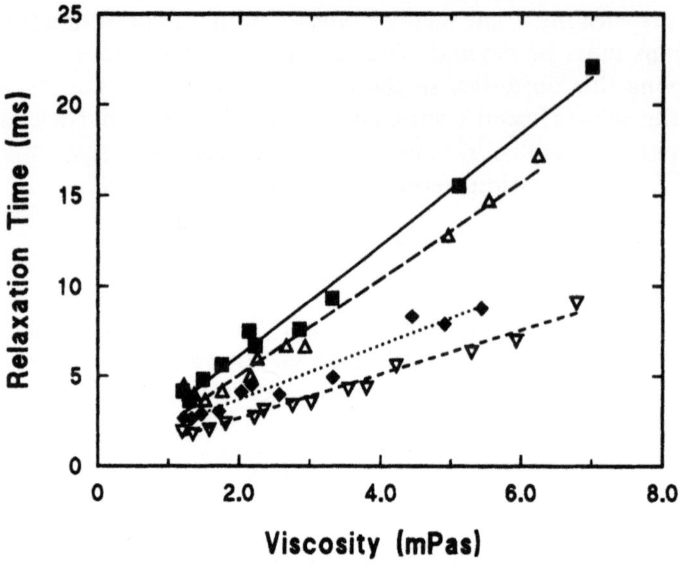

Fig. 11.10 DWS relaxation time as a function of emulsion viscosity for the diluted emulsions of Fig. 11.9. Straight lines are linear regression fits to the plotted data points.

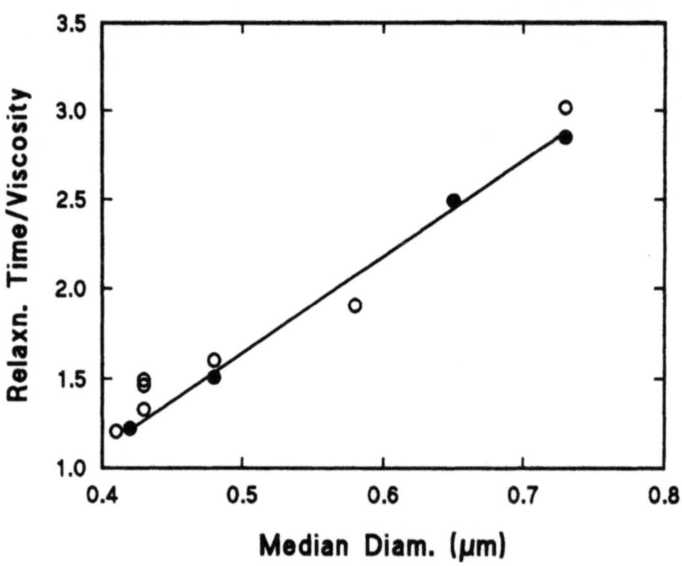

Fig. 11.11 Gradients of the linear plots of Fig. 11.10 as a function of the emulsion droplet median diameter calculated from droplet-size distributions measured by Malvern Mastersizer (●). Open circles are points derived for a second series of similar emulsions and represent the ratio, relaxation time/viscosity, *versus* median diameter. The line is a linear fit to the solid data points only.

both parameters at a single concentration will suffice. The open symbols in Fig. 11.11 demonstrate this continuing linear behaviour between the ratio τ/η and droplet diameter for a further group of protein-stabilized emulsions.

This concept of using suspension viscosity to correct concentration dependent particle mobility for interaction effects should hold true also for data obtained using single fibre techniques. There you would have the added advantage that your particle movement was measured against a known wave vector ruler, and so, provided the homodyne/heterodyne effects were not influential, the mean particle size ought to be obtainable directly.

There are also some caveats to be mentioned in relation to the DWS technique. In the first instance, in presenting this empirical approach to determining particle size using DWS, we have omitted extensive discussion of the parameter γ. In favourable circumstances, knowledge of this would allow measured relaxation times to be translated directly into estimates of particle size. A more rigorous treatment of the diffusional propagation of light predicts the value of γ to depend on both the polarization and anisotropy of the scattering (MacKintosh and John, 1989). Experiment confirms these effects (MacKintosh *et al.*, 1989), with parallel polarization showing the value of γ increasing with size and perpendicular polarization the converse. Both curves, however, asymptotically approach a value of ~2.0 as particle size increases above 0.1 μm. In the multi-fibre situation, each fibre launches light into the suspension at a random polarization angle. Parallel and perpendicular effects cancel and hence a constant value averaging to approximately 2.0 would be expected with this apparatus for all sizes.

A further problem in using DWS for particle sizing is that, unlike conventional single scattering DLS techniques, no independent information can be extracted from DWS measurements on sample polydispersity or particle size distribution. In a polydisperse system, as the photon proceeds along its diffusional pathway, it can encounter any size of particle in any random order. Each tortuous path thus reflects an average over particle size, even before the contributions over individual paths are integrated. Thus, in an individual measurement all information on polydispersity is lost, rendered inaccessible by the very nature of the experiment.

11.4 Dynamic light scattering from gelling systems

Dynamic light scattering in concentrated suspensions is sensitive to particle–particle interactions. Relaxation time is lengthened as particle motion is hindered or slowed. A logical progression from this would be towards particle entrapment and complete immobility when the particle becomes part of a gel network. At this stage, we should distinguish between true entrapment, where, say, an emulsion droplet is caught up in a biopolymeric gel of perhaps gelatin or carrageenan, and the case where the particle itself forms part of the matrix. This latter scenario is closer to the situation believed to exist when gelation of milk is induced either by acidification or enzymic treatment with chymosin. In effect, these treatments form a network of colloidal particles where particle integrity may still be assumed, at least in the short-term post-gelation period.

Pusey and van Megen (1989) pointed out that such gels whether prepared from polymers or colloids are non-ergodic media. Conventional dynamic light scattering, as emphasized earlier, measures a time-averaged property. The assumption is made that given enough time, the system will evolve through a representative fraction of all spatial configurations, that is, enough of phase space will be sampled for the measured time average to give a good estimate of the system ensemble average. Such a system is said to be ergodic. In particular, this permits the time-averaged intensity correlation function to be equated to the ensemble-averaged intensity correlation function which can then be interpreted in terms of the ensemble-averaged properties of the scattering system, the properties we can calculate. In the non-ergodic gel, however, the scatterers are localized near fixed average positions and are able to execute only limited Brownian motion about these positions. They are thus trapped in a restricted region of phase space, a sub-ensemble. A single dynamic light scattering measurement made on this scattering volume will yield a time-averaged property of only this sub-ensemble. The form of this time-averaged or sub-ensemble-averaged correlation function will depend on the particular (scattering) volume of the scattering medium being sampled, since different regions of the same overall system will contain scatterers trapped in different spatial configurations, thus describing different sub-ensembles. Thus, in a study of non-ergodic media, such as a clear gel with a low level of scatterers, a single time-averaged correlation function is unlikely to equal the intensity correlation function averaged over a set of configurations sufficient to represent the whole of phase space, i.e. over the full ensemble. In practice an ensemble average can only be obtained by moving the sample through a series of positions, illuminating in this fashion a series of scattering volumes. The ensemble average for the whole assembly is then evaluated from a summation of the time-averaged scattering behaviour at each position.

In many of the early studies of dynamic light scattering from gel systems the existence of the problem of non-ergodicity was not recognized. In spite of this, two models describing the dynamics of gels were developed. In the first of these, the gel is viewed as an elastic medium (Tanaka *et al.*, 1973). Thermally-induced fluctuations in the density of the medium are considered the source of the observed dynamic light scattering behaviour. In theory these fluctuations give a correlation function which decays to zero at long times. However, typical gels have frozen-in structures, inhomogeneities or dust particles, which give rise to static light scattering. In essence, these inhomogeneities act as local oscillators and the experiment becomes one with heterodyne rather than homodyne detection. In the second model, the fluctuations are considered to arise as the result of the motion of harmonically bound Brownian particles (Wun and Carlson, 1975). At short times the particles undergo Brownian diffusion but on departing from the equilibrium position they experience the network restoring force, restricting them to a maximum displacement which coincides with the point when the elastic energy is equal to the thermal energy, $k_B T$. Xue *et al.* (1992) showed that, after correctly ensemble averaging to overcome the ergodicity problem, the distinction between these two models appears in the wave vector dependence of their correlation functions. For clear polyacrylamide gels, they were then able to demonstrate that the dynamic light scattering behaviour was consistent with the elastic gel model with static inhomogeneities rather than with the harmonically bound particle model. This does not mean that all gels will follow this model. In particular, particle gels where the network is comprised of particles which are themselves extensive light scatterers may behave differently, or more likely, as a mixture of both types of behaviour.

The study of such particle gels will require those dynamic light scattering techniques adapted for use in concentrated opaque systems. The problem of non-ergodicity will continue to arise with the single fibre instruments. The requirement to change positions to achieve a representative sampling of phase space may present further problems with these instruments as it may not be possible to do this without disrupting the gel. Xue *et al.* (1992) have suggested that ensemble averaging can also be accomplished by moving the detector but this could also be difficult unless a custom-built apparatus is being employed. Diffusing wave spectroscopy does not have these problems. The time-averaged intensity correlation function is made up from appropriately weighted contributions from all possible diffusive light paths in the system. The scattering volume is the total volume and therefore the time-averaged correlation accumulated must always be the ensemble-average in these measurements, although experiment durations may have to be considerably extended to accommodate the much longer relaxation times being measured. Differentiating between the two models, however, is difficult because the wave vector q is no longer an experimental variable with this technique, unless a range of light sources is available.

11.4.1 Milk gels studied using diffusing wave spectroscopy

With milk as a substrate, we have monitored the changes seen in DWS relaxation time as gel formation progresses. Such data can be utilized at several levels of sophistication. Clotting by the enzyme chymosin is the first stage in cheese manufacture. (For a more complete discussion of the mechanism of this reaction, see, for example, Dalgleish (1992).) Briefly, during the primary stage, the κ-casein component of the casein micelle is attacked by the enzyme at a specific bond. The loss of the sterically stabilizing hydrophilic peptide which then diffuses away from the micelle leads to a progressive destabilization of the micelle as the proteolysis of its κ-casein coating continues. The subsequent aggregation of the destabilized micelles through to curd or gel formation constitutes the secondary stage of the process which we follow using DWS.

Typical changes in DWS relaxation time consequent on the addition of enzyme are illustrated in Fig. 11.12. The plots show a lag phase where micelle size remains essentially constant at its untreated milk value. With the single-fibre Malvern Hi-C apparatus, the apparent size shows a slight decrease as the enzyme cleaves the κ-casein hairs during this period of the reaction (Horne and Davidson, 1993a). At the end of this lag phase, the relaxation time increases rapidly by several orders of magnitude indicating massive aggregation, the onset time being designated the

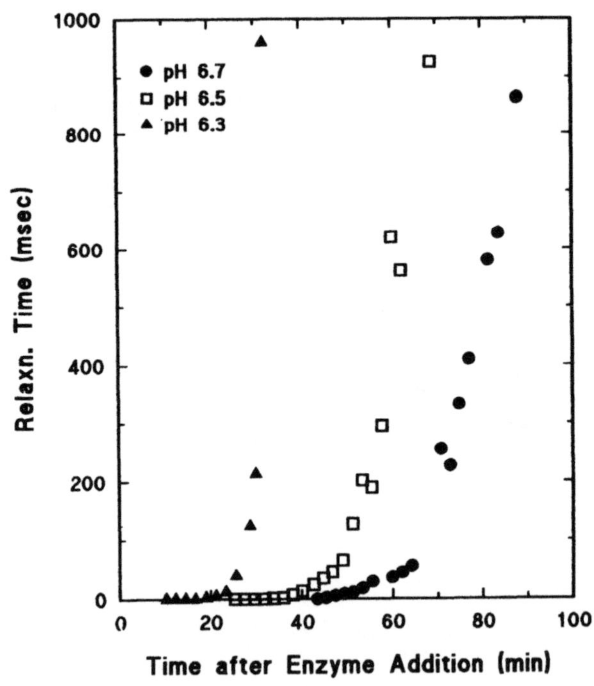

Fig. 11.12 Changes in DWS relaxation time of milk due to curd formation following the addition of chymosin at time zero. Data are shown for three samples adjusted to the pH values indicated.

coagulation time. As a critical time in the progress of the reaction, this coagulation time can be usefully related to many reaction parameters for testing of postulated mechanisms and so on. As a non-destructive, non-invasive, non-perturbing probe of the system, DWS represents an extremely useful technique for on-line monitoring of the progress of such reactions on an industrial scale. DWS is, however, intrinsically more useful than this rudimentary analysis suggests, in that as a non-destructive measurement it provides a continuity of output from which may be derived a flow of fundamental kinetic and mechanistic information from the earliest stages of the gelation process and before.

Thus, we have been exploring the link between the light scattering relaxation behaviour and the rheological properties of the developing gel. Elastic shear moduli were measured by a rheometer operating in oscillatory mode which allows estimation of the viscous (loss) and elastic (storage) components. The ratio of these two components yields the tangent of the phase angle δ, which is the angle by which the response of the viscoelastic medium lags the applied strain. For a simple mechanical model of the gel, the Maxwell model (Ferry, 1980), the reciprocal of this tangent defines the stress relaxation time of the gel. Because the rheometer applies a mechanical impulse, it requires the presence of a gel of a finite strength dictated by rheometer sensitivity to produce a measurable shear modulus. Thus, the growth of shear modulus in an enzyme-treated milk is delayed compared to the increase in relaxation time detected by DWS. Nevertheless there is a window in the reaction time profile where the responses of the two instruments overlap. In this time span, we find that DWS relaxation time and Maxwellian relaxation time are linearly correlated (Horne, unpublished observations).

It therefore appears that observation of dynamic light scattering from these opaque systems using the DWS technique, by virtue of its sensitivity to particle mobility and relaxation in the system, must eventually reflect changes in viscoelastic properties. A major advantage over traditional mechanical techniques of rheology is that light scattering is non-perturbing. Particularly with very weak gels, the very act of applying a mechanical strain to the system may destroy or distort the nascent structure, whereas DWS simply measures the effect of spontaneous thermal fluctuations present in all systems. The interpretation of this behaviour in terms of system structure and interactions is an ongoing and active area of research.

Acknowledgment

The research carried out at the Hannah Research Institute is supported by The Scottish Office Agricultural and Fisheries Department.

References

Berne, B.J. and Pecora, R. (1976) *Dynamic Light Scattering with Applications to Chemistry, Biology and Physics*, Wiley-Interscience, New York.

Bremer, L.G.B., Deriemaeker, L., Finsy, R., Gelade, E. and Joosten, J.G.H. (1993) Fibre optic dynamic light scattering, neither homodyne nor heterodyne. *Langmuir*, **9**, 2008–2014.

Brown, J.C. and Pusey, P.N. (1975) Photon correlation study of polydisperse samples of polystyrene in cyclohexane. *J. Chem. Phys.*, **62**, 1136–1144.

Brown, W. (ed) (1993) *Dynamic Light Scattering: The Method and Some Applications*, Oxford University Press, Oxford.

Chu, B. (1974) *Laser Light Scattering*, Academic Press, New York.

Cummins, P.G. and Staples, E.J. (1987) Particle size distributions determined by a 'multi-angle' analysis of photon correlation spectroscopy data. *Langmuir*, **3**, 1109–1113.

Dalgleish, D.G. (1992) The enzymatic coagulation of milk, in *Advanced Dairy Chemistry – 1: Proteins* (ed. P.F. Fox), Elsevier Applied Science, London, pp. 579–619.

Drewel, M., Ahrens, J. and Podschus, U. (1989) Decorrelation of multiple scattering for an arbitrary scattering angle. *J. Opt. Soc. Am.*, **A7**, 206–210.

Ferry, J.D. (1980) *Viscoelastic Properties of Polymers*, 3rd edn., Wiley, New York.

Horne, D.S. (1989) Particle size measurement in concentrated latex suspensions using fibre-optic photon correlation spectroscopy. *J. Phys. D., Appl. Phys.*, **22**, 1257–1265.

Horne, D.S. (1990) Influence of polydispersity on dynamic light scattering measurements in concentrated suspensions. *J. Chem. Soc., Faraday Trans.*, **86**, 1149–1150.

Horne, D.S. (1991) Diffusing wave spectroscopy of gelling systems, in *Photon Correlation Spectroscopy: Multicomponent Systems* (ed. K.S. Schmitz), Proceedings SPIE Vol. 1430, SPIE, Bellingham, Washington, USA, pp. 166–180.

Horne, D.S. and Dalgleish, D.G. (1985) A photon correlation spectroscopy study of size distributions of casein micelle suspensions. *Eur. Biophys. J.*, **11**, 249–258.

Horne, D.S. and Davidson, C.M. (1993a) Direct observation of decrease in size of casein micelles during the initial stages of renneting of skim milk. *Intern. Dairy J.*, **3**, 61–71.

Horne, D.S. and Davidson, C.M. (1993b) Application of diffusing wave spectroscopy to particle sizing in concentrated dispersions. *Colloids Surf. A*, **77**, 1–8.

Kerker, M. (1969) *The Scattering of Light and Other Electromagnetic Radiation*, Academic Press, New York.

Koppel, D.E. (1972) Analysis of macromolecular polydispersity in intensity correlation spectroscopy: the method of cumulants. *J. Chem. Phys.*, **57**, 4814–4820.

Lilge, D. and Horn, D. (1991) Diffusion in concentrated dispersions: a study with fiber-optic quasi-elastic light scattering (FOQELS). *Colloid Polym. Sci.*, **269**, 704–712.

Maret, G. and Wolf, P.E. (1987) Multiple light scattering from disordered media. The effect of Brownian motion of scatterers. *Z. Phys. B*, **65**, 409–413.

MacKintosh, F.C. and John, S. (1989) Diffusing-wave spectroscopy and multiple scattering of light in correlated random media. *Phys. Rev. B*, **40**, 2383–2406.

MacKintosh, F.C., Zhu, J.X., Pine, D.J. and Weitz, D.A. (1989) Polarization memory of multiply scattered light. *Phys. Rev. B*, **40**, 9342–9345.

Oliver, C.J. (1974) Correlation techniques, in *Photon Correlation Spectroscopy and Light Beating Spectroscopy* (eds. H.C. Cummins and E.R. Pike), Plenum, New York, pp. 151–223.

Ostrowsky, N., Sornette, D., Parker, P. and Pike, E.R. (1981) Exponential sampling method for light scattering polydispersity analysis. *Opt. Acta*, **28**, 1059–1070.

Pine, D.J., Weitz, D.A., Chaikin, P.M. and Herbolzheimer, E. (1988) Diffusing-wave spectroscopy. *Phys. Rev. Lett.*, **60**, 1134–1137.

Pine, D.J., Weitz, D.A., Maret, G., Wolf, P.E., Herbolzheimer, E. and Chaikin, P.M. (1990a) Dynamical correlations of multiply scattered light, in *Scattering and Localization of Classical Waves in Random Media* (ed. P. Sheng), World Scientific, Singapore, pp. 312–372.

Pine, D.J., Weitz, D.A., Zhu, J.X. and Herbolzheimer, E. (1990b) Diffusing-wave spectroscopy: dynamic light scattering in the multiple scattering limit. *J. Phys. (Paris)*, **51**, 2101–2127.

Provencher, S.W. (1979) Inverse problems in polymer characterization: direct analysis of polydispersity with photon correlation spectroscopy. *Makromol. Chem.*, **180**, 201–209.

Provencher, S.W. (1982) A constrained regularization method for inverting data represented by linear algebraic or integral equations. *Comput. Phys. Commun.*, **27**, 213–227.

Pusey, P.N. and Tough, R.J.A. (1981) Particle interactions, in *Dynamic Light Scattering: Applications of Photon Correlation Spectroscopy* (ed. R. Pecora), Plenum, New York, pp. 85–179.

Pusey, P.N. and van Megen, W. (1989) Dynamic light scattering by non-ergodic media. *Physica A*, **157**, 705–741.

Siegert, A.J.F. (1943) *MIT Rad. Lab. Rpt.* No. 465.

Tanaka, T and Benedek, G.B. (1975) Measurement of the velocity of blood flow (*in vivo*) using a fiber optic catheter and optical mixing spectroscopy. *Appl. Opt.*, **14**, 189–196.

Tanaka, T., Hocker, L.O. and Benedek, G.B. (1973) Spectrum of light scattered from viscoelastic gel. *J. Chem. Phys.*, **59**, 5151–5159.

Thomas, J.C. (1989) Fiber optic dynamic light scattering from concentrated dispersions. 2. Concentration dependence of the apparent diffusion coefficient for small spheres. *Langmuir*, **5**, 1350–1355.

Thomas, J.C. (1991) Photon correlation spectroscopy: technique and instrumentation, in *Photon Correlation Spectroscopy: Multicomponent Systems* (ed. K.S. Schmitz), Proceedings SPIE Vol. 1430, SPIE, Bellingham, Washington, USA, pp. 2–18.

Thomas, J.C. and Tjin, S.C. (1989) Fiber optic dynamic light scattering (FODLS) from moderately concentrated suspensions. *J. Colloid Interface Sci.*, **129**, 15–31.

van der Meeren, P., Stastny, M., Vanderdeelen, J. and Baert, L. (1993) Particle sizing of concentrated emulsions using fibre optic quasi-elastic light scattering. *Colloid Surf. A*, **76**, 125–133.

van Megen, W., Underwood, S.M., Ottewill, R.H., Williams, N.St.J. and Pusey, P.N. (1987) Particle diffusion in concentrated dispersions. *Faraday Discuss. Chem. Soc.*, **83**, 47–57.

van Veluwen, A., Lekkerkerker, H.N.W., De Kruif, C.G. and Vrij, A. (1987) Brownian diffusivities of interacting colloidal particles measured by dynamic light scattering. *Faraday Discuss. Chem. Soc.*, **83**, 59–67.

Weitz, D.A., Zhu, J.X., Durian, D.J., Gang, H. and Pine, D.J. (1993) Diffusing-wave spectroscopy: the technique and some applications. *Phys. Scri.*, **T49**, 610–621.

Wiese, H. and Horn, D. (1991) Single-mode fibres in fiber-optic quasielastic light scattering: a study of the dynamics of concentrated latex dispersions. *J. Chem. Phys.*, **94**, 6429–6443.

Wiese, H. and Horn, D. (1992) Fiber-optic quasielastic light scattering in concentrated latex dispersions: the performance of single-mode vs. multimode fibres. *Ber. Bunsenges. Phys. Chem.*, **96**, 1818–1828.

Wun, K.L. and Carlson, F.D. (1975) Harmonically-bound particle model for quasi-elastic light scattering by gels. *Macromolecules*, **8**, 190–194.

Xue, J.Z., Pine, D.J., Milner, S.T., Wu, X.-I. and Chaikin, P.M. (1992) Non-ergodicity and light scattering from polymer gels. *Phys. Rev. A*, **46**, 6550–6563.

Chapter 12

Small-Angle Neutron Scattering and Neutron Reflection

J. Eastoe
School of Chemistry,
University of Bristol, UK

12.1 Introduction

When studying liquid phase systems containing macromolecules or colloids, the first questions to be asked are concerned with the structure in the system. 'What is the particle size?' 'How thick is the adsorbed layer?' 'How do the particles (molecules) interact with one another?' Knowledge of the structure allows us to understand the system at a deeper level, and fine-tune its characteristics. As the instrumental tools required for work of this kind become increasingly sophisticated, it is becoming essential to make use of external laboratories or centralized facilities like neutron scattering institutes. We shall see, in what follows, that neutron scattering is an extremely powerful non-perturbative method for investigating structures *in situ* and is now the technique of choice for studying a wide range of phenomena. A great wealth of structural (and dynamic) information can be obtained using neutrons. But, owing to the large expense involved, governments, individually and in consortia, have recognized the need to provide central neutron sources for the scientific and industrial community. Experimental time is usually obtained through peer-review panels or it may be purchased directly for work of a confidential nature.

Consider a solution containing surfactant micelles, macromolecules or colloidal particles. Depending on the precise nature of the system, the characteristic length scales of the individual particles and interparticle interactions may range from 1 to 10^4 nm. There are a number of appropriate methods available for structural characterization, as described in detail elsewhere in this volume. Electron microscopy, and methods like the ultracentrifuge, can be used to determine particle sizes down to about 5.0 nm. However, their use will always involve some sample perturbation, or sample preparation, and this is often undesirable for obtaining reliable information on equilibrium properties. Dynamic light scattering (DLS) is a convenient laboratory-based method for accessing length scales down to ~5 nm via the diffusion coefficient. Since DLS relies on the interaction of the incident electromagnetic field with electrons in the sample, no chemical or physical perturbation is needed. However, the wavelength of laser light is long ($\lambda \sim 500$ nm) and so internal particle details cannot be resolved. Furthermore light scattering methods are of little use if (a) the sample is opaque, (b) the optical refractive indices of the particle and solvent are matched (only very weak or negligible scattering); (c) the characteristic size is less than $\lambda/100$, or (d) one is interested in the internal details of particles or of adsorbed layers at interfaces. DLS data are also liable to mis-interpretation, since for polymer solutions and colloidal particles strong interactions are often present, particularly if the system is close to a phase separation boundary; this has to be taken into account. The recent development of pulsed-field-gradient-spin-echo NMR (PFGSE-NMR) is in many ways complementary to DLS since opaque or optically absorbing samples can be investigated (Stilbs, 1987). Again, the length scale of interest is obtained by measuring the diffusion coefficient, and so, as with DLS, interactions may be extremely important.

For a wide range of systems the 'in-house' methods mentioned above are often the most appropriate. However, when the measurements become technically unfeasible (e.g. refractive index matching in light scattering), or when a finer level of detail is required (e.g. the nature of interfaces), and/or it is not desirable to perturb the structure, it is necessary to study the scattering of short wavelength radiation, such as a neutron beam or an X-ray beam ($\lambda \approx 0.10$–2.0 nm). For particles in water- or oil-continuous media, there are considerable advantages in using neutrons over X-rays, owing to the contrast afforded by isotopic H/D substitution, or contrast variation, of the various components. The contrast variation method permits a detailed investigation of the system structure. It exploits the isotope-dependent nature of physical interaction between free neutrons and bound nuclei in the sample. These interactions operate via the short range nuclear force (range ~10^{-15} m). Fortunately, for most nuclei, the neutron absorption cross-section is low, meaning that many different materials are amenable to study. With reference to food-related systems, there are three important neutron methods. This chapter concentrates on the two used for studying static properties of macromolecular and self-assembling systems: for structures and interactions in solution we have small-angle neutron scattering (SANS), and neutron reflectometry for investigating structure and properties of interfacial layers. Kinetic measurements can be made using these techniques, however, only if the relaxation time τ is long (~10–30 min). For information on molecular dynamics at the sub-nanosecond time-scale quasi-elastic neutron scattering (QENS) may be used, and details can be found elsewhere (Leadbetter and Richardson, 1979). We concentrate here on instruments and examples from the EPSRC ISIS time-of-flight spallation neutron source in Oxfordshire, UK.

It is instructive in passing to compare scattering methods with an imaging technique such as microscopy. In microscopy, the waves scattered by each point in the sample are focused to give an image of the sample structure. Scattering methods record an interference pattern, due to waves scattered by all points, which can be related by Fourier transformation to a model of the system. Obviously, data interpretation and analysis is more difficult for scattering methods.

12.2 Principles of scattering methods

Before we consider how neutrons are obtained, it is first important to define some general terms relevant to any radiation scattering process. We are interested here only in elastic scattering; the incident and final energies E_i and E_f are equal and the incident radiation only suffers a small change in direction. Furthermore, the relevant scattering events are coherent, i.e. the scattering properties of the nuclei are related to their location in the sample. Incoherent scattering processes also occur, but they convey no structural (only dynamic) information (Leadbetter and Richardson, 1979). The incident radiation of wavelength λ is directed towards the sample (Fig. 12.1(a)). Owing to the physical interaction between radiation and sample, some of the beam is scattered though an angle θ. The measured quantity $I(\theta)$ yields the desired information on the structure of the sample.

The incident beam has wave vector $k_i = 2\pi n/\lambda$ where n is the refractive index of the medium (~1.0 for neutrons). Consider two nuclei O and P that act as point scatterers located in a particle by position vector r where O represents the origin (Fig. 12.1(b)). For elastic scattering the magnitudes of the incident and scattered ($\boldsymbol{k}_s$) wave vectors are equal. For the two nuclei the wave scattered in the direction of $\boldsymbol{k}_s$ is phase shifted with respect to that scattered in the same direction from the origin O. This phase difference is equal to $\boldsymbol{Q}\cdot\boldsymbol{r}$, where the scattering vector $\boldsymbol{Q}$ is simply the difference in the incident and scattered wave vectors, $\boldsymbol{k}_i - \boldsymbol{k}_s$. The magnitude $\boldsymbol{Q}$ has dimensions of reciprocal length, and equation (12.1) may be obtained from simple geometry:

$$Q = \frac{4\pi}{\lambda}\sin\left(\frac{\theta}{2}\right). \qquad (12.1)$$

Since the incident neutron wavelength is usually quoted in units of Å (= 0.1 nm), the units of Q are commonly Å^{-1}.

In any particle there are not just two, but many, atoms; and the Q dependence of the amplitude of the scattering $F(Q)$ is given by

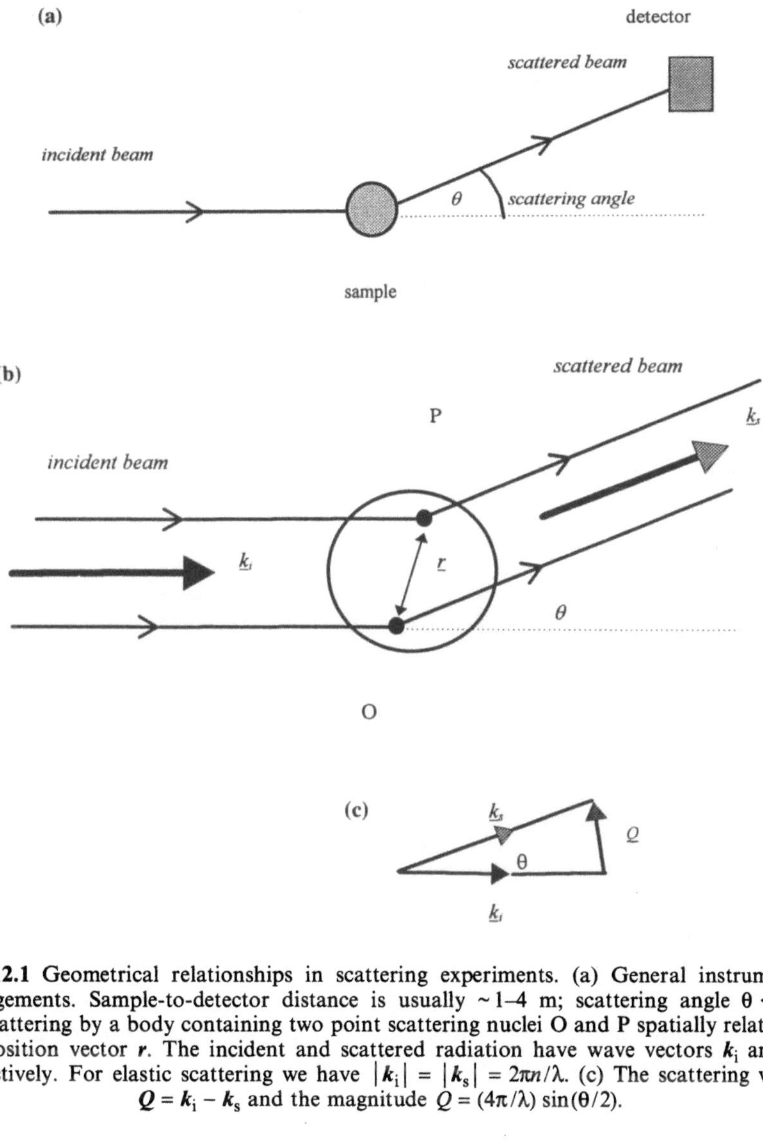

Fig. 12.1 Geometrical relationships in scattering experiments. (a) General instrumental arrangements. Sample-to-detector distance is usually ~1–4 m; scattering angle $\theta < 10°$. (b) Scattering by a body containing two point scattering nuclei O and P spatially related by the position vector $\boldsymbol{r}$. The incident and scattered radiation have wave vectors $\boldsymbol{k}_i$ and $\boldsymbol{k}_s$, respectively. For elastic scattering we have $|\boldsymbol{k}_i| = |\boldsymbol{k}_s| = 2\pi n/\lambda$. (c) The scattering vector $\boldsymbol{Q} = \boldsymbol{k}_i - \boldsymbol{k}_s$ and the magnitude $Q = (4\pi/\lambda)\sin(\theta/2)$.

$$F(\boldsymbol{Q}) = \int \rho(\boldsymbol{r}) \exp(\mathrm{i}\boldsymbol{Q}\cdot\boldsymbol{r})\,\mathrm{d}\boldsymbol{r}. \tag{12.2}$$

$F(\boldsymbol{Q})$ is the single particle form factor, and it is the Fourier transform of the scattering length density distribution of the particle, $\rho(r)$, which is related to the internal composition and structure of the particle. The origin of $F(Q)$ is the destructive (and constructive) interference between all waves scattered coherently from all nuclei in the sample, represented

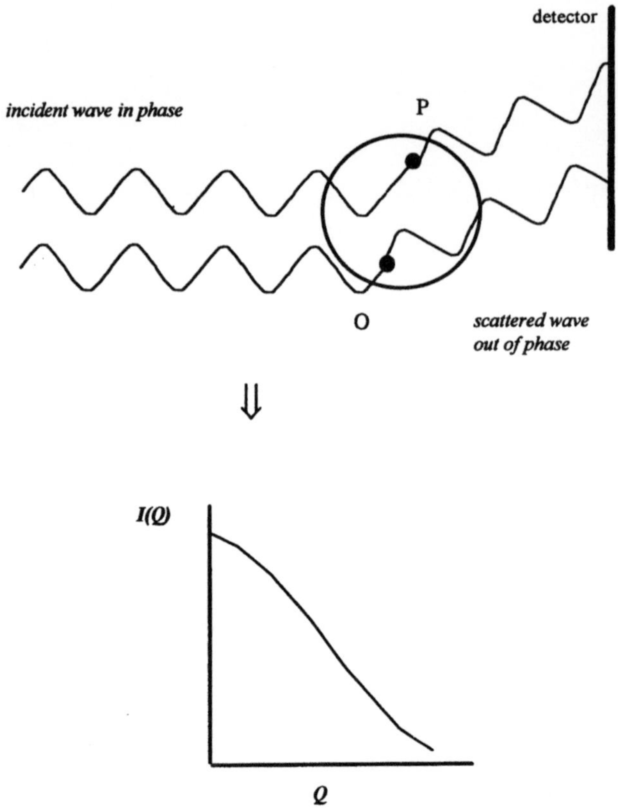

Fig. 12.2 Schematic diagram of interference between two point scatterers and the resulting decay in $I(Q)$.

by the $\exp(\mathrm{i}\boldsymbol{Q}\cdot\boldsymbol{r})$ term (shown schematically in Fig. 12.2). In the experiment, we do not measure the amplitude of the interference, since the detector is not phase sensitive; rather we measure the intensity I, which is the amplitude squared. Since the sample contains not one but n_p particles per unit volume, we can then write

$$I(Q) = n_\mathrm{p}\langle\langle\,|F(\bar{Q})|^2\rangle_\mathrm{o}\rangle_\mathrm{s}, \tag{12.3}$$

where the ensemble averages are over all orientations o and shapes s.

We now have a convenient relationship between two instrumental variables, θ and λ, and a reciprocal distance which is related to the positional correlations $\boldsymbol{r}$ between point scattering nuclei in the sample. Correlations between nuclei that are close may be probed at high Q, and long scale correlations at low Q values. Generally, for dilute samples, $I(Q)$ is a decay function like that shown in Fig. 12.2. At higher concentrations, a peak appears in the profile characteristic of the nearest-neighbour distance. $I(Q)$ from SANS contains information on intra-particle and inter-particle structure, while for neutron reflection the reflectivity $R(Q)$ depends on the thickness and composition of the interface.

12.3 Why use neutrons?

It has already been stated that there are important advantages of neutrons over X-rays in these kinds of experiments. This needs clarification, since it is relatively straightforward and cheap to produce X-rays, but very expensive, and technically more difficult, to provide neutron sources. Obviously the intensity of the signal will depend on the strength of interaction between incident radiation and sample. This may be quantified, and comparisons between different atoms made, by introducing a scattering length b. For compounds we use a more useful parameter called the scattering length density ρ, which is the scattering length per unit volume. For X-rays the interaction is with electrons in the sample, and so b scales with atomic number. The systems we are interested in contain chiefly C, H and O. This means that, when using X-rays to study hydrocarbon- and/or water-continuous samples, the values of ρ are virtually the same. In other words, when using X-rays, it is difficult to 'see' anything. However, for neutrons the interaction is with nuclei, and the parameters b (and ρ) are isotope dependent. Fortunately it turns out that the values of b for hydrogen and deuterium differ widely (see Table 12.1). Therefore it is easy to 'see' proton-containing hydrocarbon-type

Table 12.1 Selected values of coherent scattering length b and coherent scattering length density ρ at 25°C.

Nucleus		$b\,(10^{-12}\,\text{cm})$
^{1}H		−0.3741
^{2}H		0.6674
^{12}C		0.6653
^{16}O		0.5805
^{19}F		0.5650
^{23}Na		0.3630
^{31}P		0.5130
^{32}S		0.2804
Cl		0.9579
Substance		$\rho\,(10^{10}\,\text{cm}^{-2})$
Water	H_2O	−0.56
	D_2O	+6.40
Hydrocarbon	C_8H_{18}	−0.53
	C_8D_{18}	+6.43
Surfactant	$C_{12}H_{25}OSO_3^-Na^+$	+0.387
	$C_{12}D_{25}OSO_3^-Na^+$	+6.704

material dissolved in heavy water D_2O. Herein lies the importance of neutron scattering–the ability to resolve different parts of a multi-component system using contrast variation, which is achieved by controlling the level of H/D substitution. Apart from a few subtle effects (e.g. H-bonding), this isotopic exchange does not usually alter the chemical or physical properties of the system. Another benefit is that the cross-sections for absorption of neutrons are generally much lower than those for X-rays; this means that sample containers can be thicker and experimental environments more versatile. For example, using equipment available at ISIS, it is now routine to perform experiments at high pressure (< 1 kbar) (Winter *et al.*, 1991), high or low temperatures (~2–1000 K), in the presence of strong magnetic fields (< 2 T) (Groot *et al.*, 1994); or to study solutions as a function of shear ($< 10^4\,s^{-1}$) (Schurtenberger *et al.*, 1990). Many of these experiments could not even be contemplated using X-rays.

(a)

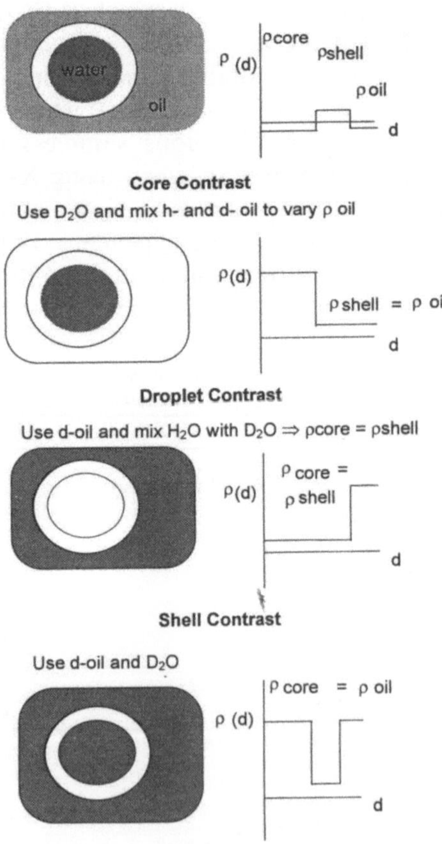

12.3.1 Scattering length densities

A useful parameter in neutron scattering is the coherent scattering length density ρ since it quantifies the strength of interaction with different materials. The value of ρ is readily calculated from

$$\rho = \frac{\sum_i b_i}{V_m}, \tag{12.4}$$

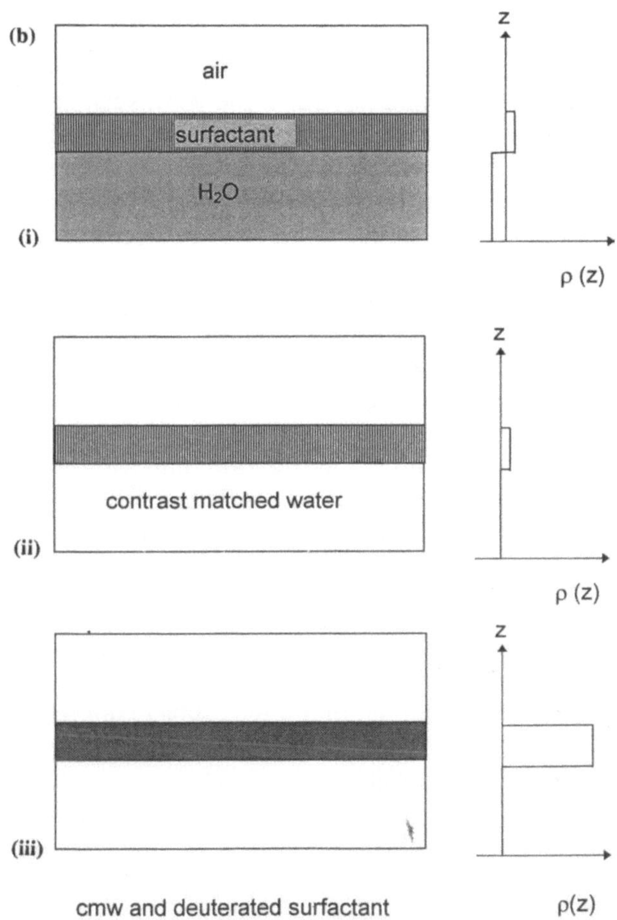

Fig. 12.3 Contrast variation in neutron scattering. (a) SANS: elucidation of the structure of water-in-oil microemulsion droplets. The scattering length density ρ depends on *d*, the distance from the centre of a droplet. (b) Neutron reflection: the structure of a sufactant monolayer adsorbed at the air–water interface can be resolved using contrast variation.

where the sum is over all i atoms in a molecule. The quantity b_i is the coherent scattering length of atom i (in cm) and V_m is the molecular volume (cm^3); the units of ρ are therefore cm^{-2}. The scattering properties of the material then depend (weakly) on temperature and pressure (since $\rho \propto 1/V_m$). Values of b_i can be found in standard texts (Lovesey, 1984). Table 12.1 lists b and ρ values for nuclei and molecules of interest. The negative b value of ^1H arises due to a phase change of the incident wave on scattering.

12.3.2 Neutron contrast variation

We can understand more about the contrast variation method if we consider a specific example of a water/oil/surfactant system where the relative volumes are approximately 10, 5 and 85%, such as a microemulsion (Clint, 1992). SANS has been instrumental in the elucidation of the phase structure of these systems (Eastoe *et al.*, 1991). Clearly the scattering length density ρ of water and oil can be readily adjusted using commercially available materials, but it is often necessary to synthesize deuterated surfactant analogues (Thomas *et al.*, 1989, 1993).

It can be seen from Table 12.1 that the scattering length densities of H_2O, n-alkanes (hydrogenated) and surfactants are all very similar. The initial situation is shown schematically at the top of Fig. 12.3(a). The distance from the droplet centre is d, and so ρ depends on d owing to the presence of the different materials. Before we introduce deuteration, the interpretation of the SANS diffraction pattern turns out to be somewhat ambiguous as structural parameters for both core and shell contribute to the scattering.

(i) *Core contrast*: If we now arrange that the oil phase is a suitable mixture of h- and d-oil, so that its scattering length density is exactly matched to that of the surfactant shell, and use D_2O rather than H_2O, then the droplet core will be contrasted. The schematic $\rho(d)$ profile shows that the SANS will yield information on the droplet core radius.
(ii) *Droplet contrast*. Similarly we can determine the overall droplet size, i.e., the core + surfactant shell. This can be done by setting $\rho_{core} = \rho_{shell}$, using an appropriate mixture of H_2O and D_2O, and using fully deuterated oil.
(iii) *Shell contrast*. Taking this idea one stage further we can study the shell only by matching $\rho_{core} = \rho_{oil}$; this is conveniently done using wholly D_2O and d-oil.

In order to extract the maximum information from the $I(Q)$ data, it must be fitted to a structural model. The validity of the model will be seriously tested in fitting the three different contrasts with the same structural parameters.

Although we shall cover the experimental technique of neutron reflectivity in more detail later, it is instructive to mention here the reflection of neutrons from material adsorbed at an interface. A monolayer adsorbed at the air–water interface serves as an excellent example (Fig. 12.3(b)). In this case it is the refractive index n of the various materials that is important. There is a simple relationship between n and ρ, and for air we have $\rho = 0$ and $n = 1.0$ (see equation (12.18)). For reflection from a surfactant monolayer, sandwiched between semi-infinite air and water phases, the dependence of ρ on the vertical distance z is of interest. The initial situation is shown in Fig. 12.3(b)(i), with H_2O, h-surfactant and air. As for SANS, before selective deuteration, we would not expect to obtain much useful information. The first change we can make is to contrast match the water sub-phase to air (Fig. 12.3(b)(ii)); this can be done by an appropriate H_2O/D_2O mixture, so-called contrast matched water (cmw). The reflectivity $R(Q)$ would then arise principally from the surfactant layer. An important advance (Fig. 12.3(b)(iii)) is the use of custom-synthesized fully and partially deuterated surfactants (Thomas *et al.*, 1993). Then the $R(Q)$ signal is maximum, and parameters such as the monolayer thickness and area per surfactant molecule at the interface can be reliable extracted from the data. Furthermore, the use of surfactants that are partially deuterated along their chain has allowed the detailed interfacial structure to be resolved (Thomas *et al.*, 1989).

12.3.3 How do we get neutrons?

There are two important ways of obtaining neutrons.

In a conventional nuclear reactor, a number of neutrons are produced per fission event, and a continuous neutron flux results. There are just a few research reactors in Europe designed for optimum neutron, rather than heat, production. The most famous is the Institut Max-von Laue Paul Langevin in Grenoble, France. This is an intergovernmental institute, and the main partners are Germany, France and UK. The thermal population of neutrons produced in the reactor core is too 'hot' (i.e. the wavelengths λ are too short) to be useful in SANS and reflection experiments. Therefore they are moderated by thermal equilibration with a cold body (often liquid hydrogen) to shift their wavelength distribution to ~ 1–10 Å. The technology of 'piping' neutrons away from the reactor to the experimental station is well developed. Neutron guides are evacuated and coated inside with a thin Ni film which is totally reflecting at small angles of incidence. At steady-state sources it is necessary to select a particular neutron wavelength before the beam is incident on the sample. This is usually done using a mechanical velocity selector that will give a typical resolution $\Delta\lambda/\lambda \sim 5 \rightarrow 10\%$, and this naturally reduces the incident flux.

In contrast, the UK neutron source ISIS at the EPSRC DRAL laboratory in Oxfordshire produces pulses of neutrons at 50 Hz. This is done by firing a pulsed beam of fast protons at a target made of a heavy metal (e.g. U, Ta or W). On impact, nuclear spallation occurs, during which some of the target neutrons are quickly 'boiled off'. Therefore, each pulse of protons results in a white beam daughter pulse of neutrons which has a Maxwellian λ-distribution. Again, for SANS and reflection experiments, the neutrons must be slowed down using a moderator. The result is that a spread of wavelengths, with a well-defined pulse structure, is incident on the sample, giving an extra experimental degree of freedom over steady-state sources. The wavelength of each neutron can be easily determined by 'time-of-flight' methods. Since none of the incident beam is 'thrown away', the spallation source is more efficient. It is now widely recognized that spallation sources offer considerable advantages and future developments will almost certainly be in this direction.

12.4 Small-angle neutron scattering (SANS)

SANS is routinely used to obtain information on the size and shape of macromolecules and colloidal particles in solution and their interactions. Interfacial structure can also be studied to determine the thickness and density of surface layers (Fig. 12.3(a)).

12.4.1 Instrumentation

Figure 12.4 shows a schematic diagram of the LOQ time-of-flight small-angle diffractometer at ISIS. The instrument control, data acquisition and normalization are all under computer control. To minimize the background, all flight paths are evacuated, although the beam travels through ~0.5 moderator are guided to a chopper which reduces the pulse rate from 50 Hz (the ISIS frequency) to 25 Hz, thus eliminating frame overlap problems (fast neutrons from the next pulse over taking slow neutrons of the first). The beam is then collimated by a series of apertures so that on arrival at the sample position it is of a well-defined size. A scintillator/photomultiplier tube monitor (not shown), placed before the sample, produces an output proportional to the incident beam flux and pulse structure. There is a range of sample environments, although for liquid phase work at around ambient temperatures, 1 or 2 mm path length quartz glass optical cells are used and a convenient computer-controlled automatic sample changer may be installed. After interaction with the sample, the beam is scattered about the forward position, and passes into an evacuated tank housing the ^{3}He—CF_4 filled detector placed 4.5 m from the sample. The incident wavelength range is $2.2 \rightarrow 10$ Å, and so the effective Q range is $0.007 \rightarrow 0.22$ Å^{-1}. The

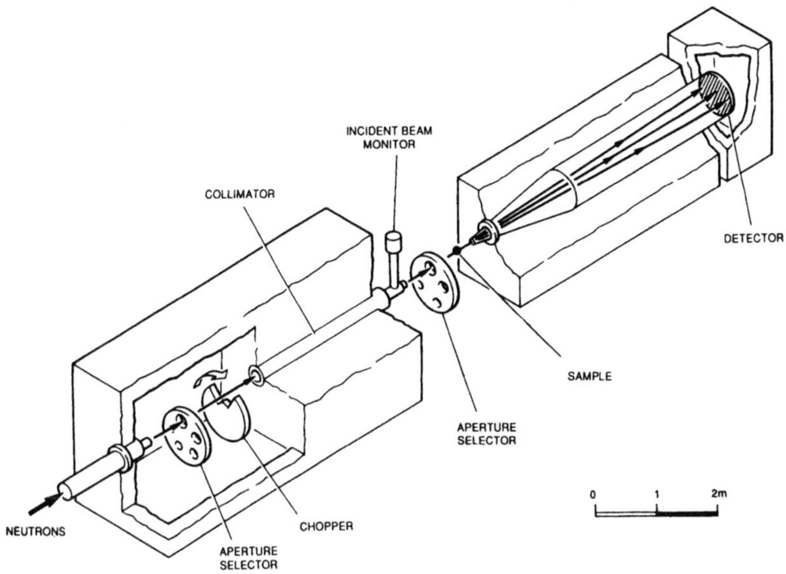

Fig. 12.4 Schematic diagram of the SANS instrument LOQ.

ionization detector is both time- and position-sensitive with an active area of $64 \times 64\ \text{cm}^2$. The scattering angle θ can be calculated from simple geometry whilst the wavelength λ is readily determined by time-of-flight (up to 40 ms for LOQ).

Software-controlled radial averaging calculates the signal intensity $s(\theta, \lambda)$ as a function of θ and λ. Counting times are usually 0.5–1 h and so $s(\theta, \lambda)$ is typically the sum of 10^4 pulses. A beam stop prevents damage by the unscattered neutrons to the central detector elements. It is also necessary to measure the λ-dependent transmission of each sample; this is done by taking the ratio of the output of two time-sensitive scintillator monitors, immediately before and after the sample.

12.4.2 Measured quantities

In order to produce the absolute scattering cross-section of the sample it is necessary to make four different measurements: (1) the sample of interest, including the background and the cell (samp); (2) the background solvent in a cell (bkd); (3) a normalization standard + background (std); and (4) the background of the standard (bkd-std). The background count arises principally from isotropic incoherent scattering of ^{1}H. The SANS signal $s(\theta, \lambda)$ and transmission factors $t(\lambda)$ must all be measured separately. The transmission is given by

$$t(\lambda) = \frac{\Sigma(\lambda)}{\Sigma_{EB}(\lambda)}, \tag{12.5}$$

where $\Sigma(\lambda)$ is the sum of counts on the post sample detector divided by that of a pre-sample monitor M for each wavelength. The subscript EB denotes the empty beam (i.e. no sample present). The relative intensities $i(\theta, \lambda)$ are then

$$i(\theta, \lambda) = \frac{s(\theta, \lambda)}{t(\lambda) M_S}. \tag{12.6}$$

The quantity $i(\theta, \lambda)$ is then corrected for the number of incident neutrons (M_S) and the proportion that actually reach the detector ($t(\lambda)$); all samples 1–4 above are treated in this way. It is a simple matter then to convert $i(\theta, \lambda)$ to $i(Q)$ using equation (12.1). Then the absolute SANS cross-section $I(Q)$ ((cm^{-1}) is

$$I(Q) = A \frac{i_{\mathrm{samp}}(Q) - i_{\mathrm{bkd}}(Q)}{i_{\mathrm{std}}(Q) - i_{\mathrm{bkd-std}}(Q)}, \tag{12.7}$$

where the constant A is the known cross-section of the standard (in cm^{-1}). Further corrections may be necessary to compensate for sample volume fraction.

12.4.3 Data interpretation

The general expression for the intensity $I(Q)$ may be found elsewhere (Chen, 1986). Under certain limiting conditions the equation reduces to a readily understood form. Consider the case of monodisperse, homogeneous spherical particles of radius R, volume V_p, number density $n_p\,(\mathrm{cm}^{-3})$, and coherent scattering length density ρ_p, dispersed in a medium of density ρ_m; the normalized SANS intensity $I(Q)$ (cm^{-1}) may be written as

$$I(Q) = n_p \Delta\rho^2 V_p^2 P(Q, R) S(Q), \tag{12.8}$$

where $\Delta\rho = \rho_p - \rho_m (\mathrm{cm}^{-2})$. The first three terms in equation (12.8) are independent of Q and account for the absolute intensity of the scattering. The last two terms are Q-dependent functions. $P(Q,R)$ is the single, dilute, particle form factor describing the angular distribution of the scattering due to the particle radius. It is the square of the modulus of $F(Q)$ (equation (12.3)). $S(Q)$ is the structure factor which arises from interparticle interactions, and it is the Fourier transform of the particle pair correlation function, and is therefore related to the interparticle interaction potential. Figure 12.5 shows how $P(Q,R)$ and $S(Q)$ combine to give the overall scattering $I(Q)$.

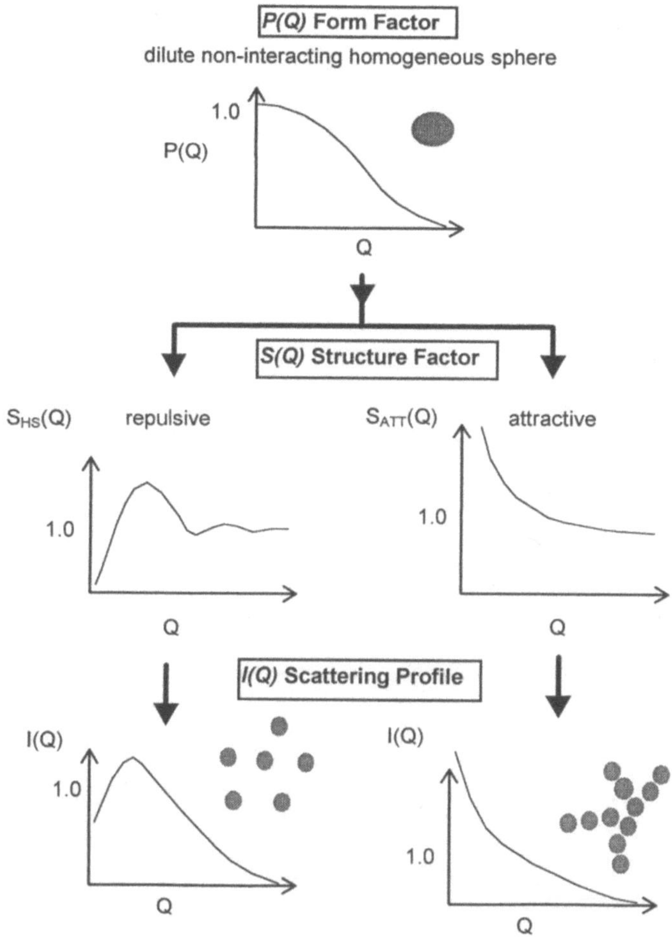

Fig. 12.5 Schematic representation of the particle form factor $P(Q,R)$ for attractive and repulsive homogeneous spheres.

(a) The single particle form factor $P(Q)$. An approximate representation of the form factor $P(Q,R)$ for spherical particles is shown in Fig. 12.5. In general, it appears as a decay, although under high resolution maxima and minima are expected at high Q values. Expressions for $P(Q)$ are known for a wide range of different particle shapes such as homogeneous spheres, spherical shells, ellipsoids, cylinders, concentric cylinders and discs (Cabane, 1987). They are often complicated functions, and there seems no need to quote them here. We will later consider Guinier approximations for $P(Q)$ of spheres, cylinders and discs (equations (12.14) to (12.17)). A polydispersity function may be introduced to account for the particle-size distribution. Then $P(Q,R)$ may be expressed as

$$P(Q, R) = [\sum_i P(Q, R)_i^{\text{sphere}} X(R)_i], \tag{12.9}$$

where $X(R)_i$ is the distribution function. The Schultz distribution is often used (Chen, 1986).

(b) The interparticle structure factor S(Q). The nature of the structure factor $S(Q)$ depends on the type of interparticle interaction in the system.

For many particles, a reasonable first approximation for the interaction is a hard sphere potential. An analytical expression for the hard-sphere structure factor $S_{HS}(Q)$ has been given by Ashcroft and Lekner (1966):

$$S_{HS}(Q) = f(\phi_{HS}, R_{HS}). \tag{12.10}$$

Here, ϕ_{HS} is the volume fraction of hard spheres, and R_{HS} the hard sphere radius. As can be seen in Fig. 12.5, the main features of $S_{HS}(Q)$ are to reduce the scattering at low Q values and to produce a peak in the profile. The position of the peak at $Q = Q_{MAX}$ is related to the nearest neighbour distance in the sample D, where

$$D \approx \frac{2\pi}{Q_{MAX}}. \tag{12.11}$$

For dilute systems we have $\phi_{HS} \rightarrow 0$, in which case the structure factor disappears, i.e. $S(Q) \rightarrow 1$. Electrostatic interactions also result in a repulsive $S(Q)$ with similar features although the mathematical expression is very different (Hayter and Penfold, 1981).

For systems close to a phase boundary that is driven by attractive interactions (i.e. in the proximity of a critical point), a very different form of $S(Q)$ is observed. There is an increase in intensity at low Q, and there is no 'nearest neighbour' peak (Fig. 12.5). An appropriate from for $S_{ATT}(Q)$ is given by the Ornstein–Zernike expression

$$S_{ATT}(Q) = 1 + \left[\frac{S(0)}{1 + (Q\xi)^2}\right], \tag{12.12}$$

where $S(0) = n_p k_B T \chi$, with k_B the Boltzmann constant, T the temperature, and χ the isothermal compressibility. The quantity ξ is a correlation length. Far from the phase boundary we have $S(0) \rightarrow 0$, and so $S_{ATT}(Q)$ disappears, i.e. $S(Q) \rightarrow 1$.

An important use of the absolute cross-section scale is for model checking. In the dilute case, and far from phase boundaries, we have $S(Q) \rightarrow 1$. Since $P(Q,R)$ at $Q = 0$ is defined as 1.0, the observed scale factor A $(= n_p \Delta\rho^2 V_p^2)$ provides a consistency check on the model fitting since the data are supplied in absolute units (cm^{-1}). It is important to note that the volume fraction ϕ $(= n_p V_p)$ and thus the expected scale factor A, are known from the sample composition.

(c) Interfacial Porod scattering. At high Q values, the intensity from SANS is sensitive to the scattering from local interfaces and the asymptotic intensity may be analysed using the Porod equation

$$I(Q) = 2\pi\Delta\rho^2\left(\frac{S}{V}\right)Q^{-4}, \tag{12.13}$$

where S/V is the total interfacial area per unit volume of solution (cm^{-1}). Recently this approach has been successfully adapted to determine volume fraction profiles of polymer layers adsorbed at the interface of large (~ 1 μm radius) colloidal particles (Auvray and Auroy, 1991). For surfactant systems the average area per surfactant molecule at the interface, a_S, can be estimated from $a_S = ((S/V)/N_S)$ where N_S is the number density of surfactant molecules.

For a homogeneous, monodisperse sphere of radius R, a plot of $[I(Q)\cdot Q^4]$ *versus* Q gives a first maximum at $\sim 2.7/R$ and a minimum at $\sim 4.5/R$. A plot of this kind is useful for measuring particle radii, particularly when the low Q regime (i.e. $QR << 1$, the so-called Guinier regime) is experimentally inaccessible.

12.4.4 Examples – surfactant micelles

The SANS profile $I(Q)$ is particularly sensitive to different particle shapes. In the dilute limit it is a relatively simple matter to analyse $I(Q)$ to extract the radius of a sphere or cylinder, or the thickness of a disc particle. The examples used here are micelles of a homologous series of non-ionic surfactants in D_2O (Eastoe *et al.*, 1995). There are no temperature-induced phase boundaries for these systems and the micellar volume fraction is $\phi < 0.05$. Therefore, to a first approximation, we can ignore $S(Q)$ effects, and so $I(Q)$ is a direct measure of $P(Q)$. As the hydrocarbon chain length increases, significant changes in the form of $I(Q)$ are obvious, and an initial analysis can be performed using the Guinier approximation. In the absence of micelle interactions (i.e. $S(Q) \to 1$), and over the appropriate range of Q, the scattering from the micelles can be accounted for by this simple relationship (Cabane, 1987):

$$I(Q) \propto Q^{-D}\exp(-Q^2R^2/K). \tag{12.14}$$

The proportionality constant in equation (12.14) depends on the concentration and isotopic composition. The exponent D is 1 for cylinders, 2 for discs, and 0 for spheres. R is the characteristic dimension of the particle, i.e. the cross sectional radius for cylinders, the thickness for discs, and the radius for spheres. K is an integer of value 4 for cylinders, 12 for discs, and 5 for spheres. Depending on the geometry, the dimension R can be obtained by plotting different quantities against $\boldsymbol{Q}^2$:

$$\ln(I(Q) \cdot Q) \textit{ versus } Q^2\text{: cylinder radius} = \surd(\text{slope} \times 4), \quad (12.15)$$

$$\ln(I(Q) \cdot Q^2) \textit{ versus } Q^2\text{: disk thickness} = \surd(\text{slope} \times 12), \quad (12.16)$$

$$\ln(I(Q)) \textit{ versus } Q^2 \ (QR < 1)\text{: sphere radius} = \surd(\text{slope} \times 5). \quad (12.17)$$

Figure 12.6 demonstrates this simple data treatment for surfactants from the homologous series at various concentrations. Since the Q-dependence of the scattering is essentially concentration (ϕ) independent, the approximation $S(Q) \rightarrow 1$ holds. The micelle dimensions derived from the ana-

(a)

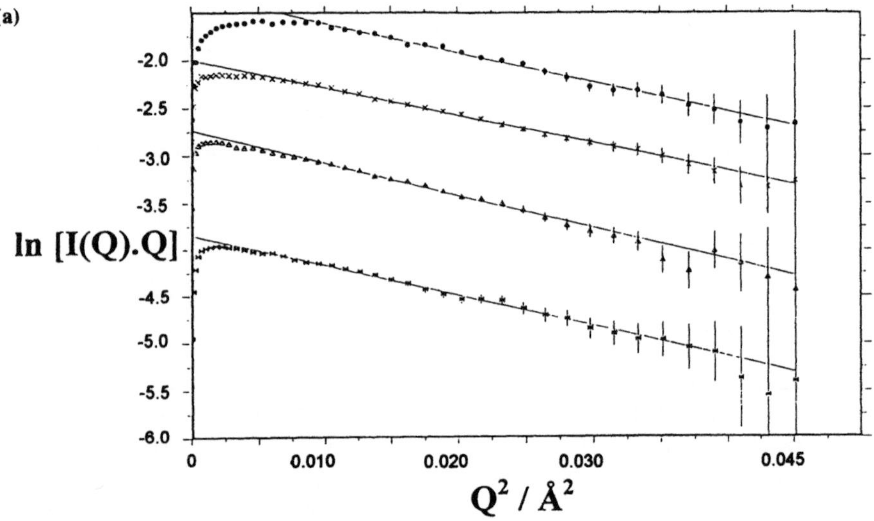

(b)

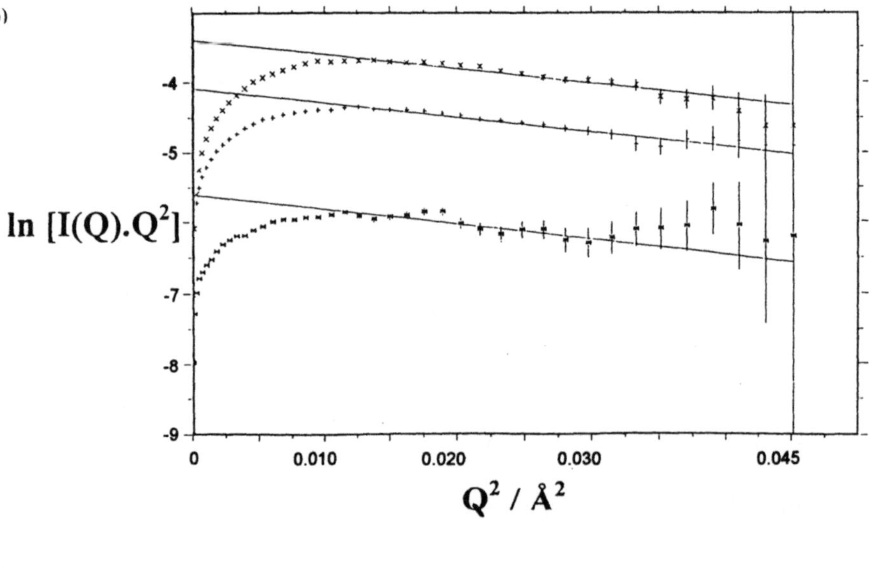

(c)

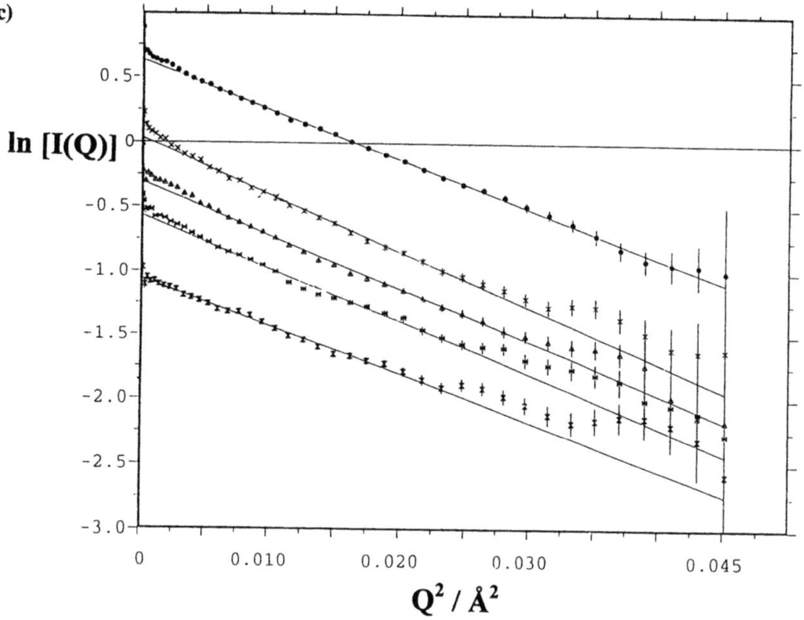

Fig. 12.6 Treatment of SANS data from non-ionic micelles at various surfactant concentrations. Data are consistent with presence of cylinder (radius = 11 ± 2 Å), disk (thickness = 16 ± 5 Å) and sphere (radius = 14 ± 2 Å) micelles respectively.

lysis are given in Fig. 12.6. In order to confirm the presence of different micelle shapes, the data must be fitted to the full model for $I(Q)$ (i.e. an expression like equation (12.8) containing an explicit expression for $P(Q)$).

12.4.5 Examples – food and biological science

SANS is an important method for studying biological structures such as enzymes, vesicles, ribosomes, nucleosomes and viruses, and a number of excellent reviews exist (e.g. see Jacrot, 1976; Perkins, 1988, 1994). The LOQ spectrometer has been used to study a range of systems of this type. The structure of novel microemulsion-based organogels has been elucidated by contrast variation experiments (Robinson *et al.*, 1991). SANS has shown the effects of the local anaesthetic tetracaine on the phase behaviour and structure of model biological membranes, and the data indicate that an order–disorder transition occurs at a critical pressure (Winter *et al.*, 1991). Using a magnetic field sample environment, the effect of field strength and direction, as well as electrolyte concentration, on the alignment of cholesteric liquid crystals of DNA fragments has been investigated (Groot *et al.*, 1994). Using contrast variation methods, the internal structure of waxy maize starch granules, and the changes that accompany gelatinization, have recently been studied (Jenkins *et al.*,

1994). The alignment of anisotropic reversed micelles of soybean lecithin in a shear field has been investigated using a Couette shear cell (Schurtenberger *et al.*, 1990). Other workers have investigated lipoproteins (Perkins *et al.*, 1993, 1994) and poly(γ-benzyl-L-glutamate) gels (Cohen, 1993).

12.5 Neutron reflectometry

The recent development of the technique of specular reflection of neutrons from interfaces has been an important advance in surface and colloid science (Thomas and Penfold, 1990; Penfold 1991). Neutron reflectometry is fast becoming the method of choice for the investigation of adsorbed layers of thickness in the 0.5–100 nm range, such as surfactants, phospholipids, proteins and enzymes at air–water or water–solid interfaces. The reflectivity profile $R(Q)$ gives information on the structure normal to an interface, and, as with the reflection of light, it is the refractive index normal to the surface that is important. For any material the refractive index to neutrons n is wavelength dependent, i.e.

$$n = 1 - \lambda^2 A + i\lambda C, \qquad (12.18)$$

where $A = Nb/2\pi$ and $C = N\sigma_{abs}/4\pi$ are constants, with N the atomic number density, b the bound coherent scattering length, and σ_{abs} the absorption cross-section (which may often be ignored).

12.5.1 Instrumentation

Figure 12.7 shows a schematic diagram of the CRISP reflectometer at ISIS. The beam line is inclined at 1.5° to the horizontal, and most of the components can be easily accessed for sample alignment and instrumental changes. The incident angle θ can be varied down to about 0.2° by use of a 'supermirror'. Neutrons are guided from the liquid H_2 moderator, and they pass through a single disk chopper which defines the wavelength range. After collimation, a set of frame overlap mirrors eliminates all wavelengths greater than 13 Å. The incident wavelengths are 0.5 → 13 Å, and the accessible Q-range extends from ~4×10^{-3} Å^{-1} to 0.65 Å^{-1}. Final collimation before the sample gives a ribbon-like beam of ~40 mm width and between 0.25 and 5.0 mm in height, depending on sample conditions and experimental set-up. Therefore the beam illuminates a 'footprint' of length L on the sample surface. Since the total number of neutrons incident on the surface depends on L, which obviously depends on θ/2, great care is needed in alignment. Immediately after the final collimation is a scintillator/PMT monitor M which has an output proportional to the incident beam flux, recorded by time-of-

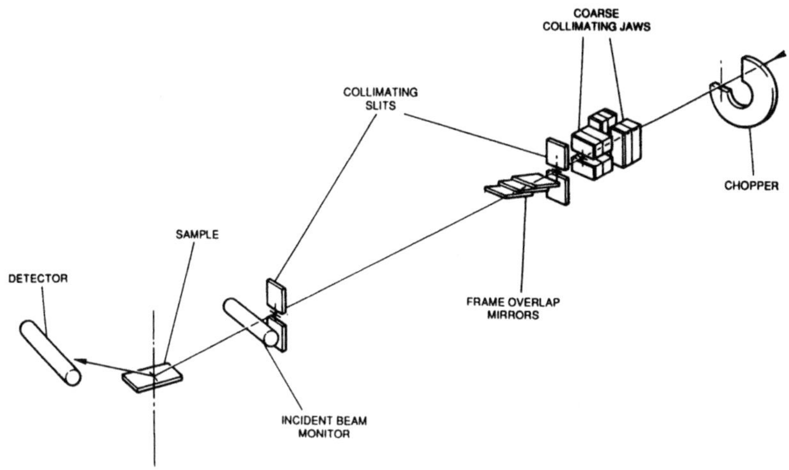

Fig. 12.7 Schematic diagram of the CRISP reflectometer at ISIS.

flight. There is a range of sample environments and goniometers, and for liquid surfaces there is the convenience of a four-position computer-controlled automatic sample changer comprising hermetically sealed and efficiently thermostatted PTFE-based troughs or a Langmuir-type film balance. The intensity of the reflected beam is measured using a detector D, which may be either a single element ^{3}He tube or a linear position-sensitive detector. The measurement is carried out at a fixed angle $\theta/2$ and the reflectivity is determined as a function of wavelength $R(\lambda)_\theta$. Since the distances from the moderator to M and D are known, λ can be determined by the time-of-flight. The collimation slits, the sample position, and the data acquisition and reduction and are all under computer control. A second reflectometer SURF is due in service at ISIS in 1995.

Although it is possible to perform surface reflection experiments with a steady-state source (e.g. at the ILL), it is preferable experimentally to use a white incident beam. This is because a range of Q values is accessed in a single measurement, and the sample illumination length L is constant. In contrast, on a reactor-based reflectometer, a single wavelength is selected, and the angle is scanned; in doing so, L varies and this has to be taken into account when normalizing the data.

12.5.2 Measured quantities

Owing to the direction of the scattering vector $\boldsymbol{Q}$ (Fig. 12.8), information on interfacial structure is obtained normal to the interface (Fig. 12.3). In order to produce the reflectivity profile $R(Q)$ on an absolute scale, it is necessary to measure (at fixed $\theta/2$) the wavelength dependencies of the

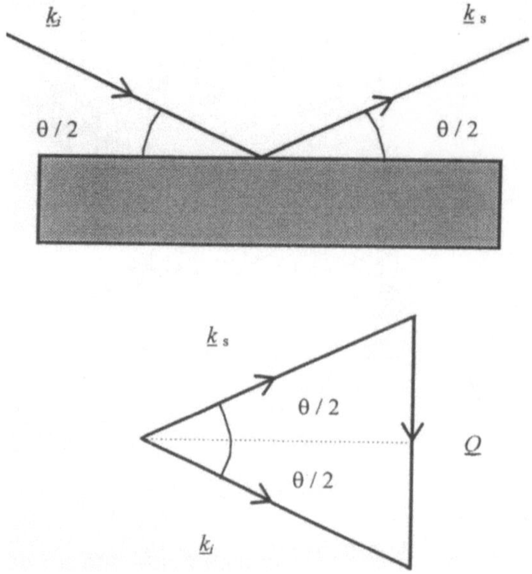

Fig. 12.8 Geometry of a neutron reflection experiment and direction of the scattering vector $\boldsymbol{Q}$, where $\boldsymbol{k}_i$ and $\boldsymbol{k}_s$ are the incident and scattered wave vectors, and θ is the scattering angle.

intensity in the incident beam monitor $i(\lambda)_M$ and at the detector $i(\lambda)_D$. The wavelength efficiencies of the monitor and detector, $E(\lambda)_M$ and $E(\lambda)_D$, which have to be pre-determined, are also required, as are the 'dark count' backgrounds, b_M and b_D, recorded with the neutron shutter closed.

Then the reflectivity is given by

$$R(\lambda, \theta) = K\frac{[i(\lambda)_D - b_D]E_M(\lambda)}{[i(\lambda)_M - b_M]E_D(\lambda)}, \tag{12.19}$$

which can easily be converted to $R(Q)$ using equation (12.1). The pre-factor K is a scale factor which for liquid surfaces at $\theta/2 = 1.5°$ is usually obtained by normalization to a clean D_2O standard surface. At a sufficiently low Q value, the critical value Q_c (defined by the wavelength λ_c or angle θ_c) may be reached. Below Q_c there will be total reflection and so we have $R(Q) \equiv 1$ (Figs 12.9 and 12.10); in this case it is a trivial matter to determine K. For monolayers on water subphases, however, it is usually necessary to make measurements at $\theta/2 < 1.5°$ in order to reach Q_c. Below Q_c we have $R(Q) = 1.0$, but it falls off quickly, as Q^{-4}, for a perfect interface such as clean D_2O (Figs 12.9 and 12.10). Clearly the absolute value of the background will limit the maximum value of Q. The instrumental background is normally about $\sim 10^{-7}$ in

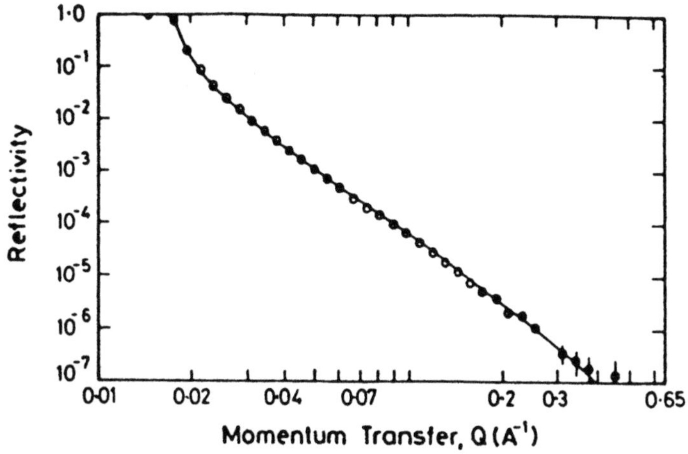

Fig. 12.9 Reflectivity profile $R(Q)$ for a pure D_2O surface, measured at three different angles on CRISP, $\theta/2 = 0.4$, 0.8 and 1.5°. A background of 1.5×10^{-6} has been subtracted from the data. The continuous line is a least-squares fit to the Fresnel law reflectivity with a surface roughness $\sigma = 2.8$ Å (equations (12.18) and (12.22)). (With permission from Elsevier.)

reflectivity, but for the D_2O standard surface this increases to $\sim 2 \times 10^{-6}$.

12.5.3 Data interpretation

As with SANS, a full analysis is best performed by least-squares model fitting. There are two approaches: an exact optical matrix method, and the kinematic approximation method (Thomas and Penfold, 1990). To date, the former method has been most widely used, and its basic principles are described below.

Fresnel's law describes the specular reflectivity R_F at the interface between two bulk media; for $\theta < \theta_c$, we have $R_F \equiv 1$, whilst for $\theta > \theta_c$ we have

$$R_F = \left| \frac{n_0 \sin \theta_0 - n_1 \sin \theta_1}{n_0 \sin \theta_0 + n_1 \sin \theta_1} \right|^2, \qquad (12.20)$$

with n_0 and n_1 the refractive indices of the two media, θ_0 the angle of incidence, and θ_1 the angle of refraction (Fig. 12.10) (Born and Wolf, 1970). (Note that the angle of incidence θ_0 (Fig. 12.10) is the same as half the scattering angle $\theta/2$ in Fig. 12.8.)

Equation (12.20) describes the reflectivity of a perfect interface. But if now there is a single uniform film **1**, of thickness d, introduced between bulk media **0** and **2** (Fig. 12.10), then we have

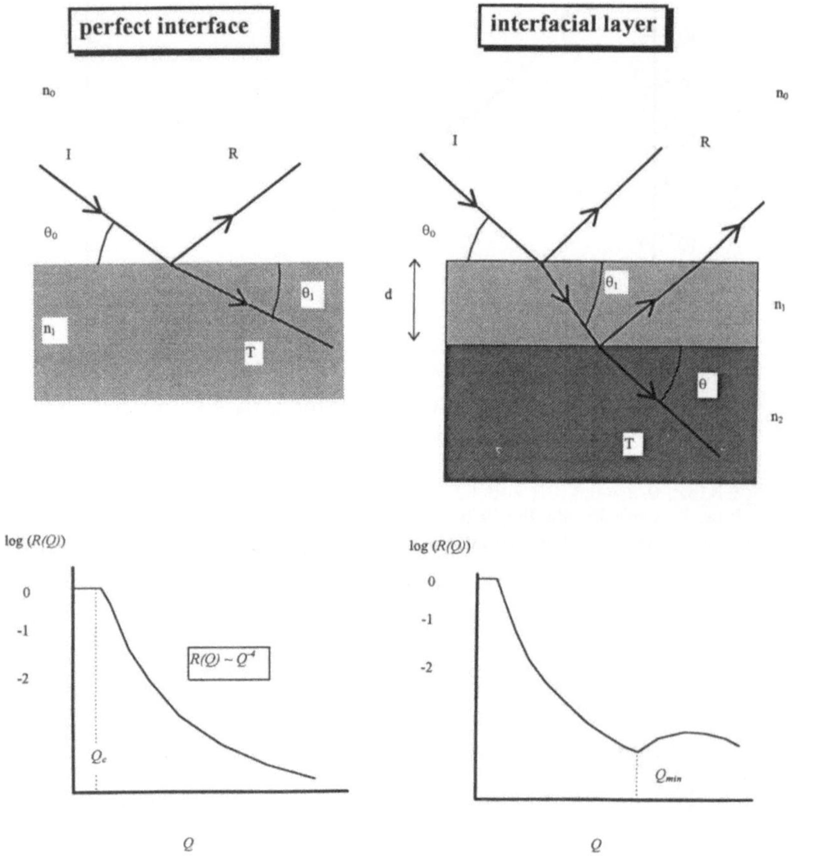

Fig. 12.10 Specular reflection at the sharp interface between two bulk media, and for a thin interfacial layer of thickness d sandwiched between two media. I is the incident beam, R is the reflected beam, and T is the transmitted beam. Other symbols as defined in the text. Also illustrated are schematic $R(Q)$ profiles for the two situations.

$$R_{\mathrm{F}} = \left| \frac{r_{01} + r_{12}\exp(-2i\beta)}{1 + r_{01}r_{12}\exp(-2i\beta)} \right|^2, \tag{12.21}$$

where r_{lm} is the Fresnel coefficient at the lm interface defined as

$$r_{lm} = \frac{(p_l - p_m)}{(p_l + p_m)}, \tag{12.22}$$

where $p_m = n_m \sin\theta_m$, and $\beta = (2\pi/\lambda)n_1\, d \sin\theta_1$ (the optical path length of the film). In addition to the beam reflected from the top **(0–1)** interface, the introduction of the thin interfacial film results in a reflection from the **1–2** interface. There is then an interference between these two reflected beams, resulting in the appearance of a 'fringe' in the $R(Q)$ profile (Fig. 12.10). The position of the first minimum is related to the film thickness d by $Q_{\min} \sim 2\pi/d$.

So far, we have dealt with perfect interfaces only. It is unlikely that on a 1–20 Å length scale (probed at high Q in the measurements) that this is actually the case, especially for liquid surfaces where thermally excited capillary waves may also be present. These 'roughness' imperfections will modify the reflectivity, and this effect may be taken into account by damping the reflectivity R_F, i.e.

$$R = R_F \exp(-q_0 q_1 \langle\sigma\rangle^2), \tag{12.23}$$

where $\langle\sigma\rangle$ is the root-mean-square roughness of the interface, and $q_i = (4\pi/\lambda) \sin \theta_i$.

The above expressions have proved extremely useful in describing the reflectivity of a molecular monolayer adsorbed at, say, the air–water interface (Thomas and Penfold, 1990). The approach can be extended to compute $R(Q)$ of any number of discrete layers; this involves the solution of optical matrix equations (Thomas and Penfold, 1990). The dependence of the refractive index on incident wavelength λ for any phase n_1 is known (equation (12.18)), as are the refractive indices n_0 and n_2 (say air and cmw), and so equations (12.21) and (12.23) can be recast in terms of $R(Q)$. The $R(Q)$ data are routinely analysed by model fitting using this approach to obtain, say, n_1, d and σ (the film roughness). The refractive index n_1 is related to the film composition via the parameter A in equation (12.18) ($A = Nb/4\pi$). Obviously, the parameters derived from fits to a number of isotopic variants must be consistent for the model to be valid (Thomas *et al.*, 1989). More recently, data modelling has advanced to a much more sophisticated stage using a partial structure factor analysis (Thomas *et al.*, 1993). For adsorbed surfactants, the chain and solvent distributions in the monolayer can be determined, and this provides a detailed picture of the interfacial structure.

12.5.4 Examples – surfactants at the air–water interface

An excellent example of the power of this technique is the study of the adsorption of the cationic surfactant decyltrimethylammonium bromide (DTAB) at the air–water interface (Thomas *et al.*, 1989). For these measurements three isotopic variants were synthesized: (i) dDdTAB with a C_{10} chain and three head-group methyls fully deuterated, (ii) dDhTAB with a fully deuterated C_{10} chain and proteinated head-group, and (iii) hDdTAB where only the head group is deuterated. The $R(Q)$ data were analysed using the method outlined in the previous section. Figure 12.11

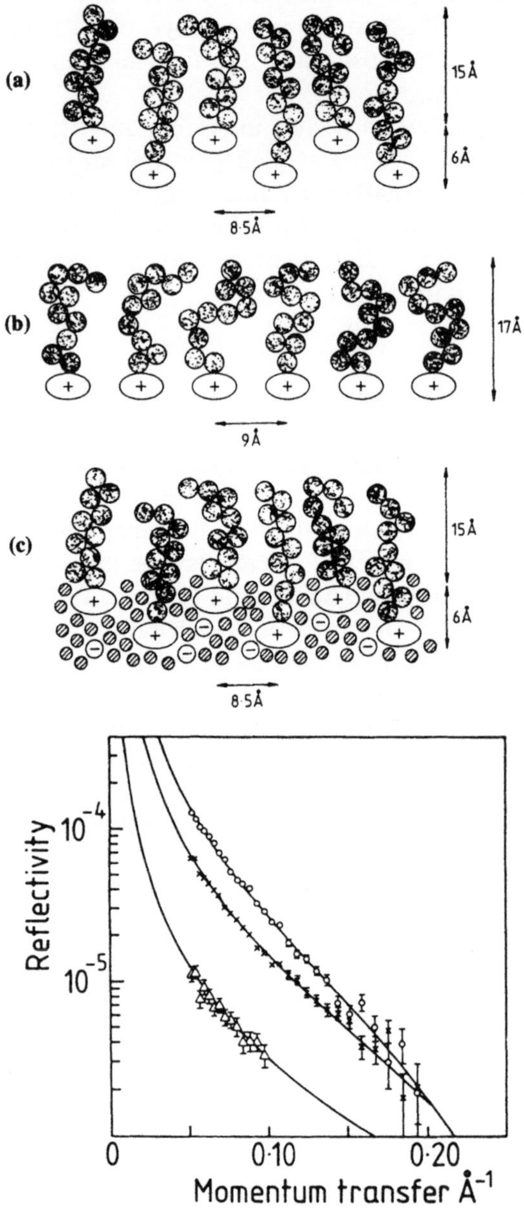

Fig. 12.11 Results from neutron reflectivity experiments of DTAB monolayers at the air-water interface. Top: interfacial structure derived from data analysis of isotopic variation experiments. Configuration of DTAB molecules in the adsorbed layer at a bulk concentration of (a) 0.05 and (b) 0.01 mol dm^{-3}; (c) full structure of the layer at 0.05 mol dm^{-3}, including interfacial water molecules. Bottom: observed and calculated $R(Q)$ profiles from different isotopic variants of DTAB adsorbed on cmw. The reflectivity decreases in the order dDdTAB > dDhTAB > hDdTAB. The fit parameters are given in Table 12.2. (Used with permission from the American Chemical Society.)

Table 12.2 Values derived from neutron reflectivity measurements on dDhTAB monolayers at the air–water interface (Thomas *et al.*, 1989)

Bulk concentration (mol dm^{-3})	0.05	0.01	0.005
Film thickness d (Å)	17 ± 1	15 ± 2	15 ± 2
Area per molecule a_S (Å^2)	58	82	100

shows, schematically, the structure of the monolayer derived from these experiments at bulk surfactant concentrations of 0.05 and 0.01 mol dm^{-3}. The headgroup and tail regions were modelled as two separate layers by a simultaneous least-squares fit to the three sets of reflectivity data. The technique is sensitive enough to resolve an increase in head group staggering (roughness) with concentration, i.e. (b) → (a) in Fig. 12.11. This subtle structure change must occur to reduce mutual head-group repulsion as the surface density increases. Furthermore, for 0.05 mol dm^{-3} surfactant concentration, the 6 Å thick, staggered, head-group region was found to contain on average 6–7 D_2O molecules per surfactant and about 10% C_{10} alkyl chains. The values obtained for film thickness, d, and area per surfactant headgroup, a_s for the dDhTAB species are given in Table 12.2. (Note that in this case d corresponds to the thickness of the alkyl chain region only.) It should be appreciated that for cationic surfactants, like DTAB, it is extremely difficult to obtain reliable values for a_s using conventional methods such as du Noüy tensiometry, owing to specific interactions between the tension ring and the surfactant (Thomas *et al.*, 1992). Nonetheless, the level of structural detail that can be obtained is matched by no other method.

12.5.5 Examples – food and biological science

Although to date the majority of reflectivity work has concentrated on model surfactant systems, there has been a number of studies of relevance to biological and food science. Surface pressure-induced phase transitions of phospholipid monolayers at the air–water interface, and the incorporation of spectrin and polylysine, have been investigated (Bayerl *et al.*, 1990, 1991). The structure of spread monolayers of bovine serum albumin has been determined (Eaglesham *et al.*, 1992). Monolayers of the disordered milk protein β-casein, the segment density profile, and competitive adsorption with a surfactant have also been studied (Dickinson *et al.*, 1993a–c). In a recent advance, the structure of fatty acids and triblock co-polymer monolayers adsorbed at the water–hexane interface has been determined (Cosgrove *et al.*, 1992).

12.6 Outlook

Using the technique of neutron scattering, the structure of self-assembling surfactant systems, macromolecular and colloidal solutions, and surfactant and polymer-containing interfaces can be determined in detail, and in many instances unique information can be obtained. Recent advances in neutron source technology and instrumentation, combined with advances in isotope enrichment chemistry and data modelling, mean that the investigation of complex realistic systems is now feasible. The continual development of the these flexible neutron methods suggests that they will become increasingly important in the study of material structures and properties.

Acknowledgements

I would like to thank my co-workers, Dr David Steytler (School of Chemistry, University of East Anglia, UK) and Dr Richard Heenan (ISIS, Oxfordshire, UK), for their continuing support and encouragement. Kodak Limited Research Division UK are thanked for supplying the surfactants used for the experiments reported in Fig. 12.6. Figures 12.4 and 12.7 were kindly provided by EPSRC DRAL ISIS.

References

Auvray, L and Auroy, P. (1991) Scattering by interfaces: variations on Porod's law, in *Neutron, X-Ray and Light Scattering* (eds. P. Lindner and Th. Zemb), Elsevier, Amsterdam, pp. 199–223.

Ashcroft, N.W. and Lekner, J. (1966) Structure and resistivity of liquid metals. *Phys. Rev.*, **145**, 83–90.

Bayerl, T.W., Thomas, R.K., Rennie, A., Penfold J. and Sackmann, E. (1990) Specular reflection of neutrons from phospholipid monolayers at the transition from the expanded to the condensed state. *Biophys. J.*, **57**, 1095–1098.

Bayerl, T.W., Thomas, R.K., Rennie, A., Penfold J. and Sackmann, E. (1991) Coupling of spectrin and polylysine to phospholipid monolayers studied by the specular reflection of neutrons. *Biophys. J.*, **60**, 1017–1025.

Born, M. and Wolf, E. (1970) *Principles of Optics*, Pergamon, Oxford.

Cabane, B. (1987) Small-angle scattering methods, in *Surfactant Solutions – New Methods of Investigation* (ed. R. Zana), Marcel Dekker, New York pp. 58–145.

Chen, S.-H. (1986) Small-angle neutron scattering studies of the structure and interaction in micellar and microemulsion systems. *Annu. Rev. Phys. Chem.*, **37**, 351–399.

Clint, J. (1993) *Surfactant Aggregation*, Blackie, Glasgow.

Cohen, Y. (1993) *ISIS Annual Report RAL*, SERC, Chilton, UK, pp. A194.

Cosgrove, T., Phipps, J.S. and Richardson, R.M. (1992) Neutron reflection from a liquid/liquid interface. *Colloids Surf.*, **62**, 199–206.

Dickinson, E., Horne, D.S., Phipps, J. and Richardson, R.M. (1993a) A neutron reflectivity study of the adsorption of β-casein at fluid interfaces. *Langmuir*, **9**, 242–248.

Dickinson, E., Horne, D.S., Phipps, J. and Richardson, R.M. (1993b) Interfacial segment density profile of adsorbed β-casein from neutron reflectivity measurements, in *Food*

Colloids and Polymers: Stability and Mechanical Properties (eds. E. Dickinson and P. Walstra), Royal Society of Chemistry, Cambridge, UK.

Dickinson, E., Horne, D.S. and Richardson, R.M. (1993c) Neutron reflectivity study of the competitive adsorption of β-casein and water-soluble surfactant at the planar air–water interface. *Food Hydrocolloids*, **7**, 497–505.

Eaglesham, A., Herrington, T. and Penfold, J. (1992) A neutron reflectivity study of a spread monolayer of bovine serum albumin, *Colloids Surf.*, **65**, 9–16.

Eastoe, J., Steytler, D., Robinson, B.H. and Thorn-Leeson, D. (1991) Structural studies of microemulsions stabilised by Aerosol-OT, *Adv. Colloid Interface Sci.*, **36**, 1–31.

Eastoe, J, Rogueda, P., Howé, A.M., Pitt, A. and Harrison, W. (1995) Properties of a dichained sugar surfactant. *Langmuir*, **10**, 4429–4433.

Groot, L.C.A., Kuil, M.E., Van der Maarel, J.C.R., Leyte, J.C., King, S.M.K., Heenan, R.K. and Jannik, G. (1994) Neutron scattering experiments on magnetically aligned liquid-crystalline DNA fragment solutions. *Liquid Crystals*, **17**, 263–276.

Hayter, J. and Penfold, J. (1981) An analytic structure factor for macroion solutions. *Mol. Phys.*, **42**, 109–118.

Jacrot, B. (1976) The study of biological structures by neutron scattering from solution. *Rec. Prog. Phys.*, **39**, 911–53.

Jenkins, P.J., Donald, A.M., Cameron, R.E., Bras, W., Derbyshire, G.E. and Ryan, A.J. (1994) *In-situ* simultaneous wide-angle X-ray scattering. A new technique to study starch gelatinization. *J. Polym. Sci. B, Polym. Phys.*, **32**, 1579–1583.

Leadbetter, A.J. and Richardson, R.M. (1979) in *The Molecular Physics of Liquid Crystals*, (eds. G.R. Luckhurst, and G.L. Gray), Academic Press, New York, pp 451–483.

Lovsey, S.W. (1984) *Theory of Neutron Scattering From Condensed Matter*, Clarendon Press, Oxford.

Penfold, J. (1991) Data interpretation in specular neutron reflection, in *Neutron, X-Ray and Light Scattering* (eds. P. Lindner and Th. Zemb), Elsevier, Amsterdam pp. 223–236.

Perkins, S.J. (1988) X-ray and neutron scattering, in *Modern Physical Methods in Biochemistry, Part B* (eds. A. Neuberger, and L.L.M. VanDeenen) Elsevier, Amsterdam, pp. 143–165.

Perkins, S.J. (1993) *ISIS Annual Report RAL* SERC, Chilton, pp. A222

Perkins, S.J. (1994) Physical methods of analysis, in *Methods in Molecular Biology*, (eds. C.Jones, B. Mulloy and A.H. Thomas), Humana Press, New Jersey Vol. 22, pp. 39–60.

Robinson, B.H., Atkinson, P.J., Howe, A.M. and Heenan R.K. (1991) Structure and stability of microemulsion-based organogels. *J. Chem. Soc., Faraday Trans.*, **87**, 3389–3397.

Schurtenberger, P., Magid, L., Penfold, J. and Heenan, R.K. (1990) Shear-aligned lecithin reverse micelles: a SANS study of the anomalous water-induced growth. *Langmuir*, **6**, 1800–1803.

Stilbs, P. (1987) Fourier transform pulsed-field gradient spin-echo studies of molecular diffusion. *Prog. Nucl. Magn. Reson. Spectrosc.*, **19**, 1–45.

Thomas, R.K. and Penfold, J. (1990) The application of the specular reflection of neutrons to study surfaces and interfaces, *J. Phys.: Condens. Matter*, **2**, 1369–1412.

Thomas, R.K., Lee, E.M., Penfold, J. and Ward, R.C. (1989) Structure of aqueous decyltrimethylammonium bromide solutions at the air/water interface studied by the specular reflection of neutrons. *J. Phys. Chem.*, **93**, 381–388.

Thomas, R.K., Simister, E.A., Penfold, J. Aveyard, R., Binks, B.P., Cooper, P., Fletcher, P.D.I., Lu, J.R. and Sokolowski, A. (1992) Comparison of neutron reflection and surface tension measurements of the surface excess of tetradecyltrimethylammonium bromide layers at the air/water interface. *J. Phys. Chem.*, **96**, 1383.

Thomas, R.K., Lee, E.M., Lu, J.R., Penfold, J. and Flitsch, S.L. (1993) Direct determination by neutron reflection of the structure of triethylene glycol monododecyl ether layers at the air/water interface. *Langmuir*, **9**, 1352–1360.

Winter, R., Christmann, M.T., Bottner, M., Thiyagarajan, P. and Heenan, R.K. (1991) The influence of the local anaesthetic tetracaine on the temperature and pressure dependent phase behaviour of model biomembranes. *Ber. Bunsenges. Phys. Chem.*, **95**, 811–820.

Chapter 13

High Resolution Nuclear Magnetic Resonance Spectroscopy of Solid and Semi-Solid Food Components

M.J. Gidley, A.J. McArthur, A.H. Darke and S. Ablett
Unilever Research Laboratory, Colworth House, Bedford, UK

13.1 General considerations

Nuclear magnetic resonance (NMR) is a versatile spectroscopic technique which can be used to probe food systems in a wide variety of ways (Ablett, 1992). Conventionally, four types of NMR experiment are carried out: pulsed (low resolution), high resolution solution state, high resolution solid state, or spatial localization (imaging). Increasingly, however, hybrid techniques and methods drawing on more than one conventional strand of NMR are being applied to complex systems such as foods. This review focuses on information obtainable by high resolution NMR from solid and semi-solid foods and their components, and explores overlaps in technique application between 'solution' and 'solid' state spectroscopy and low resolution relaxation measurements. Following a superficial introduction to the relevant materials, NMR background, and experimental approaches, examples are discussed of systems exhibiting either a high degree of solid state order (e.g. semi-crystalline and crystalline systems) or a low degree of solid state order (e.g. gel networks and plant cell walls). Examples from the carrageenan family of food polymers are used to illustrate that the labels 'solution state' and 'solid state' NMR are misnomers under certain circumstances.

More comprehensive introductions to NMR are given by Sanders and Hunter (1987) and Hemminga (1992), with various applications in food science covered by Ablett (1992), Gribnau (1992), Gidley (1992), Berendsen (1992) and McCarthy and McCarthy (1994).

13.1.1 Solid and semi-solid food components

Virtually all foods contain at least some components which have the structural permanence characteristic of solid-like materials. Examples of non-obvious solid-like components would be hazes and sediments in many beverages, and arguably even 'liquid' emulsion systems. Most foods are more obviously solid, and are amenable to direct analysis by 'solid state' techniques.

Many quality parameters of foods can be related to solid-like structures. This not only includes texture and mouthfeel but also processing and packaging, flavour and aroma delivery, and nutritional status/benefit. NMR spectroscopy is well suited to probe the origins of these quality parameters at a molecular level as well as providing insight into molecular/segmental mobility. This arises from the combination of chemical and physical information which can be obtained from appropriate high resolution NMR experiments as outlined in the following section.

13.1.2 Relevant NMR concepts for application to solid food components

The NMR technique is based on the absorption of radiation energy by certain nuclei which induces transitions between (at least) two energy states defined by different values for the spin quantum number (for most common NMR nuclei these are the $+\frac{1}{2}$ and $-\frac{1}{2}$ states). These two states have different energies as they align differently with an imposed magnetic field. The energy required for the transition is

$$\Delta E = h\mu, \tag{13.1}$$

where h is Planck's constant and μ is the radiation frequency. Because ΔE is also proportional to the strength of the applied magnetic field, for a given nucleus μ will be proportional to the applied field. This energy/frequency is called the chemical shift and is expressed as a fraction (parts per million) of the applied field in order to be independent of the strength of the magnetic field. The reason why NMR is such a powerful probe of chemistry is that the chemical shift is determined by the electronic environment of the nucleus and therefore reports on both covalent and 'through space' atomic connections. General 'rules of thumb' for interpreting chemical shifts are well known, based upon analogies between similar systems, but *ab initio* calculation of shifts from quantum chemical basis sets is a daunting task for all but relatively simple systems.

Although the chemical (electronic) environment of a nucleus at any position/conformation within a molecule determines the chemical shift, there are two common mechanisms which can cause broadening of signals (resonances) such that chemical shifts are obscured in solid materials. These are dipole–dipole interactions and chemical shift anisotropy. In nuclei within applied magnetic fields, individual nuclear dipoles can interact through space with a strength dependent on inter-nuclear distance and mutual orientation of nuclei with respect to the magnetic field. In solution, the rapid tumbling of molecules usually leads to an averaging of dipolar interactions to zero. In solids, reduced molecular motion means that dipolar effects are significant; in principle they can be averaged to zero by rapid sample spinning, but in practice most solids have such strong dipolar interactions that unrealistically rapid spinning would be required. The more usual way of removing dipolar resonance broadening is by irradiation at frequencies corresponding to the chemical shift(s) of interacting nuclei ('dipolar decoupling'). Chemical shift anisotropy arises from the fact that electron densities are not arranged spherically around nuclei. In an applied magnetic field, the chemical shift of an individual nucleus will be influenced by molecular orientation as well as chemistry. In solutions, rapid motion averages orientation effects to zero, revealing the underlying chemical information. In systems with constrained motion, incomplete motional averaging results in a range of chemical shifts for nuclei from equivalent sites in different molecules oriented differently with respect to the magnetic field, leading to a chemical shift anisotropy broadening effect. As the orientational dependence of chemical shift anisotropy (and dipolar interactions) contains the term $(3\cos^2\theta - 1)$, spinning of the sample at an angle ($\sim 54° 44'$) for which $\cos^2\theta = 1/3$ with respect to the applied magnetic field and at speeds in excess of the frequency spread in chemical shift anisotropy results in line narrowing. This is conventionally referred to as 'magic-angle' spinning. Spinning speeds of several kiloHertz are usually used to average anisotropy effects.

Another major feature of NMR spectroscopy is that, because the energy level difference between the lower and upper spin states is very small, there is only a slight excess population in the lower energy state (*ca.* 0.005% for protons). This leads to a relatively insensitive response requiring multiple repeats of experiments to obtain good signal-to-noise ratios in many cases. The small energy gap also means that many spins migrate from one energy level to the other as if they were resonating, and hence the technique is referred to as a resonance spectroscopy.

13.1.3 Choice of NMR nucleus

The prerequisite for atomic nuclei to exhibit NMR effects is that the nuclear spin quantum number is non-zero. This rules out ^{12}C and ^{16}O, two of the most abundant nuclei in food systems. The other common nucleus (^{1}H) is, however, the archetypal and most widely studied NMR nucleus. Other abundant nuclei found in food include ^{23}Na, ^{31}P and ^{39}K. The other class of useful NMR nuclei are stable isotopes such as ^{13}C, ^{15}N and ^{17}O which represent low abundance (1.1%, 0.4% and 0.04%, respectively) forms of important food nuclei.

Table 13.1 Properties of NMR nuclei of relevance to foods

Nucleus	Spin	Relative sensitivity[a]	Natural abundance (%)
^{1}H	1/2	100.00	99.98
^{13}C	1/2	1.59	1.11
^{15}N	1/2	0.10	0.37
^{17}O	5/2	2.91	0.04
^{23}Na	3/2	9.25	100.00
^{31}P	1/2	6.63	100.00
^{39}K	3/2	0.05	93.10

[a]Determined by the magnetogyric ratio and expressed relative to ^{1}H.

In addition to abundance, three other factors determine the utility of nuclei for NMR studies. These are the spin quantum number, the magnetogyric ratio which determines relative sensitivity, and the ability to remove the line broadening effects described above. Spin quantum numbers greater than 1/2 result in more complex spectra than those for spin 1/2 nuclei and, although interpretable, tend to obscure chemical information. The magnetogyric ratio is a fundamental constant for each nucleus which determines the strength of interactions with magnetic fields and hence the inherent NMR sensitivity (Table 13.1). In order to determine relative molar sensitivities for different nuclei, the relative absolute sensitivity is multiplied by the natural abundance of each nucleus. For organic materials, ^{1}H is clearly the most attractive nucleus from a sensitivity viewpoint, being nearly four orders of magnitude more sensitive than, say, ^{13}C (Table 13.1). However, this very abundance means that there is a multitude of potential dipolar interactions which are very difficult to remove by decoupling and (magic angle) spinning techniques. In practice, at the current time, the resolution achievable for solid state ^{1}H NMR spectroscopy is only useful for identifying general chemical classes rather than details of solid state structures or conformations. The relatively narrow chemical shift range for ^{1}H (~10 ppm) places great demands on line narrowing techniques, as resolution to at least 0.1 ppm is required for useful chemical information. Whilst ^{1}H NMR spectroscopy will remain a powerful probe of soluble components in foods, there is currently very limited value in its use as a probe of more solid components.

Nearly all solid state NMR studies of food components use the ^{13}C nucleus. Despite the relative insensitivity (Table 13.1), there are several benefits of NMR studies using ^{13}C. Firstly, the low (*ca.* 1%) isotope abundance means that there are negligible dipole–dipole interactions with other nuclei. Instead dipolar effects involve close (typically covalently attached) ^{1}H nuclei. These effects can be efficiently minimized by irradiation ('decoupling') at the ^{1}H NMR frequency window. The power required for proton irradiation in dipolar decoupling must be greater than that associated with ^{1}H–^{13}C interactions in order to effectively 'short circuit' the line broadening effect. This power is typically at least an order of magnitude greater than that used in solution state ^{13}C spectroscopy to remove ^{1}H J couplings (known as broadband decoupling). Secondly, the chemical shift window for ^{13}C is large at *ca.* 200 ppm resulting in a much more useful dispersion of signals from different chemical environments than, say, ^{1}H (10 ppm). Thirdly, there is a choice of magnetization strategies for ^{13}C, primarily between direct magnetization (or single pulse) methods as used in 'solution state' spectroscopy, and cross polarization in which magnetization is transferred from ^{1}H to ^{13}C (typically covalently connected) provided that the ^{1}H–^{13}C dipole is not motionally averaged over the millisecond timescale. This approach in combination with dipolar decoupling and magic angle spinning is normally referred to as CP/MAS, CP/DD/MAS or 'solid state' high resolution spectroscopy. Fourthly, in addition to control of magnetization pathways, the resulting ^{13}C spectra can be 'edited' in several ways in order to enhance or suppress signals which correspond to different relaxation rates through either ^{1}H or ^{13}C pathways. This approach of mobility-resolved spectroscopy is amplified in later sections.

13.2 Chemical information: molecular conformation and order

High resolution solid state NMR is rarely the most effective means of determining chemical compositions of foods, as the bulk materials and major components which would be identifiable are usually known, and minor components are not amenable to study by the relatively insensitive ^{13}C NMR method. Although such primary structure effects are not best studied by NMR, certain aspects of molecular conformation and order are uniquely studied by high resolution solid state ^{13}C NMR. As an example, Fig. 13.1 shows spectra for sucrose in solution (single pulse excitation) and in solid glassy (amorphous) and crystalline states (cross polarization excitation). Chemical shift differences between solution and crystalline states indicate changes in electronic environment for nuclei at the same molecular sites within the sucrose molecule for the different states. The solution spectrum has 12 resolved lines corresponding to each of the 12 carbon atoms in the sucrose molecule. Only 11 signals are resolved in the crystalline spectrum, indicating that one of the signals corresponds to 2 atomic sites (most likely at ~73 ppm). The solution state spectrum corresponds to motionally averaged chemical shifts with weighted contributions from all significantly populated conformations. This NMR average conformation is therefore a virtual conformation which may or may not correspond to any real conformation. In contrast, the absence of significant motional averaging in the solid state means that observed chemical shifts contain true conformational information. Unfortunately, at the present time, our ability to understand the

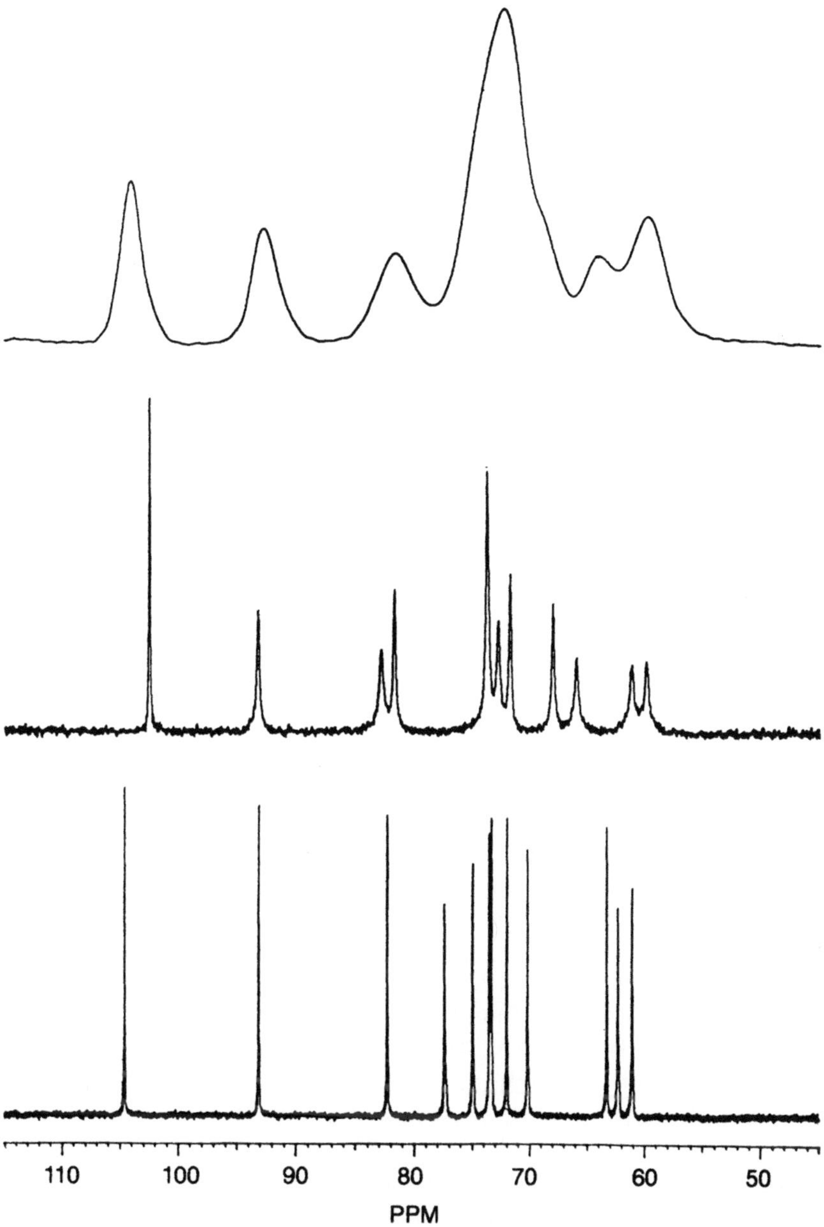

Fig. 13.1 ^{13}C NMR spectra for sucrose, illustrating information available from high resolution solid state spectra. The CP/DD/MAS spectrum of the glassy state (50% solution lyophilized below 245 K) exhibits broad peaks indicative of a range of local conformations which are averaged to sharp lines in solution (298 K). The crystalline state has relatively narrow peaks in the CP/DD/MAS spectrum at chemical shifts different from those for glass or solutions, which is consistent with a conformationally well-defined crystalline state.

conformational origins of chemical shifts (Saitô, 1986) is in its infancy, and so the main approach which has been adopted to date has been to compare 'fingerprint' spectra, for which independent structural information (from, say, X-ray diffraction) is available, with experimental spectra. The spectrum of glassy amorphous sucrose (Fig. 13.1) has broader lines than the crystalline solid. This is not considered to be due to inefficient removal of dipole–dipole or chemical shift anisotropy effects, but rather to result from a dispersion of chemical shifts resulting from a dispersion of conformations, as would be expected to be present in an amorphous material. As illustrated by glassy and crystalline sucrose, line widths in CP/DD/MAS ('solid state') ^{13}C spectra are normally interpreted in terms of well-defined (sharp) or heterogeneous (broad) local conformations. Glassy sucrose is prepared by rapid quenching (to prevent crystallization) of sucrose solutions/melts; so the fact that the average chemical shifts for each of the broad lines in the spectrum of glassy sucrose match closely to those in the solution spectrum would be expected. Similar effects have also been seen in comparisons between ambient and frozen solutions of starches (Gidley and Bociek, 1988a). It is therefore a tenable hypothesis that the dispersion of chemical shifts observed in glassy solids or frozen solutions represents the range of conformations adopted in solution. This provides an additional stimulus for uncovering the conformational origins of chemical shift effects.

13.2.1 Triglycerides and other lipids

The application of solid state NMR methods to food lipids (fats and oils) has been surprisingly infrequent (Eads and Croasmun, 1988). Bulk structuring by the model triglycerides, tristearin and tripalmitin, has been investigated with information on polymorphic structure and internal mobility (Bociek *et al.*, 1985) and relationships with thermodynamic properties (Norton *et al.*, 1985) derived. Solid state high resolution techniques have also been found to be useful for observing spectra for margarine oil (Eads and Croasmun, 1988) and source oil seeds (Rutar, 1989). The unique contribution of NMR to these studies is in the assessment of relative mobilities at resolved atomic sites in the solid and semi-solid materials. Identification of polymorphic form can also be achieved by X-ray diffraction, as it seems likely that molecular order in the absence of crystalline order (i.e. giving rise to specific NMR shifts without showing diffraction effects) is not significant in these systems.

Lipids present in various forms of bilayers and monolayers within foods (e.g. membranes, micelles, emulsions) can also be studied by high resolution solid state NMR methods. Conventionally, 'solution state' NMR spectra of membranes are obtained following sonication; however, this could lead to structural disruption. A combination of magic angle spinning and either broadband or dipolar ^{1}H decoupling, however, yields good quality high resolution spectra (Oldfield *et al.*, 1987; Forbes *et al.*, 1988), which have been used to study egg yolk lecithin and the effect of added cholesterol. The wealth of spectral information obtained could be exploited in studies of molecular architecture in important bilayer/membrane systems, but applications have so far not been forthcoming. Many important food lipids are phosphorylated, and so ^{31}P NMR becomes a possible tool. Odahara *et al.* (1990) showed how cross-polarization from ^{1}H to ^{31}P could be used to study (phosphorylated) lipid bilayers in a cell system. In this example, chemical shift anisotropy was not reduced as this was the feature which was interpreted in terms of phase characteristics. This illustrates that scientific opportunities may be lost in a headlong rush for narrower signals, as the signal shapes and widths in the broad form contain significant structural information, particularly on molecular alignment.

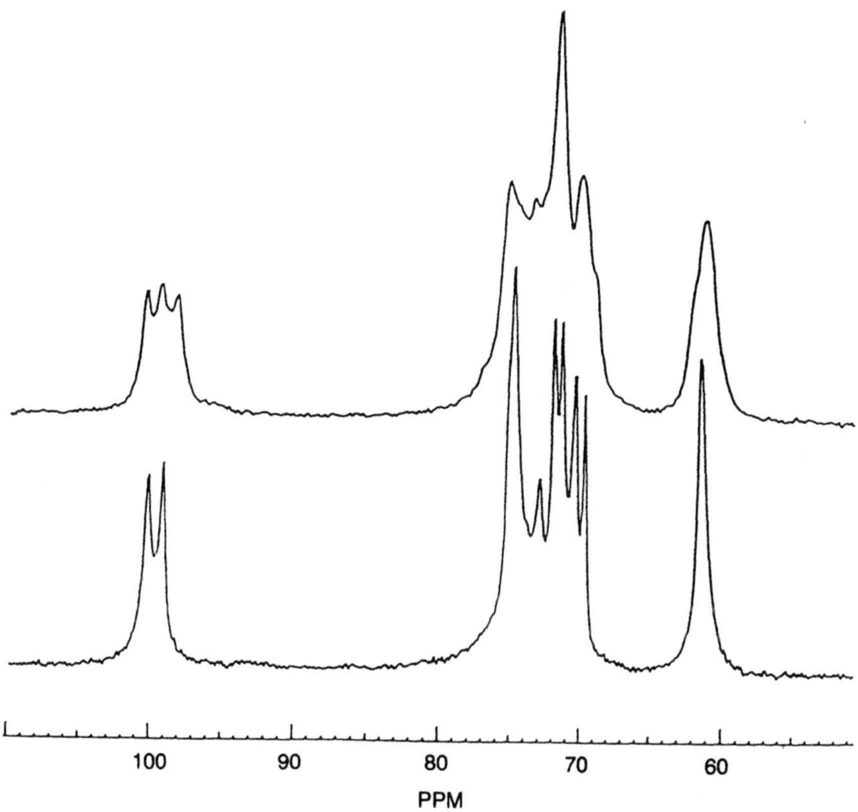

Fig. 13.2 ^{13}C CP/DD/MAS spectra of crystalline double helical polymorphs of α-(1 → 4)-D-glucan (starch) polysaccharides. Chemical shift ranges are similar in the two spectra consistent with similar double-helical conformations. The multiplicity observed at C-1 (~ 100 ppm) matches the number of non-equivalent sites in the relevant crystal structures.

13.2.2 Starch

Starch has been one of the most successfully studied food systems by solid state NMR technology. Two limiting crystalline polymorphs (A and B) based on similar double helical structures have been characterized (Fig. 13.2). The two polymorphs exhibit different multiplicity of signals, particularly pronounced for the anomeric carbon at *ca.* 100 ppm. The two peaks observed for the B type and the three peaks observed for the A type correlate with the number of non-equivalent carbon atoms in the relevant crystal structures (Gidley and Bociek, 1985; Marchessault *et al.*, 1985); but whether the multiplicity is due to different conformations within the double helix (Veregin *et al.*, 1986) or to inter-helix shift effects (Gidley and Bociek, 1985) remains to be determined unambiguously.

The ^{13}C CP/DD/MAS spectra of dry granular starch can be simulated by simple addition of spectral features due to the relevant crystalline polymorph and solid amorphous starch. From quantitation of crystalline and amorphous features required to match spectra of granular starches, the percentage of amorphous and ordered chains can be derived (Gidley and Bociek, 1985). Powder X-ray diffraction patterns can similarly be interpreted in terms of crystalline and amorphous fractions. Table 13.2 shows a comparison between NMR and X-ray measurements of starch granule order. The significantly greater values from NMR are interpreted (Gidley and Bociek, 1985; Cooke and Gidley, 1992) in terms of a fraction of double helices which are not present in sufficiently ordered arrays to diffract X-rays. This illustrates the complementarity between measurements of local molecular order by ^{13}C NMR and extended crystalline order by X-ray diffraction. Quantitation similar to that shown in Table 13.2 has been used to deduce that the enthalpy of starch gelatinization is due predominantly to melting of double helices with no significant perturbation by crystalline order (Cooke and Gidley, 1992). The levels of double helical order are in line with estimates of outer chain lengths of amylopectin by β-amylase digestion, and suggest that virtually all amylopectin segments not topologically constrained by adjacent branch points are involved in double helix formation.

Table 13.2 Comparison of crystalline (from X-ray diffraction) and double helical (from ^{13}C NMR) contents in a range of granular starches

Botanical origin	% Crystallinity	% Double helix
Potato	25	50
Maize	30	42
High amylose maize	22	40
Waxy maize	35	53
Wheat	30	48
Rice	35	49

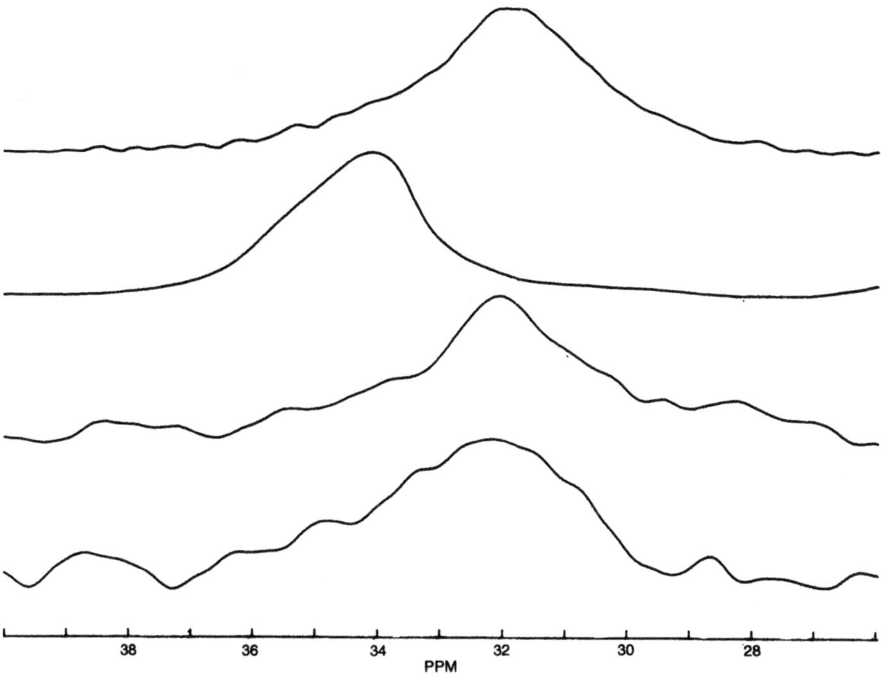

Fig. 13.3 The alkyl CH_2 region of the ^{13}C CP/DD/MAS spectra for stearic acid, either free or in an amylose complex, in comparison with two native cereal starches, illustrating the use of 'fingerprint' spectra for diagnosing the local environment of minor components within complex structures such as lipids in starch granules. (a) Amylose/stearic acid complex; (b) stearic acid; (c) maize starch; (d) barley starch.

Recently, a further question concerning starch granule structure has been resolved by ^{13}C CP/DD/MAS NMR, namely, the presence of lipids as inclusion complexes in cereal starches. Indirect evidence for and against inclusion complexes within granules had been obtained by extraction chemistry and DSC studies (Morrison and Karkalas, 1990). The ^{13}C NMR spectra of model amylose/lipid complexes, in comparison with granular starch and non-complexed lipid, provides clear evidence in support of complexes within granules (Morrison *et al.*, 1993). This is based on chemical shifts of lipid methylene carbons as illustrated in Fig. 13.3. The close match between spectral patterns for granular starches and the model complex, and their differentiation from the non-complexed lipid, provides the necessary evidence. Physical mixing of lipid and amylose or starch in the dry state yields spectra only of non-complexed lipid.

The wide range of well-defined crystal structures for starch-related materials, has allowed an approach to evaluating possible chemical shift/conformation relationships (Gidley and Bociek, 1988b). The relative orientation of adjacent glucose residues (as expressed in glycosidic torsion angles) appears to dominate chemical shifts. However, empirical relationships differ depending on the data set used and the degree of rigour of shift assignments (Veregin *et al.*, 1987; Gidley and Bociek, 1988b; Jarvis, 1994). *Ab initio* chemical shift calculations are currently being used in our laboratory to evaluate possible conformational origins for chemical shift effects in polysaccharides.

13.2.3 Polysaccharide gels

The ability to detect molecular order in the absence of crystalline packing makes solid state NMR spectroscopy a method of choice for characterizing polysaccharide gels and other networks. As the cross-linking in food polysaccharide gels is non-covalent, many contiguous residues need to be involved in order to confer stability. This results in regions of polymer chains being involved in junction zones (Rees *et al.*, 1982; Gidley and Robinson, 1990) with defined conformations compatible with inter-polymer association. In many cases junction zones are thought to involve lateral alignment of several polysaccharide chains mediated by either hydrogen bonding or charge interactions (Rees *et al.*, 1982). However, rarely are sufficient chains aligned to give X-ray diffraction patterns which could be directly compared with those for previously characterized conformations. In addition, as detailed secondary/tertiary structures from diffraction techniques tend to be obtained from extensively annealed and sometimes chemically treated fibres, it cannot be directly assumed that such structures are present in aqueous gels/networks. Furthermore, quantities of fibres produced for X-ray diffraction are very small, making acquisition of ^{13}C NMR data currently impractical. In favourable circumstances, however, CP/DD/MAS NMR can be used to identify conformations in gels by comparison with spectra for previously characterized 'crystalline' material (Fig. 13.4). This approach has been

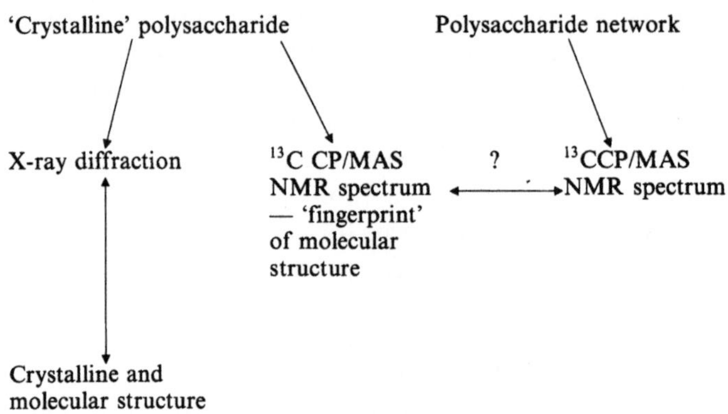

Fig. 13.4 Schematic illustration of the use of ^{13}C CP/DD/MAS spectroscopy for deducing conformational states in non-crystalline systems provided suitable model crystalline materials are available.

used to show that junction zones in amylose gels have the B-type conformation (Gidley, 1989) and that those in galactomannan and glucomannan gels have the mannan II conformation (Gidley *et al.*, 1991) rather than the mannan I form found by X-ray diffraction studies of low moisture fibres (Song *et al.*, 1989). For other food polysaccharides such as carrageenans, agarose and pectins, ^{13}C CP/DD/MAS spectra of gels have been recorded (Saitô *et al.*, 1990; Gidley, 1992), but the absence of NMR spectra for crystalline models means that detailed conformational deductions are problematic. For non-bulk-crystallizing systems, progress must await developments, either in the acquisition of ^{13}C spectra from X-ray fibre diffraction specimens, or greater understanding of the conformational dependence of chemical shifts.

13.3 Physical information: mobility-resolved spectroscopy

As introduced in Section 13.1.3, the NMR operator has a choice of experiments based on different excitation strategies and a potentially infinite range of detection strategies. This element of control over the experiment is a major facet of the utility of NMR spectroscopy, particularly in the study of heterogeneous systems where different components or regions within the system have different NMR properties. As the interaction and relaxation parameters which determine NMR responses to different excitation and detection strategies are heavily influenced by segmental (for polymers) molecular mobility, NMR spectra can be tailored to enhance or suppress signals from components differing in mobility. This general approach is termed 'mobility-resolved' spectroscopy (Gidley, 1992). There are two commonly used excitations (direct polarization, or single pulse excitation, and cross polarization); information obtainable from each is discussed below in the context of both practical experience and the theoretical framework shown in Fig. 13.5.

The dependence of the key relaxation parameters, T_1, T_2 and $T_{1\rho}$ on correlation time T_c (Fig. 13.5) can be predicted theoretically (Bloembergen *et al.*, 1948; Harris, 1983). The correlation time can be considered to represent the time required for the equivalent of rotation through 1 radian by a segment within the NMR sample, i.e. longer correlation times correspond to less mobile segments. Determination of correlation times for real systems requires measurement of relevant relaxation properties as a function of magnetic field strength and comparison with various models (correlation times are independent of magnetic fields). An example of this approach for a relatively mobile polymer is in the study of glycogen by Zang *et al.* (1990). This approach is not practical for many materials due to the requirement for multiple spectrometers, and the inability to study less mobile segments using the same method.

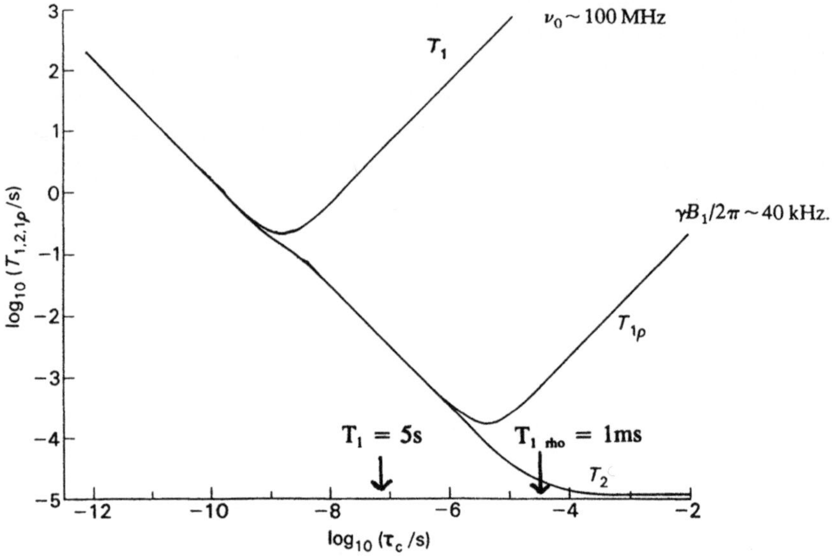

Fig. 13.5 Graphical representation of the effect of (segmental) correlation times on relaxation processes relevant to the ability to observe spectra following various acquisition strategies. Values are for 100 MHz fields (i.e., ^{13}C on a 400 MHz spectrometer) and a typical ^1H rotating frame frequency of 40 kHz for T_1 and $T_{1\rho}$, respectively. For typical cross-polarization contact times of 1 ms and an effective maximum recycle time of 5 s in single pulse experiments on dilute systems, the indicated $T_{1\rho}$ and T_1 values represent the maximum and minimum correlation times, respectively, for significant spectral observability. For correlation times between 10^{-5} and 10^{-7} s, spectral observability problems can be anticipated. (Adapted from Harris, 1983.)

13.3.1 Cross polarization

In order for effective cross polarization from ^1H to ^{13}C, the relative positions of the two nuclei should remain nearly constant during the cross polarization process. Under magic angle spinning conditions, it is the ^1H $T_{1\rho}$ value which reports on modulation of intensity due to internal motion. Typical cross-polarization times for food components are *ca.* 1 ms (Gidley and Bociek, 1985). If $T_{1\rho}$ values are less than 1 ms, then magnetization will be ineffective (i.e. lost as soon as it is generated). From Fig. 13.5 it can be seen that 1 ms $T_{1\rho}$ corresponds to correlation times between 10 and 100 μs. For materials with longer correlation times (more 'solid'), it is predicted that ^1H–^{13}C cross polarization should be effective and hence relevant signals will be observed in ^{13}C CP/DD/MAS experiments. It is interesting that this cut-off also corresponds to the correlation-time independence region ('rigid lattice') for T_2 of *ca.* 10 μs. If the idealized situation shown in Fig. 13.5 exists in reality, it would be predicted that material showing rigid lattice T_2 behaviour should correspond to intensity observable in ^{13}C CP/DD/MAS spectroscopy. This was indeed found to be the case for amylose gels (Gidley, 1989), but may not be generally true as shown below for carrageenans.

Within the CP-observable range of correlation times, decreased molecular mobility is expected to result in an increase in $^{1}H\ T_{1\rho}$ values. This can be exploited to generate spectra corresponding to more and less mobile 'solid' components as recently demonstrated by Newman *et al.* (1994), who showed how apple cell wall ^{13}C CP/DD/MAS NMR spectra could be manipulated to produce sub-spectra corresponding to cellulose (less mobile, $T_{1\rho}$ 9.7 ms) and pectin and hemicellulose (more mobile, $T_{1\rho}$ 4.0 ms). This effect has also been shown in studies of other cell walls using a variable contact time approach (Foster *et al.*, 1995). For nearly dry systems, there is much less mobility discrimination; e.g. in dry granular starches cross polarization is approximately twice as efficient for methylene compared with methyl groups, but $^{1}H\ T_{1\rho}$ values are all similar (Gidley and Bociek, 1985).

13.3.2 Single pulse excitation

There is no significant barrier to single pulse excitation as direct polarization times ('pulse lengths') are typically a few microseconds, i.e. shorter than the most rapid relaxation process (Fig. 13.5). The ability to observe a signal is, however, dependent on the T_1 and T_2 values for sites of interest. Line widths at half height ($\Delta\nu_{1/2}$) are determined by the T_2 value

$$\Delta\nu_{1/2} = \frac{1}{\pi T_2}. \qquad (13.2)$$

That is, shorter T_2 values lead to broader lines. For the case of ^{13}C relaxation, T_2 is difficult to measure due to low abundance, but assuming compliance with Fig. 13.5, correlation times of 1 μs and 100 ns correspond to T_2 values which translate into linewidths of *ca.* 1000 and 100 Hz, respectively. Correlation times longer than this are therefore likely to lead to lines so broad (via for example strong dipole–dipole interaction) as to be indistinguishable from the baseline. $^{13}C\ T_1$ processes lead to a return to the pre-excitation ground state; if this is incomplete before the next excitation is applied, there will be a reduced response due to a saturation effect. Due to the low sensitivity of ^{13}C NMR (Table 13.1), there is a desire to repeat experiments as often as possible particularly for systems of low carbon content such as gels. A typical recycle time would be up to 4 s; any T_1 values of this order or longer would give severely attenuated signal intensity. From Fig. 13.5, a T_1 value of 5 s corresponds to a correlation time of approximately 100 ns. Both T_1 and T_2 considerations, therefore, indicate a maximum correlation time of *ca.* 100 ns for visibility in single pulse excitation experiments. As cross polarization excitation is indicated to be efficient only for correlation times greater than 10–100 μs, there is the possibility that, for correlation times intermediate between these values, NMR signal visibility may not be possible (a 'black hole'). One way to influence correlation times is to manipulate sample temperatures; in particular, we have found it useful to acquire CP/DD/MAS spectra in the frozen state to be certain of seeing full signal intensity (Gidley, 1989, 1992); see following discussion on carrageenan.

In conjunction with single pulse excitation, a number of other experimental conditions can be varied to produce mobility-resolved spectra. At the mobile extreme, standard 'solution state' conditions of scalar decoupling and no magic angle spinning will reveal signals from segmentally mobile components which can be sub-discriminated by line width (T_2) or variable acquisition times/sequences (T_1). Addition of magic angle spinning in principle will reveal highly mobile segments with slow reorientation times. In practice, the best way to reveal slightly less mobile signals is to move to high power (dipolar) ^{1}H decoupling, and then to apply magic angle spinning (SP/DD and SP/DD/MAS experiments). Further discrimination within any of these spectra is in theory possible through changes in recycle times which suppress signals from sites with T_1 values of the order of the recycle time and longer via saturation.

13.3.3 Practical applications

In practice, different groups with different materials under study choose selected experiments to obtain mobility-resolved spectra. As the link to theory (Fig. 13.5) is not established directly for systems of interest, the semi-empirical lines of reasoning outlined earlier are employed in experiment selection. Examples of systems which have been studied are deoxyhaemoglobin gelation (Sutherland *et al.*, 1979), bovine eye lens (Morgan *et al.*, 1989), glycogen hydration (Jackson and Bryant, 1989) and polysaccharide ion-exchange resins (Morgan *et al.*, 1994).

For semi-solid food components such as polysaccharide gels and plant cell walls, practical experience has identified a particularly useful suite of experiments, listed below in order of decreasing mobility of observed features:

(i) single pulse with scalar decoupling;
(ii) single pulse with dipolar decoupling;
(iii) single pulse with dipolar decoupling and magic angle spinning;
(iv) CP/DD/MAS at ambient temperature; and
(v) CP/DD/MAS of frozen system.

Experiment (v) serves the purpose of identifying all carbon sites within the material and is useful for assessing whether correlation times in the 'black hole' region (Fig. 13.5) are present. This experiment does not, however, give a spectrum composed of a linear combination of the other experiments, as single pulse observable sites are motionally averaged to give narrower lines than is observed upon freezing where the full dispersion of chemical shifts is displayed (Gidley and Bociek, 1988a). Examples of this approach have been in the study of gels from amylose (Gidley, 1989), glucomannans and galactomannans (Gidley *et al.*, 1991) and κ-carrageenan (Gidley, 1992), leading to the identification of three categories (so far) of molecular structures within polysaccharide gels (Gidley, 1992):

1. ordered junction zones different from solution conformations with little mobile inter-junction material (κ-carrageenan in 0.1 M KCl);
2. ordered junction zones different from solution conformations with significant mobile inter-junction material (amylose); and
3. ordered junction zones with similar conformations to solutions enriched in certain primary structural features compared with mobile inter-junction material (glucomannan and galactomannan).

Other classes may emerge from further study. The challenge now is to relate these molecular characteristics to the macroscopic features which underlie rheological and food textural properties. For plant cell walls, the same approach yields much novel information on the segmental dynamics for onion (McCann *et al.*, 1995) and several other cell wall systems in the hydrated state (Foster *et al.*, 1995). This approach builds on previous high resolution ^{13}C NMR studies of cell walls in more dehydrated states (Jarvis and Apperley, 1990; Jarvis, 1990; Newman *et al.*, 1994).

13.4 Agarose, ι-carrageenan and λ-carrageenan

To further probe the utility of mobility-resolved ^{13}C NMR spectroscopy, we now describe results obtained for agarose, ι-carrageenan in 0.2% $CaCl_2$ solution, and λ-carrageenan. For 5% agarose gels (in deionized water), there was no detectable intensity in single-pulse excitation experiments, with significant amounts (most if not all of the carbons) visible in a CP/DD/MAS experiment as evidenced by thė same signals at similar signal-to-noise ratios in ambient and frozen systems. In terms of Fig. 13.5, this places correlation times in excess of 10 μs. In line with this interpretation, ^1H T_2 measurements using a combination of decay fitting and direct 'CPMG' determination (Ablett *et al.*, 1982) show *ca.* 90% of the material to have a rigid lattice value. Chemical shifts observed for the gel state (Fig. 13.6) are very similar to those reported previously (Rochas and Lahaye, 1989; Saitô *et al.*, 1990) and are compared with the solution state spectrum (Fig. 13.6). Although many chemical shifts are similar between sol and gel states, spectral patterns in the region 75–85 ppm

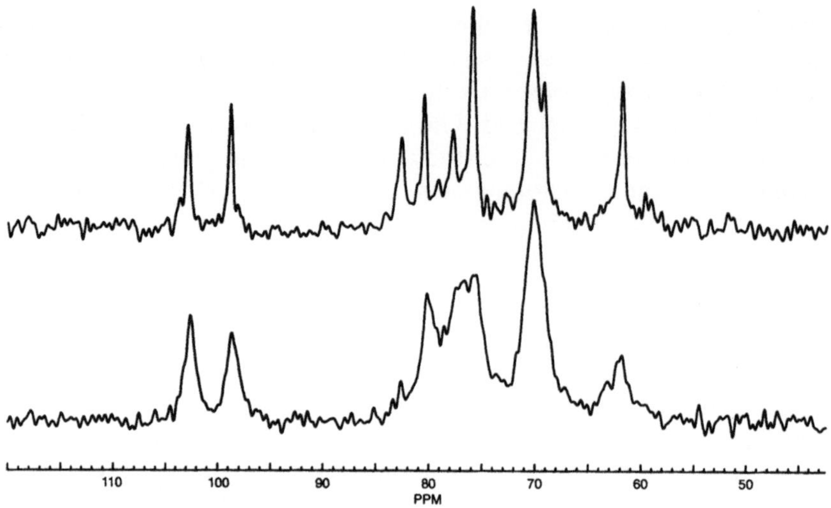

Fig. 13.6 ^{13}C spectra for 5 wt% agarose (Oxoid L11) in solution at 338 K (single pulse, 5 s recycle time) and in the gel state at 297 K (CP/DD/MAS, 1 ms contact time, 3 s recycle time). Spectral patterns differ in the region 75–85 ppm but are similar in other chemical shift regions.

differ significantly. On the basis of chemical shift similarity and the effect of hydration on dried gels, Saitô *et al.* (1990) suggested that a single helical structure was the most likely conformation, in contradiction with the received wisdom (Rees *et al.*, 1982) that a double-helical structure (Arnott *et al.*, 1974) underlies junction zone formation. The relatively small changes between the motionally averaged solution state spectrum and the conformational dispersion of shifts in the gel state spectrum cannot, however, be used as evidence for similar conformations, as the conformational origin of shifts is unknown (i.e. a lack of conformational sensitivity cannot be ruled out). An over-interpretation of the data in Fig. 13.6 might suggest that the small peak at 83 ppm in the gel represents residual single chain forms in the presence of a dominant amount of a different conformation. Evidence from optical rotation appears to be less ambiguous and strongly favours a double-helical structure in the gel state (Arndt and Stevens, 1994).

For 5% ι-carrageenan in 0.2% calcium chloride solution, significant intensity for the gel state is seen in both single pulse (with dipolar decoupling and magic angle spinning) and CP/DD/MAS spectra (Fig. 13.7). At 299 K, the SP/DD/MAS response is greater than the CP/DD/MAS response with the latter enhanced at lower temperature. Chemical shift patterns are different between gel and solutions states; for example, the peak at *ca.* 102 ppm in the sol is shifted to *ca.* 105 ppm in the gel (Fig. 13.7). Spectra for gels obtained by either single pulse or

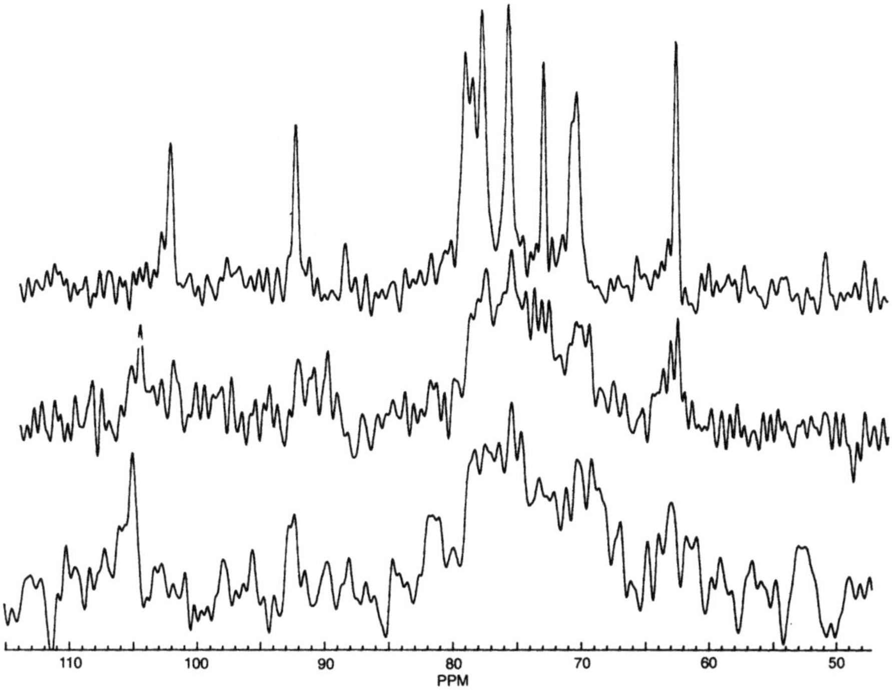

Fig. 13.7 ^{13}C spectra for 5 wt% ι-carrageenan (Sigma) in 50 mM calcium chloride. In solution at 338 K there is high signal intensity (90° pulse, 3 s recycle time, 4000 scans) which is much reduced in the gel state either at 298 K (90° pulse/DD/MAS, 3 s recycle time, 4000 scans) or at 278 K (CP/DD/MAS, 1 ms contact time, 3 s recycle time, 27 000 scans). Despite the difficulties in observing spectra, there is evidence for a predominant conformation in the gel state which differs from the solution average (e.g. 102.5 → 105.0 ppm).

cross polarization routes are similar and are indicative of a single conformational state; on previous evidence (Rees *et al.*, 1982) this would be the double helix (Millane *et al.*, 1988). ^{13}C spectra of solid ι-carrageenans have been reported previously (Rochas and Lahaye, 1989; Saitô *et al.*, 1990), but we are not aware of previously published gel state spectra. Chemical shifts for the gel (Fig. 13.7) and previously characterized solids are very similar indicating that the ordered gel state (presumed double helix) was dominant in the materials examined by Rochas and Lahaye (1989) and Saitô *et al.* (1990). ^{13}C NMR therefore indicates for this system that segmental mobility of the gel state conformation is significantly greater than that for κ-carrageenan or agarose. In further support of this deduction, ^1H T_2 measurements on ι-carrageenan gels indicate an absence of significant 'rigid lattice' relaxation (10 μs) with the dominant T_2 value being *ca.* 40 μs. In Fig. 13.5, this places the average correlation time at *ca.* 10 μs: presumably correlation times around this average extend into both the CP-observable and SP-observable regimes. The structural origin for the relatively mobile segmental behaviour of

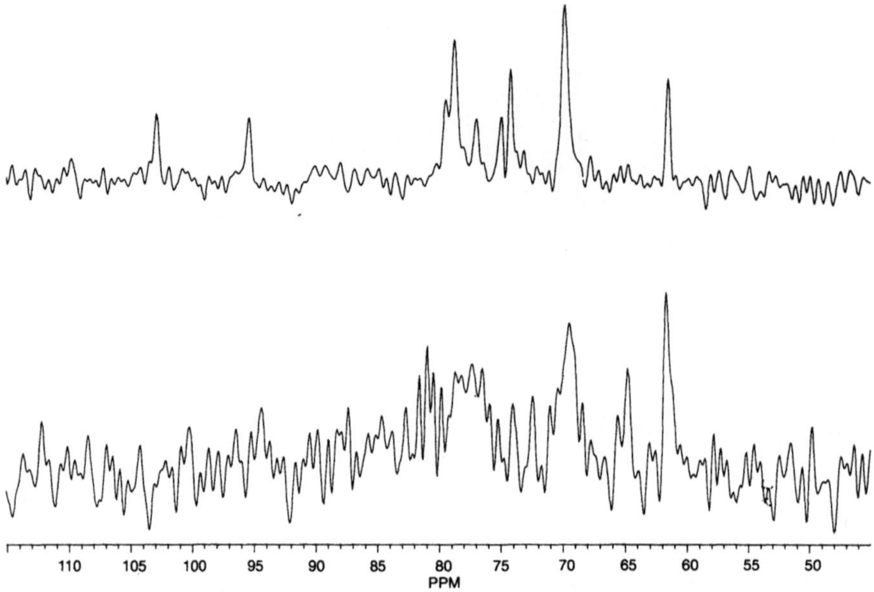

Fig. 13.8 Comparison of signal observability for 2 wt% solutions of κ- and λ-carrageenans (Sigma) at 338 K under identical acquisition conditions (90° pulse/DD/MAS, 3 s recycle time, 10 000 scans).

ι-carrageenan gels probably lies in the thinner network strands present in the gel compared with both κ-carrageenan and agarose (Brigham *et al.*, 1994). A solution of 5% ι-carrageenan in 0.2% $CaCl_2$ represents a clearly 'solid' system which is partially observable by both 'solid state' and 'solution state' NMR spectroscopy.

It is known that λ-carrageenan is a non-gelling viscous polysaccharide for which high resolution NMR spectra are notoriously difficult to observe (Rochas and Lahaye, 1989; Stortz *et al.*, 1994). We have confirmed this (Fig. 13.8) by being able to observe only the weakest of signals at 338 K in a SP/DD/MAS spectrum compared with the same concentration of κ-carrageenan. On lowering the sample temperatures to 278 K, comparably weak intensity is observed in a CP/DD/MAS spectrum (Fig 13.9). However, on freezing the system and re-recording the CP/DD/MAS spectrum, greater signal is observed after 3600 scans than after 10 000 scans for either of the previous two spectra (Fig. 13.9). This indicates that there is ^{13}C material in the sample: chemical shifts (e.g., *ca.* 103 and *ca.* 92 ppm) are in line with previous studies of λ-carrageenan (Rochas and Lahaye, 1989; Stortz *et al.*, 1994). The ^1H T_2 measurements at 338 K and 298 K suggested average values of 7 and 3 ms, respectively, which according to Fig. 13.5 would correspond to a correlation time of *ca.* 100 ns. Figure 13.9 suggests that, within this average, there may be

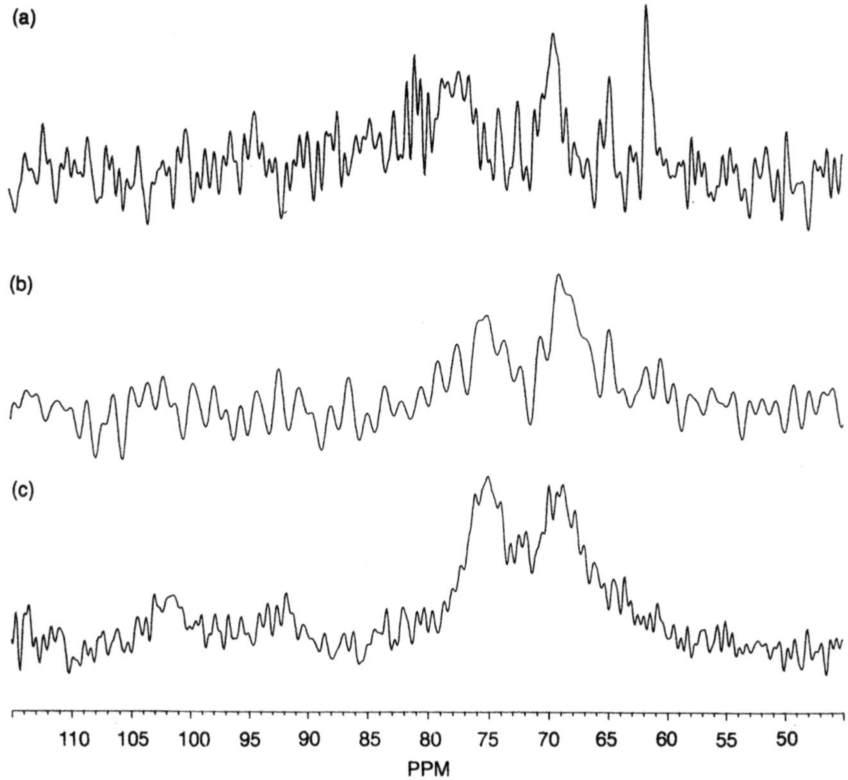

Fig. 13.9 ^{13}C spectra for 2% λ-carrageenan (Sigma) solutions at (a) 338 K under SP/DD/MAS conditions with 3 s recycle time and 10 000 scans, (b) 278 K under CP/DD/MAS conditions with 1 ms contact time, 3 s recycle time and 10 000 scans and (c) 253 K under CP/DD/MAS conditions as for (b) but only 3600 scans. The observation of significant signal in a frozen (long correlation time) 'solid state' spectrum contrasts with the limited signal intensity found for the non-frozen solution, either at low temperatures by 'solid state' or at high temperatures by 'solution state' spectroscopy.

effective correlation times both less than 10–100 ns (SP observable) and greater than 10 μs (CP observable), but that a significant fraction of material may not be observable by either SP or CP techniques for this sample. This would support the notion of a 'black hole' of difficult spectral observability. The system containing 2% λ-carrageenan (ex. Sigma) represents a solution which is only partially observable by either 'solid state' or 'solution state' NMR spectroscopy.

13.5 Summary and future challenges

High resolution (^{13}C) NMR of solid and semi-solid food components provides important information on the chemical and physical environment of molecules. In terms of chemical information, it is most useful as a probe of molecular order, conformations and interactions in relatively simple systems through observable chemical shift effects. Better understanding of the relationships between local conformations and chemical shifts is a future challenge which should increase its applicability. It is not the method of choice for molecularly complex mixtures, nor for investigating minor components. Mobility-resolved spectroscopy provides molecular descriptions of components in motionally heterogeneous states. For polymer-based systems, segmental mobility/rigidity determines, in a reasonably predictive way, observability by either 'solution state' (single pulse) or 'solid state' (cross polarization) spectroscopy. A semi-empirical link to effective correlation times suggests that some segmentally rigid solutions and some segmentally mobile solids may be difficult to observe by high resolution ^{13}C NMR at ambient temperatures, being intermediate between 'solution state' and 'solid state' conditions. A challenge for the future is to describe the molecular/segmental motion responsible for different signal observabilities in mobility resolved spectroscopy in terms of rheological properties.

References

Ablett, S. (1992) Overview of NMR applications in food science. *Trends Food Sci. Technol.*, **3**, 246–250.

Ablett, S., Clark, A.H. and Rees, D.A. (1982) Assessment of the flexibilities of carbohydrate polymers by ^{1}H NMR relaxation and line shape analysis. *Macromolecules*, **15**, 597–602.

Arndt, E.R. and Stevens, E.S. (1994) A conformational study of agarose by vacuum ultraviolet circular dichroism. *Biopolymers*, **34**, 1527–1534.

Arnott, S., Fulmer, A., Scott, W.E. *et al.* (1974) The agarose double helix and its function in agarose gel structure. *J. Mol. Biol.*, **90**, 269–284.

Berendsen, H.J.C. (1992) Rationale for using NMR to study water relations in foods and biological tissues. *Trends Food Sci. Technol.*, **3**, 202–205.

Bloembergen, N., Purcell, E.M. and Pound, R.V. (1948) Relaxation effects in nuclear magnetic resonance absorption. *Phys. Rev.*, **73**, 679–712.

Bociek, S.M., Ablett, S. and Norton, I.T. (1985) A ^{13}C NMR study of the crystal polymorphism and internal mobilities of the triglycerides, tripalmitin and tristearin. *J. Am. Oil Chem. Soc.*, **62**, 1261–1266.

Brigham, J.E., Gidley, M.J., Hoffmann, R.A. and Smith, C.G. (1994) Microscopic imaging of network strands in agar, carrageenan, locust bean gum and kappa carrageenan/locust bean gum gels. *Food Hydrocolloids*, **8**, 331–344.

Cooke, D. and Gidley, M.J. (1992) Loss of crystalline and molecular order during starch gelatinization: origin of the enthalpic transition. *Carbohydr. Res.*, **227**, 103–112.

Eads, T.M. and Croasman, W.R. (1988) NMR applications to fats and oils. *J. Am. Oil Chem. Soc.*, **65**, 78–83.

Forbes, J., Husted, C. and Oldfield, E. (1988) High field, high resolution proton magic angle sample-spinning NMR spectroscopic studies of gel and liquid crystalline lipid bilayers and the effects of cholesterol. *J. Am. Chem. Soc.*, **110**, 1059–1065.

Foster, T.J., Ablett, S., McCann, M.C. and Gidley, M.J. (1995) Mobility-resolved NMR spectroscopy of primary plant cell wall material. Submitted for publication.

Gidley, M.J. (1989) Molecular mechanisms underlying amylose aggregation and gelation. *Macromolecules*, **22**, 351–358.

Gidley, M.J. (1992) High-resolution solid-state NMR of food materials. *Trends Food Sci. Technol.*, **3**, 231–236.

Gidley, M.J. and Bociek, S.M. (1985) Molecular organisation in starches: a ^{13}C CP/MAS NMR study. *J. Am. Chem. Soc.*, **107**, 7040–7044.

Gidley, M.J. and Bociek, S.M. (1988a) ^{13}C CPMAS NMR studies of frozen solutions of (1 → 4)-α-D-glucans as a probe of the range of conformations of glycosidic linkages: the conformations of cyclomaltohexaose and amylopectin in aqueous solution, *Carbohydr. Res.*, **183**, 126–130.

Gidley, M.J. and Bociek, S.M. (1988b) ^{13}C CPMAS NMR studies of amylose inclusion complexes, cyclodextrins, and the amorphous phase of starch granules: relationships between glycosidic linkage confromation and solid state ^{13}C chemical shifts, *J. Am. Chem. Soc.*, **110**, 3820–3829.

Gidley, M.J. and Robinson, G. (1990) Techniques for studying interactions between polysaccharides, in *Methods in Plant Biochemistry* (ed. P.M. Dey), Academic Press, London, Vol. 2, pp. 607–642.

Gidley, M.J., McArthur, A.J. and Underwood, D.R. (1991) ^{13}C NMR characterisation of molecular structures in powders, hydrates and gels of galactomannans and glucomannans. *Food Hydrocolloids*, **5**, 129–140.

Gribnau, M.C.M. (1992) Determination of solid/liquid ratios of fats and oils by low-resolution pulsed NMR. *Trends Food Sci. Technol.*, **3**, 186–190.

Harris, R.K. (1983) *Nuclear Magnetic Resonance Spectroscopy: A Physicochemical View*, Pitman, London.

Hemminga, M.A. (1992) Introduction to NMR. *Trends Food Sci. Technol.*, **3**, 179–186.

Jackson, C.L. and Bryant, R.G. (1989) Carbon-13 NMR of glycogen:hydration response studied by using solids methods. *Biochemistry*, **28**, 5024–5028.

Jarvis, M.C. (1990) Solid state ^{13}C NMR spectra of *Vigna* primary cell walls and their polysaccharide components. *Carbohydr. Res.*, **201**, 327–333.

Jarvis, M.C. (1994) Relationship of chemical shift to glycosidic conformation in the solid-state ^{13}C NMR spectra of (1 → 4) linked glucose polymers and oligomers: anomeric and related effects. *Carbohydr. Res.*, **259**, 311–318.

Jarvis, M.C. and Apperley, D.C. (1990) Direct observation of cell wall structure in living plant tissues by solid state ^{13}C NMR spectroscopy. *Plant Physiol.*, **92**, 61–65.

McCann, M.C., Roberts, K., Wilson, R.H., Gidley, M.J., Gibeaut, D.M., Kim, J.-B. and Carpita, N.C. (1995) *Can. J. Bot.*, *in press.*

McCarthy, M.J. and McCarthy, K.L. (1994) Quantifying transport phenomena in food processing with NMR imaging, *J. Sci. Food Agric.*, **65**, 257–270.

Marchessault, R.H., Taylor, M.G., Fyfe, C.A. and Veregin, R.P. (1985) Solid-state ^{13}C-CPMAS NMR of starches, *Carbohydr. Res.*, **144**, C1–C5.

Millane, R.P., Chandrasekaran, R., Arnott, S. and Dea, I.C.M. (1988) The molecular structure of kappa-carrageenan and comparison with iota-carrageenan, *Carbohydr. Res.*, **182**, 1–17.

Morgan, C.F., Schleich, T., Caines, G.H. and Farnsworth, P.N. (1989) Elucidation of intermediate (mobile) and slow (solid-like) protein motions in bovine lens homogenates by ^{13}C NMR spectroscopy. *Biochemistry*, **28**, 5065–5074.

Morgan, K.R., Hemmingson, J.A., Furneaux, R.H. and Stanley, R.A. (1994) A ^{13}C solid state NMR study of ion-exchange resins derived from natural polysaccharides. *Carbohydr. Res.*, **262**, 185–194.

Morrison, W.R. and Karkalas, J. (1990) Starch, in *Methods in Plant Biochemistry 2* (ed. P.M. Dey), Academic Press, London, pp. 323–352.

Morrison, W.R., Tester, R.F., Snape, C.E., Law, R. and Gidley, M.J. (1993) Swelling and gelatinisation of cereal starches IV. Some effects of lipid-complexed amylose and free amylose in waxy and normal barley starches. *Cereal Chem.*, **70**, 385–391.

Newman, R.H., Ha, M.-A. and Melton, L.D. (1994) Solid-state ^{13}C NMR investigation of molecular ordering in the cellulose of apple cell walls. *J. Agric. Food Chem.*, **42**, 1402–1406.

Norton, I.T., Lee-Tuffnell, C.D., Ablett, S. and Bociek, S.M. (1985) A calorimetric, NMR and X-ray diffraction study of the melting behaviour of tripalmitin and tristearin and their mixing behaviour with triolein, *J. Am. Oil. Chem. Soc.*, **62**, 1237–1244.

Odahara, T., Akutsu, H. and Kyogoku, Y. (1990) Direct observation of the phase behaviour of the lipid bilayers of phage PM2 and the intact host cells by ^{1}H-^{31}P cross-polarisation NMR. *Biochemistry*, **29**, 5968–5974.

Oldfield, E., Bowers, J.L. and Forbes, J. (1987) High-resolution proton and carbon-13 NMR of membranes: why sonicate? *Biochemistry*, **26**, 6919–6923.

Rees, D.A., Morris, E.R., Thom, D. and Madden, J.K. (1982) Shapes and interactions of polysaccharide chains, in *The Polysaccharides* (ed. G.O. Aspinall), Academic Press, Orlando, Vol. 1, pp. 195–290.

Rochas, C. and Lahaye, M. (1989) Solid state ^{13}C-NMR spectroscopy of red seaweeds, agars and carrageenans. *Carbohydr. Polym.*, **10**, 189–204.

Rutar, V. (1989) Magic-angle sample spinning NMR spectroscopy of liquids as a non-destructive method for studies of plant seeds. *J. Agric. Food Chem.*, **37**, 67–70.

Saitô, H. (1986) Conformation-dependent ^{13}C chemical shifts: a new means of conformational characterisation as obtained by high resolution solid state ^{13}C NMR. *Magn. Reson. Chem.*, **24**, 835–852.

Saitô, H., Yokoi, M. and Yamada, J. (1990) Hydration-dehydration induced conformational changes of agarose, and kappa and iota-carrageenans as studied by high-resolution solid-state ^{13}C NMR spectroscopy. *Carbohydr. Res.*, **199**, 1–10.

Sanders, J.K.M. and Hunter, B.K. (1987) *Modern NMR Spectroscopy*, University Press, Oxford.

Song, B.K., Winter, W.T. and Taravel, F.R. (1989) Crystallography of highly substituted galactomannans: fenugreek and lucerne gums. *Macromolecules*, **22**, 2641–2644.

Stortz, C.A., Bacon, B.E., Cherniak, R. and Cerezo, A.S. (1994) High-field NMR spectroscopy of cystocarpic and tetrasporic carrageenans from *Iridea undulosa*. *Carbohydr. Res.*, **261**, 317–326.

Sutherland, J.W.H., Egan, W., Schechter, A.N. and Torchia, D.A. (1979) Carbon-13-proton nuclear magnetic double-resonance study of deoxyhemoglobin S gelation. *Biochemistry*, **18**, 1797–1803.

Veregin, R.P., Fyfe, C.A., Marchessault, R.H. and Taylor, M.G. (1986) Characterization of the crystalline A and B starch polymorphs and investigation of starch crystallisation by high resolution ^{13}C CP/MAS NMR. *Macromolecules*, **19**, 1030–1034.

Veregin, R.P., Fyfe, C.A., Marchessault, R.H. and Taylor, M.G. (1987) Correlation of ^{13}C chemical shifts with torsional angles from high resolution, ^{13}C-CPMAS NMR studies of crystalline cyclomalto-oligosaccharide complexes, and their relation to the structures of the starch polymorphs. *Carbohydr. Res.*, **160**, 41–56.

Zang, L.-H., Laughlin, M.R., Rothman, D.L. and Shulman, R.G. (1990) ^{13}C NMR relaxation times of hepatic glycogen *in vitro* and *in vivo*. *Biochemistry*, **29**, 6815–6820.

Chapter 14

Magnetic Resonance Imaging in Food Science

B.P. Hills
Institute of Food Research,
Norwich, UK

14.1 Introduction

The last decade has seen a rapid increase in the application of magnetic resonance imaging (MRI) to food science, as the powerful imaging techniques developed for observing anatomical details in medical applications have been applied to foods. Indeed, the first images of fruit and vegetables demonstrated the same beautiful anatomical detail familiar in clinical imaging. Nevertheless, with most solid foods there is little advantage in non-invasive imaging of structure over invasive techniques such as light or electron microscopy since, unlike human bodies, foods can be physically sliced with impunity. Moreover, spatial resolutions obtainable with MRI can never compete with resolutions routinely available in light and electron microscopy. The true potential of food imaging lies not in static structure determination but in following, non-invasively, in real time, dynamic changes such as mass and heat transport as foods are processed, stored, packaged and distributed. This dynamic information is contained in real-time changes in maps of the NMR parameters causing image contrast. Depending on the imaging protocol, these are spatial maps of spin number density, relaxation times, magnetization transfer rates, chemical shifts, self-diffusion coefficients and velocity flow rates, each of which will be considered in subsequent sections.

The time-dependent maps of mass and heat transport obtained by imaging can be used to develop theoretical computer models provided the appropriate transport mechanism has been identified. The theoretical model is perhaps the most valuable end-product of the whole imaging process because it enables food processing conditions to be optimized numerically without repetitive trial-and-error experimentation.

14.2 Principles of magnetic resonance imaging

14.2.1 Spatial imaging

The basic idea underlying MRI is easy to understand. The high resolution proton spectrum of a tube of water in a spatially uniform magnetic field of strength B_0 consists of a single spectral line situated at a frequency ν given by $\nu = \gamma B_0/2\pi$, where γ is the gyromagnetic ratio of the proton. If a linear magnetic field gradient G is now applied across the tube of water, each part of the tube will be associated with a different resonance frequency according to the formula, $\nu = \gamma[B_0 + Gx]/2\pi$. The water spectrum therefore gets 'smeared out' into an image (actually a one-dimensional projection) of the tube. The application of gradients in two orthogonal directions will generate a two-dimensional spectrum which is the NMR image of the tube. Usually a third gradient is applied along the axis of the tube so that 'slices' of the sample can be taken in that direction. Of course, numerous imaging protocols have been developed based on this concept, and these are well described in textbooks such as that of Callaghan (1991) or McCarthy (1994).

14.2.2 Preparation sequences

The usual imaging experiment uses a pulse sequence consisting of two parts. The first part, called the preparation pulse sequence, creates image contrast by weighting the total signal intensity by an NMR parameter such as the longitudinal relaxation time T_1 or the transverse relaxation time T_2. For T_1-weighted images, the preparation sequence consists of the standard inversion recovery or saturation recovery pulse sequences. For T_2-weighted contrast, the preparation sequence would typically be the usual CPMG sequence. Diffusion weighting can be introduced using a standard pulsed field gradient sequence such as that of Stejskal and Tanner (1965). Once the signal has been weighted by the preparation sequence, an image is created by the application of the external field gradients. Figure 14.1 shows an example of a simple one-dimensional imaging sequence and in Fig. 14.2 maps of different NMR parameters have been obtained from the same sample, in this case raw carrot, by using the appropriate preparation sequence. Such images serve to emphasize that the major challenge in food imaging lies in relating the parameter maps to physico-chemical characteristics of the food samples.

14.2.3 Digital image resolution

An image voxel (volume element) is defined by the slice width and the lengths in the read-out and phase-encoding directions. In clinical

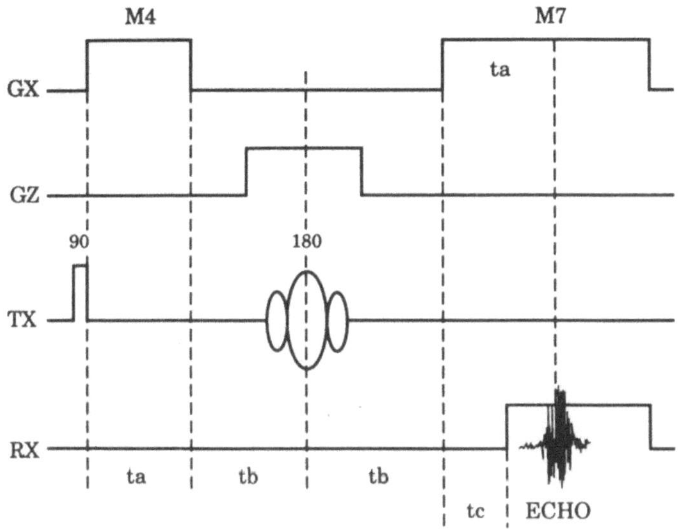

Fig. 14.1 A simple slice-selective, one-dimensional spin-echo imaging pulse sequence. TX and RX denote the transmitter output and receiver input, respectively. GX and GZ denote gradients in the x- and z-directions, respectively. The echo is Fourier transformed to obtain the image.

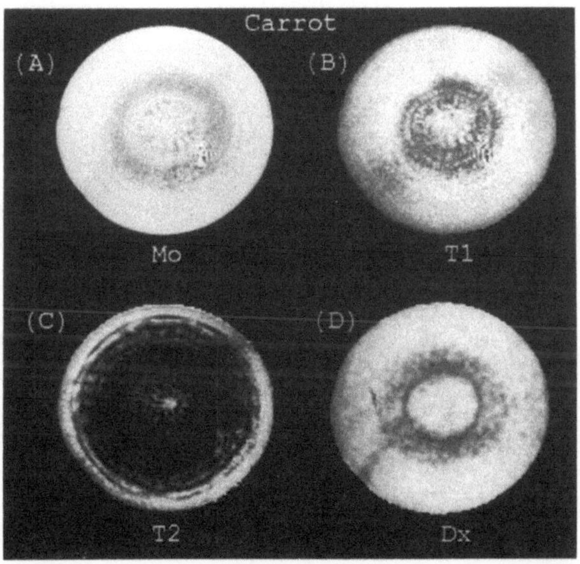

Fig. 14.2 Two-dimensional images of fresh carrot. (A) Spin density map. (B) T_1-map. (C) T_2-map. (D) Self-diffusion along the transverse direction, D_x-map. Images by courtesy of S.L. Duce and L.D.Hall of the Herchel-Smith Laboratory, Cambridge, U.K.

imaging, resolutions are quite low, with voxel dimensions of several millimetres, and this is also the case with large food samples such as whole biscuits. However, commercial imaging units are available for use with conventional high resolution NMR spectrometers using samples less than 2–3 cm in diameter. In this case voxel dimensions of a few 100 μm are not unusual, and the technique is then denoted 'NMR microimaging'. In specially designed high-field NMR microscopes, resolutions down to *ca.* 5 μm have been obtained, but such high resolutions are not routinely available.

14.3 Overview of applications

The applications of MRI in food science are discussed here according to the type of NMR parameter used to weight the image. These include (1) spin density maps, (2) relaxation time maps, (3) diffusion maps, (4) temperature mapping, (5) chemical shift imaging, and (6) velocity flow maps. These are briefly reviewed in turn below.

14.3.1 Spin density maps

The proton spin density is usually proportional to the amount of water or mobile lipid per unit volume. The time dependence of the spin density map can therefore be used to study the transport and redistribution of water or lipid during food processing or storage. Relevant processes include drying, rehydration, heating, frying, microwaving, curing, extrusion, mixing, drainage and syneresis. Figure 14.3 shows a series of moisture profiles during the drying of a bed of glass microspheres obtained by one dimensional projection imaging by the author in collaboration with the Herchel-Smith Laboratory, Cambridge (Hills *et al.*, 1994). Such profiles can be used to develop and test models for fluid transport. Similar drying profiles have been reported for potato (Ruan *et al.*, 1991) and apple (McCarthy *et al.*, 1991). MRI can also be used to study the mechanics of liquid drainage in foams such as egg white foam (McCarthy and Heil, 1990). The changing height of the foam column is related to bubble collapse and the spin density profile can be used to monitor the drainage pattern in real time. The example of the rehydration of extruded pasta is discussed in some detail in a later section.

14.3.2 Relaxation time maps

Water or lipid proton longitudinal and transverse relaxation times depend on such quantities as molecular mobility, solute concentration

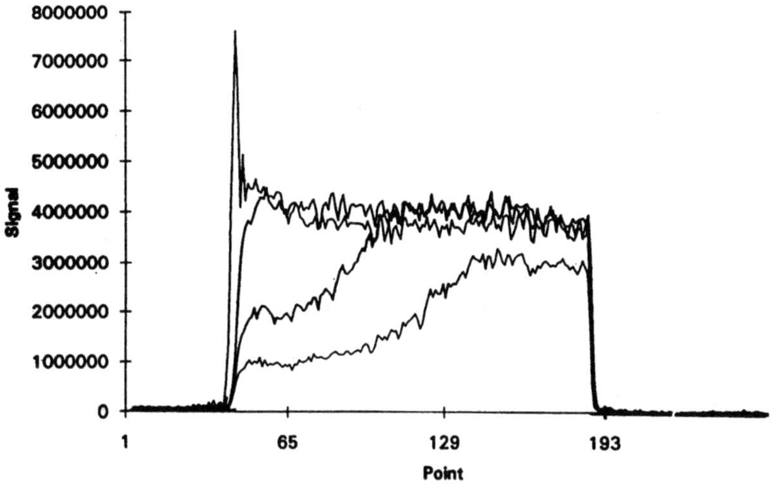

Fig. 14.3 Unprocessed image profiles during the drying of a 6.5 cm deep bed of randomly packed glass beads. Air of controlled humidity was blown over the top of the bed which corresponds to the left hand side of the profiles. The four profiles correspond to drying times of 0, 100, 300 and 500 minutes. The sharp peak in the initial profile arises from a surface meniscus. Taken from Hills *et al.* (1994).

and degrees of polymerization or cross-linking. Relaxation time maps can therefore be used to follow the kinetics of phase changes in food components, induced by crystallization of water and/or lipid, phase separation, gelation, macromolecular aggregation and denaturation. Examples include imaging the freezing of fruit and vegetables (McCarthy and Kauten, 1990), the syneresis of cheese curds (McCarthy and Kauten, 1990), and the determination of fat and lipid distributions in meat, fish and emulsions (Kauten *et al.*, 1991). The last application is based on the fact that oil and water usually have substantially different relaxation times; so by careful choice of time delay either the fat or water can be imaged. Relaxation times can also change if fruit is bruised or undergoes deterioration in storage. Thus imaging can be used to monitor bruising in apples, onions and pears, and watercore distribution in apples (Belton *et al.*, 1993). Low concentrations of paramagnetic ions dramatically shorten water proton relaxation times, and so the transport of these ions in, for example, plant tissue can be monitored (Gassner, 1989).

Ice has a very short proton transverse relaxation time of a few microseconds, which means that it is invisible in conventional imaging experiments on partly frozen foods. This permits imaging studies of the kinetics of ice formation in cryobiology since the ice appears as a black region surrounded by the high intensity unfrozen regions. Somewhat surprisingly, however, the frozen core of raw potatoes undergoing freeze-drying appears as a bright, high intensity region inside the potato. This apparent contradiction arises from the presence of unfrozen water associated with the cell walls and starch granules inside the frozen potato cells. It is therefore possible to follow freeze-drying kinetics with MRI whenever there are significant amounts of unfrozen water (Rutledge *et al.*, 1994).

When the inversion recovery pulse sequence is used as a preparation sequence to weight images by their longitudinal relaxation times, there is the additional possibility of using T_1-null imaging. Here the signal from one component can be completely eliminated by choosing the inversion recovery time to correspond to the point at which its longitudinal magnetization has recovered from its initial negative value to zero. Other components with different relaxation times will still generate a signal. This type of imaging has found wide use in clinical applications, but, as yet there are few examples in the food literature.

14.3.3 Diffusion-weighted imaging

Maps of the effective water self-diffusion coefficient can be obtained using a standard pulsed gradient spin-echo (PGSE) sequence such as that of Stejskal and Tanner, either as a preparation sequence or by inserting gradient pulses into the spin-echo imaging sequence, which is the protocol used in 'dynamic NMR microscopy'. The principles underlying the PGSE method have been discussed in a number of texts (Callaghan, 1991) and reviews (Belton *et al.*, 1993). The observed echo amplitude $S(q, \Delta, \tau)$ in the conventional (i.e. non-imaging) two-pulse Stejskal-Tanner sequence is a function of the three independent variables: q, the wave vector corresponding to the pulsed gradient area; Δ, the diffusion time (pulsed gradient separation); and τ, the 90–180° pulse spacing. The functional form of $S(q, \Delta, \tau)$ for water in a food material depends in a complicated way on both the microstructure of the food and on the microdynamics of the water transport. For present purposes it is sufficient to note that the sequence can be used to define 'effective' water self-diffusion coefficients by assuming that relaxation and diffusion are simply factorizable and that the expression for diffusive attenuation for unbounded diffusion can be applied, with an effective diffusion coefficient, in situations where diffusion may well be restricted or hindered. An example is shown in Fig. 14.2(D) where the apparent transverse water diffusivity D_x of raw carrot is imaged.

The maximum diffusion time attainable in PGSE experiments is limited by the water proton relaxation time to about 1–2 s. More usually, relaxation and dephasing in susceptibility gradients limits practical diffusion times to only a few tens of milliseconds. Even in a time of two seconds, water molecules undergoing unrestricted diffusion can diffuse an average distance of *ca.* 100 μm, so that the PGSE method measures effective diffusivity on a microscopic distance scale. The resulting effective PGSE diffusion coefficients are not therefore necessarily the same as the macroscopic diffusion coefficient measured over distances of millimetres or more, especially if there is heterogeneity over distance scales exceeding 100 μm. Fortunately, truly macroscopic water diffusion coefficients can be conveniently measured by imaging the diffusive exchange of D_2O and H_2O between two layers consisting of the sample containing D_2O and H_2O, respectively. In proton imaging the deuterated water gives no signal; so the interdiffusion is readily imaged. This method has been applied to a variety of heterogeneous model food materials including gels and cellular tissue (Hills *et al.*, 1993b). Comparison of the PGSE and the macroscopic water self-diffusion coefficient can give useful information about the system microstructure (Hills *et al.*, 1993a).

14.3.4 Temperature mapping and heat transport

Besides mass transport and phase changes, MRI can, in principle, be used to map temperature changes in food materials undergoing processes such as freezing, drying, sterilization and pasteurization. The time dependence of the temperature maps can then be modelled in terms of the intrinsic thermal properties of foods such as thermal conductivities, specific heats and latent heats. Thermal boundary conditions at the food surface can also be monitored in an analogous way to the surface moisture content.

In MRI, temperature maps are measured indirectly from the temperature dependence of other NMR parameters, usually the longitudinal relaxation time or the effective water self-diffusion coefficient discussed above. The water proton longitudinal relaxation time was found to vary roughly linearly with temperature in human tissue (Dickinson *et al.*, 1986) and in model agarose gels (Howe, 1988) which permits temperature maps to be measured. On the other hand, since, according to the Stokes–Einstein relation, the water self-diffusion coefficient is inversely proportional to the shear viscosity, an exponential temperature dependence is expected for the water self-diffusion coefficient, i.e. $D \propto \exp(-E_a/kT)$, where E_a is the activation energy and kT is the thermal energy. This is observed in MRI studies of a polyacrylamide gel and in a model food gel (Sun *et al.*, 1993) for which temperature maps have been reported. In these reports MRI temperature studies have been made in the absence of mass transport. When there is simultaneous mass and heat transport, it is necessary carefully to calibrate the system since relaxation times and the effective diffusion coefficient depend on both temperature and water content.

14.3.5 Chemical shift imaging

The principles of chemical shift imaging have been described in a number of reviews and books (Callaghan, 1991). In clinical practice chemical shift imaging can be used to resolve fat and water signals. Figure 14.4 shows chemical shift resolved images of the water and fat components in a cocoa bean taken in the Herchel-Smith Laboratory in Cambridge (Wright and Hall, 1994). Similar protocols have been used to follow the ingress of oil into water-rich foods during frying. With heterogeneous food materials, problems can arise from the broadening due to a wide distribution of relaxation times and strong internal susceptibility gradients. Methods have been proposed to alleviate these problems and have been used successfully to image oil and water distributions in cheeses. The Dixon chemical shift resolved imaging sequence has been used to map separately the spatial distribution of the liquid-like lipids and water in milk and the separation of cream from milk over a 9-h period (Duce *et al.*, 1994).

14.3.6 Flow imaging

Flow imaging has immediate relevance to problems in the food manufacturing industry where flow characteristics of complex liquid food suspen-

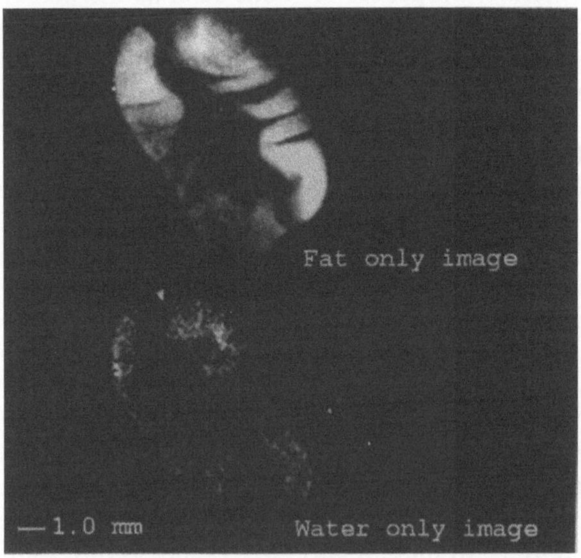

Fig. 14.4 Chemical shift resolved image of a cocoa bean. T_R = 1500 ms, TE = 12 ms, slice thickness = 1 mm, 256 × 256 data matrix. Courtesy of J.J. Wright and L.D. Hall of the Herchel-Smith Laboratory, Cambridge, U.K.

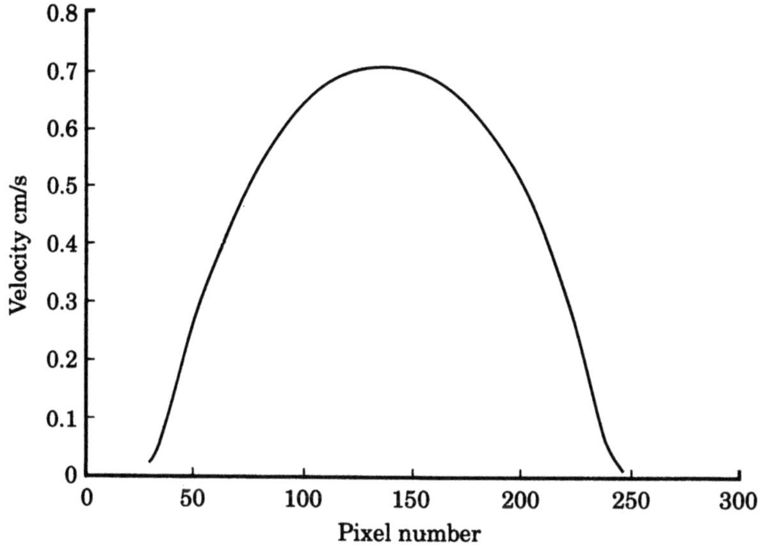

Fig. 14.5 MRI velocity profile for laminar flow of a 1.5 wt% guar gum solution flowing at 8 ml/min. Courtesy of S. Gibbs and L.D. Hall of the Herchel-Smith Laboratory, Cambridge, U.K.

sions and slurries need to be understood and monitored. As many of these complex food liquids are opaque, other flow monitoring methods are difficult or impossible to apply. Moreover, MRI is non-invasive, which means that there is no problem of disturbing the flow by inserting flow sensors. Flow imaging is based on the phase shifts induced in a PGSE experiment by coherent flow of the magnetization. These phase shifts can be translated into velocity maps (Callaghan, 1991). The (non-Newtonian) rheology of the complex liquids, suspensions and pastes used in food manufacture can be systematically studied by imaging laminar flow down a cylindrical pipe. Figure 14.5 shows an example of this type of experiment where the velocity profile across the tube has been plotted for a 1.5 wt% guar gum solution flowing at 8 ml/min. Similar protocols can be used to test models for liquid flow in complex geometries such as extruders (McCarthy *et al.*, 1992). In plant physiology flow imaging allows vascular flow in the plant at various stages of development to be monitored *in vivo*. The potential of diffusion and velocity-weighted microimaging has been illustrated beautifully by Jenner *et al.* (1988), who studied the circulation of water within a wheat grain. The transport of water in food materials can be determined, in principle, by imaging the transport of D_2O, either using the loss of the water proton signal as H_2O is replaced by D_2O, or directly using deuterium imaging. Both methods have been used to study water flow in plant materials.

Many of the applications referred to above have been in the nature of qualitative feasibility studies. We now turn to some of the challenges which arise when attempts are made to use imaging in a quantitative way to follow mass and heat transport.

14.4 Quantitative imaging protocols

Although qualitatively impressive images of food materials, weighted with a preparation sequence, are relatively straightforward to obtain, a number of difficulties arise when attempts are made to interpret these images and extract quantitative information about mass or heat transport. One problem is shrinkage/expansion of the food during a process such as drying. This volume change destroys the simple linear relationship expected between the proton spin density (signal per unit volume) and the gravimetric moisture content (w/w), which is usually the variable of interest in food manufacture. Another difficulty is the existence of large internal field gradients generated by susceptibility discontinuities across microscopic phase boundaries in the food. This is especially observed during drying operations when air replaces water in the food matrix. These air–water interfaces distort the image and cause an unpredictable loss of signal intensity. Finally there is the usual loss of signal-to-noise at low moisture contents (unless mobile lipids are present). This sets a lower limit on the range of moisture contents that can be reliably imaged.

In the following sections, the problems associated with sample shrinkage and susceptibility discontinuities are first illustrated for a well-characterized model sample consisting of Sephadex microspheres containing various amounts of water. A fast T_2-weighted radial imaging protocol is then proposed for overcoming these problems, and the technique is used to follow the ingress of water during the rehydration of extruded pasta.

14.4.1 Single-echo imaging of foods with varying water content

The image intensity I observed with the spin-echo sequence in Fig. 14.1 without a preparation sequence can be written as

$$I = M_0[1 - \exp(-T_R/T_1)]\, R(t_a, t_b) \times \mathrm{DIFF}(t_a, t_b), \qquad (14.1)$$

where M_0 is the initial water proton magnetization and $R(t_a, t_b)$ and $\mathrm{DIFF}(t_a, t_b)$ are the attenuating effects of transverse relaxation and diffusion, respectively. The time delays t_a and t_b are identified in Fig. 14.1. The repetition time T_R is usually chosen to be at least five times the longitudinal relaxation time T_1, and so the factor in square brackets can be neglected. For a homogeneous liquid like pure water, equation (14.1) can be made more explicit using

$$R(t_a, t_b) = \exp[-2(t_a + t_b)/T_2^*] \tag{14.2}$$

and

$$\mathrm{DIFF}(t_a, t_b) = \exp[-\gamma^2 G^2 D t_a^2 (2(t_a + t_b) - 4t_a/3)]. \tag{14.3}$$

Equation (14.3) expresses signal attenuation due to diffusion between the two read-out gradient pulses of amplitude G. Examination of equations (14.2) and (14.3) shows that simple variation of t_b at fixed t_a will give an exponential decay but with a time constant that is neither T_2 nor T_2^*, since the DIFF term will cause diffusive attenuation from the read-out gradient pulses. This fact is often overlooked when T_2-weighted images are generated with conventional spin-warp imaging. In heterogeneous food materials where diffusion is restricted, in general, it is no longer valid to assume that relaxation and diffusion are simply factorizable. Nevertheless, for simplicity, this complication is ignored here, and it is assumed that equation (14.1) remains applicable at all water contents, with M_0, $R(t_a, t_b)$ and $\mathrm{DIFF}(t_a, t_b)$ dependent on water content W, defined as weight of water per unit mass of dry food sample. We first test the conventional assumption that the initial magnetization, M_0, provides a useful measure of water content by considering a model sample consisting of packed beds of Sephadex G25–300 with varying water content from the (almost) dry powder to the fully saturated suspension. This sample was chosen because Sephadex exhibits many of the features typical of real foods. For example, there is volume shrinkage at lower water contents, and the system is compartmentalized on a microscopic distance scale of one to several hundred micrometres. Like many food biopolymer systems, the intrinsic water proton relaxation mechanism is dominated by proton exchange for which quantitative theoretical models are available (Hills, 1992). At lower water content the presence of air gaps in the sample creates large local internal susceptibility gradients which can cause severe dephasing.

14.4.2 Initial magnetization for Sephadex beds with single-echo imaging

Figure 14.6 shows the log–linear dependence of the signal intensity I on the increment in t_b at fixed t_a and field gradient G for beds of Sephadex G25–300 at several water contents W. The intercepts in Fig. 14.6 can be equated with $M_0(W)$, and Fig. 14.7 shows a plot of the $M_0(W)$ against water content W. It is clear that $M_0(W)$ is not a simple linear function of water content, but shows some unusual departures from linearity, arising from changes in bead radius and packing density as the water content is changed. This dependence of bead radius on water content has important implications for the interpretation of M_0 maps in slice-selective

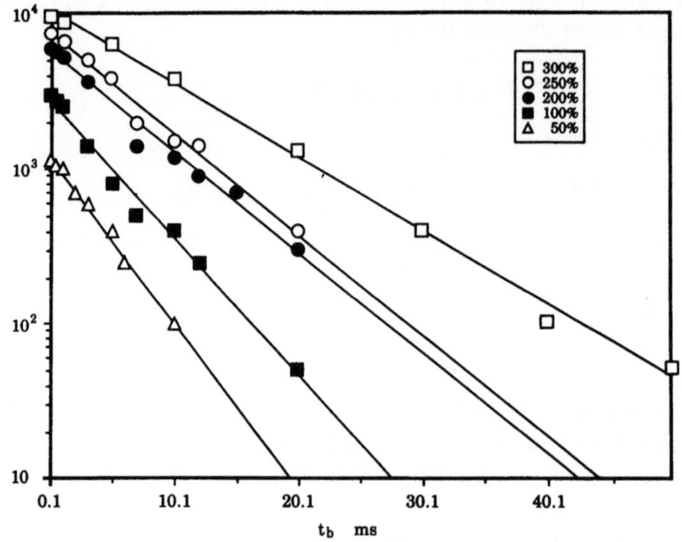

Fig. 14.6 The log–linear dependence of the average radial profile intensity (arbitrary units) obtained with the pulse sequence in Fig. 14.1 for packed beds of Sephadex G25–300 containing the indicated percent (w/w) water. Intensities are plotted against t_b at fixed t_a (2.1 ms) and gradient strength. (Taken from Hills and Babonneau (1994).)

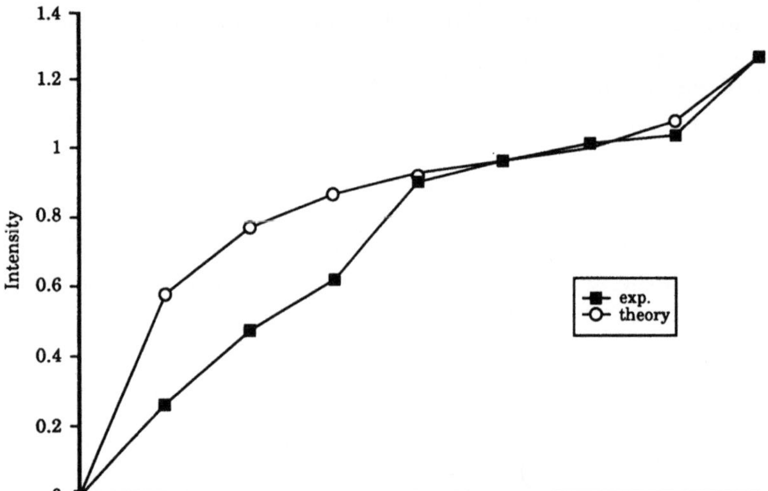

Fig. 14.7 The dependence of the initial water proton magnetization M_0 on water content W in packed beds of Sephadex G25–300. Intensities have been normalized to unity for $W = W_{in}$. The theoretical curve is calculated with equations (14.4) and (14.5). (Taken from Hills and Babonneau (1994).)

imaging because loss of water is partly compensated by the larger number of beads in the image sampling volume. For the model Sephadex system this 'compensation effect' can be calculated (Hills and Babonneau, 1994), and it can be shown that

$$M_0(W)/M_0(W_{in}) = (W/W_{in})[(1+nW_{in})/(1+nW)] \quad (0 \leq W \leq W_{in}) \quad (14.4)$$

and

$$M_0(W)/M_0(W_{sat}) = W/W_{in} \quad (W_{in} \leq W \leq W_{sat}). \quad (14.5)$$

Here W_{sat} is the water content of the saturated, packed suspension and W_{in} is the critical water content when the beads are fully hydrated but there is no water outside them. The dependence of $M_0(W)$ on water content for Sephadex G25–300 predicted by equations (14.4) and (14.5) for the experimental values of W_{in} and W_{sat} (Hills and Babonneau, 1994) is shown in Fig. 14.7. The theory accounts reasonably well for moderate water contents ($W > 2$ w/w), but overestimates M_0 at water contents below 2 w/w. A possible reason for this is seen in the light microscope, which shows that at lower water contents the beads not only shrink but tend to clump together in aggregates. If these aggregates were to pack together randomly as large 'pseudo-spheres', the packing density would be considerably lower than the regular close-packed array assumed when deriving equation (14.4). Because of these 'compensation' and 'aggregate packing' effects, the initial magnetization M_0 is clearly not a reliable measure of water content. This is a significant point, because similar effects can occur when using slice-selective pulse sequences with real food systems such as cellular tissue, pastas and gels, all of which undergo shrinkage and possibly structural reorganization as the water content is reduced.

14.4.3 Relaxation and diffusive terms for Sephadex beds with single-echo imaging

In Table 14.1 the experimental slopes derived from Fig. 14.6 for water contents W less than W_{in} are compared with those calculated with equations (14.1)–(14.3) for variation of t_b at fixed t_a and G using bulk measurements of $T_2(W)$ and $D(W)$. It can be seen that the experimental slopes are very roughly twice the theoretical values and that the difference

Table 14.1 Theoretical and experimental slopes for single-echo imaging of Sephadex G25–300 beds of varying water content W

Water content W	Theoretical slope without susceptibility gradients (s^{-1})	Experimental slope (s^{-1})
1.0	77.7	203.5
1.5	67.6	172.7
2.0	62.5	149.9
2.5	59.4	150.2
3.0	57.4	109.5

increases with lower water contents. This discrepancy is undoubtedly a consequence of large internal magnetic field gradients generated across the Sephadex–air interface, which are not included in the theoretical calculation. These gradients would be expected to increase as the beads shrink at lower water contents. Similar susceptibility effects are expected in most heterogeneous food materials.

14.4.4 The T_2-weighting imaging protocol

One method for overcoming the difficulties of determining gravimetric moisture images when there is sample shrinkage/expansion and internal susceptibility gradients is to measure T_2-weighted images using a CPMG preparation sequence and to relate the T_2 maps to gravimetric water profiles using a T_2 *versus* W calibration curve obtained from bulk, uniformly hydrated samples. The dephasing effects of inhomogeneous background gradients and internal susceptibility gradients can be minimized by keeping the echo time short (e.g. 200 μs). It is important to note that, because of proton exchange and susceptibility discontinuities, T_2 in most foods depends on the CPMG pulse spacing τ. The image intensity therefore needs to be measured as a function of increasing numbers of echoes in the CPMG preparation sequence, while keeping the pulse spacing fixed at a short value. T_1-weighting could also be considered, but T_1 is usually found to be less dependent on moisture content than T_2.

14.4.5 The radial imaging protocol

Most food samples can be cut or moulded into a cylindrical geometry. This is the most convenient geometry for MRI purposes since magnet bores, probeheads and NMR sample tubes are usually cylindrically symmetric. Provided the sample retains cylindrical symmetry, one-dimensional projections across and along the cylinder axis contain all the spatial information, and the phase encoding gradients in Fig. 14.1 can be removed. If the read-out gradient is applied across the axis of the tube, the one-dimensional Fourier transformation $F\{\ \}$ of the echo generates an image projection (a semi-ellipsoid for a uniform sample) rather than the radial profile which is of interest. In principle, Hankel transformation $H\{\ \}$ of the echo yields the radial profile directly, but it is considerably easier first to Fourier transform the echo and then to convert the projection into the radial profile with an inverse Abel transform, $IA\{\ \}$ (Hills and Babonneau, 1994). These relationships can be summarized as

$$\text{image}(r) = H\{\text{fid}(k)\} = IA\{\text{proj}(x)\}, \qquad (14.6)$$

where

$$\text{proj}(x) = F\{\text{fid}(k)\}. \qquad (14.7)$$

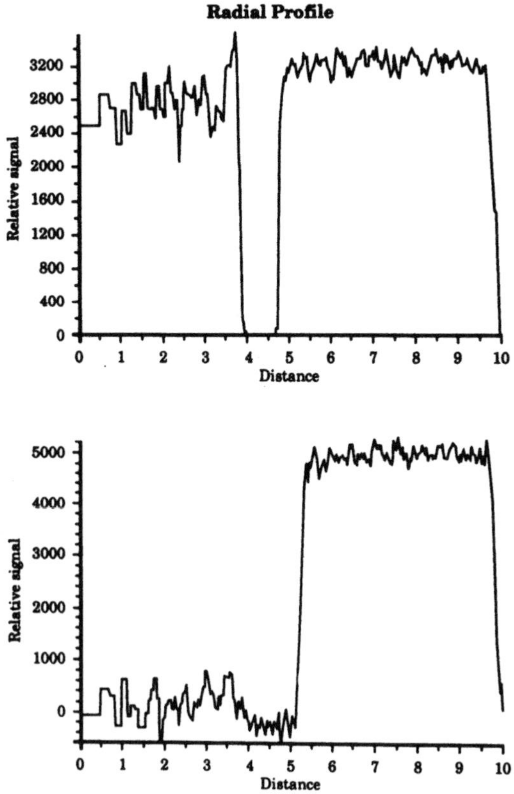

Fig. 14.8 Radial profiles of two concentric NMR tubes of 5 and 10 mm diameter. The inner tube contains a packed bed of Sephadex G25–300 either water saturated (top profile) or with 100 wt% water (bottom profile). The outer tube contains pure water. The distance scale is in arbitrary units from the centre. The central dip is the glass wall of the inner tube. (Taken from Hills and Babonneau (1994).)

Here fid(k) is the free induction decay or echo acquired in the field gradient, proj(x) is the image projected along the x axis, image(r) is the desired radial profile, and k is the wave vector γGt. This radial protocol increases the speed of image acquisition since no phase encoding gradients are needed.

Figure 14.8 shows the experimental radial profiles derived from two concentric tubes, the outer one containing water as an intensity standard, and the inner one Sephadex G25–300 at two water contents.

14.5 Application of T_2-weighted radial imaging protocol to rehydration of extruded pasta

In this section T_2-weighted radial imaging will be used to investigate the rehydration kinetics of three types of extruded pasta (spaghetti) differing in the percentage of hard and soft wheat used in their extrusion. This specific application serves to illustrate some of the difficulties which arise in quantitative imaging.

14.5.1 Sample arrangement during processing

The first practical difficulty is marrying the sample and the food processing operation to the design of the imaging probehead. In these experiments the probehead consists of a vertical 2 cm wide coil. Clearly, the rehydration cannot be done inside this probe since the cooking water will swamp the image. Moreover, pasta expands during rehydration, and so it is important that the sample can change its length inside the probehead without bending or breaking. This can be achieved by attaching the bottom of the pasta to a small cylindrical plastic cap with a diameter slightly smaller than the vertical RF coil. The sample and cap are then suspended down the centre of the coil. The lower half of the sample (including the cap) is first immersed in water at 83°C for a known time period outside the probe, then removed, cooled in cold water to stop further rehydration, and transferred to the imaging probehead.

14.5.2 Radial and length expansion rates

Figure 14.9 shows the time dependence of the radius and length of the 50:50% hard:soft wheat sample after immersion in water at 83°C for the specified periods. These measurements were made with a travelling light microscope, although MRI could also have been used. It can be seen that the length expansion is delayed compared to radial expansion because of the existence of a hard inner glassy core of unhydrated pasta, which is only rehydrated in the final stages of cooking. This difference implies the existence of stresses inside the pasta during rehydration, and it suggests that the kinetics may well depend on the sample aspect ratio. In many food samples such stresses cause cracking and shape distortion during the processing operation and considerably complicate the theoretical modelling of the images. Stress cracks appear as dark voids in the image, so that, in principle, MRI can be used to monitor the internal structure of the stress cracks non-invasively. Fortunately, no stress-induced cracking or distortion of the sample was found in the cylindrically extruded pasta samples, although it has been observed when rehydrating pastas of other morphologies. The increase in sample length during rehydration means that the total integrated signal intensity in the absence of dephasing by relaxation and diffusion in the slice-selective imaging volume is no longer proportional to the water content (by weight). Moreover, the changes in signal intensity at any fixed radial distance are caused both by water penetration and by radial expansion which must be taken into account in any theoretical interpretation of the imaging results.

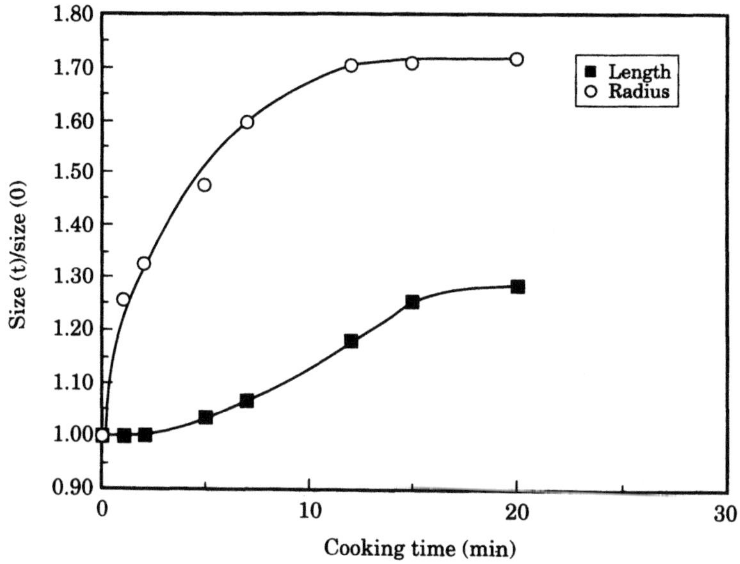

Fig. 14.9 The dependence of the radius and length of 50:50% hard:soft wheat pasta on rehydration time, measured by light microscopy.

14.5.3 Non-spatially resolved transverse relaxation

Figure 14.10 shows a typical relaxation time radial profile for the rehydration of the pasta. To convert this to a radial moisture profile, it is necessary to measure the dependence of relaxation time on moisture

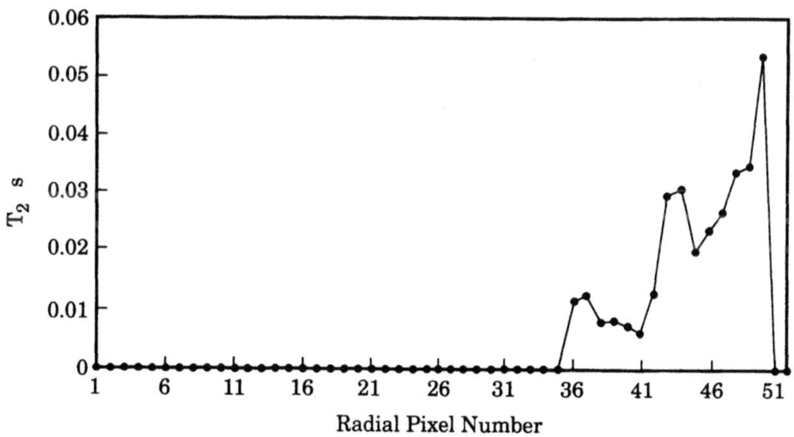

Fig. 14.10 The radial profile of the water proton transverse relaxation time T_2 observed for the 100% hard wheat sample after 2 minutes rehydration. Pixel number 1 corresponds to the centre of the pasta; pixel number 51 to the pasta surface.

content in bulk, uniformly hydrated samples using the same fixed CPMG pulse spacing. For the pasta samples, the transverse relaxation rate of the main water peak measured with a 200 μs pulse spacing was found to have a log–linear relationship with solid content, $(1 - W)$, such that $\log(1/T_2) = \alpha + \beta(1 - W)$, and this relationship was used to convert the T_2 profiles to radial moisture profiles. It is important to note that there is always a lower limit to the moisture content and that this can be determined by MRI. For the pasta the minimum echo time in the radial imaging sequence is *ca.* 8 ms, which sets a lower limit of *ca.* 2–3 ms on the shortest T_2 value that can be detected in the imaging experiment, which corresponds to a minimum water content of *ca.* 10–15% by weight.

14.5.4 Radial moisture profiles

Figure 14.11 shows an example of the moisture profile obtained using the T_2 versus moisture content calibration data and the radial expansion data in Fig. 14.9. Note that, because of the lower limit of *ca.* 8 ms in the T_2 measurements, points having $S = 0$ are actually points with $S < 0.1$. The radial profiles all show considerable 'noise'. In part it is caused by the 'graininess' of the pasta samples which can be seen in the dry pasta samples and probably arises from wheat fibres and grains that have not been evenly gelatinized by the extrusion process. The other factor is the poor signal-to-noise at low water contents, which is exacerbated by the use of the inverse Abel transformation, especially at small radial distances.

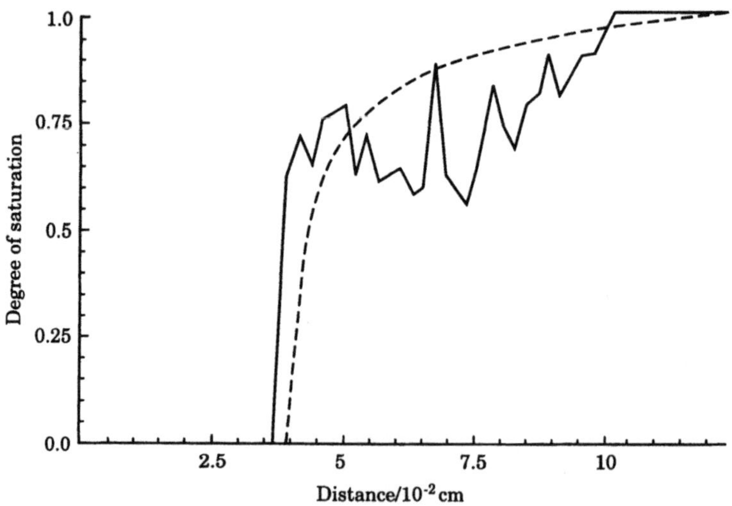

Fig. 14.11 The radial moisture profile for the hard wheat pasta after 7 minutes rehydration. The theoretical profile (dashed curve) is calculated numerically using the parameters in Table 14.2 for hard wheat.

14.6 Theoretical modelling of moisture profiles

The quantitative interpretation of MRI images represents one of the remaining outstanding challenges for the next decade. Although it is often assumed that moisture transport in foods is a diffusion process, this is not always the case. If the grain size in the food lies in the range 1–500 μm, then capillary forces will dominate the liquid phase transport. The moisture profiles in such grainy materials can then show unusual transient irregularities (Hills *et al.*, 1994). In the case of the extruded pasta, the gelatinized starch matrix is sufficiently fine-grained that the moisture profiles can be modelled using numerical, finite-element solutions of the diffusion equation in a cylindrical geometry, provided the radial expansion of the sample and dependence of the diffusion coefficient on local water content are taken into account.

14.6.1 Modelling the radial expansion

A simple model for radial expansion can be introduced by course-graining the calculated moisture profile into groups of 10 mesh points in the simulation program. Explicitly, this means that if the radial profile is characterized by a total of 1000 mesh points in the simulation then this profile is course-grained into 100 units of 10 mesh points. If the average degree of saturation $S(= W/W_{sat})$ of a 10-mesh point group exceeds some critical value S_{crit}, then the number of mesh points in the group is increased by an integral number calculated from the experimental increase in radius (R_{max}/R_{min}), and all mesh points in the group are assigned the average degree of saturation. Once a mesh point group has been expanded it cannot expand again. This ensures that the final, fully hydrated pasta has a radius equal to the measured R_{max} after full rehydration. After an expansion step the new moisture profile is used as input for the diffusional part of the calculation. To ensure consistency the number of expansion steps is increased until there is no significant change in the final moisture profile or radius.

14.6.2 Modelling non-Fickian diffusion

The effect of varying moisture content on the local self-diffusion coefficient can be incorporated by writing

$$D[S(r, t)] = D_0 \exp[A\, S(r, t)], \tag{14.8}$$

where $S(r, t)$ $(= W(r, t)/W_{sat})$ is the local degree of saturation at time t and radial position r, D_0 is the hypothetical water self-diffusion coefficient in the 'dry' pasta $(S = 0)$, and A is a constant which can be determined by fitting the experimental profiles. The two parameters D_0

and A are also related to the water self-diffusion coefficient in the fully hydrated pasta, D_{sat}, since $A = \ln(D_{sat}/D_0)$. Figure 14.11 includes the theoretical profile used to fit the moisture profile and the radial expansion.

As well as fitting the moisture profiles and the radial expansion data, the same model has been used to calculate the exponent n characterizing the increase in the total mass of water according to

$$M(t)/M(t=\infty) = K\,t^n. \tag{14.9}$$

The theoretical data points in Fig. 14.12 were calculated in this way from the integrated areas of the theoretical moisture profiles. Table 14.2 lists the sets of parameters D_0, A and n (theory) obtained from the model for three pasta types differing in the amount of hard and soft wheat used in their manufacture. It is worth noting that although the n-values derived from bulk weighing experiments are widely quoted in the literature, they do not have the significance that is often attached to them. They depend on the morphology of the sample (whether rectangular, cylindrical or spherical) and are usually calculated assuming constant radii and constant diffusion coefficients. Neither of these assumptions is valid for the

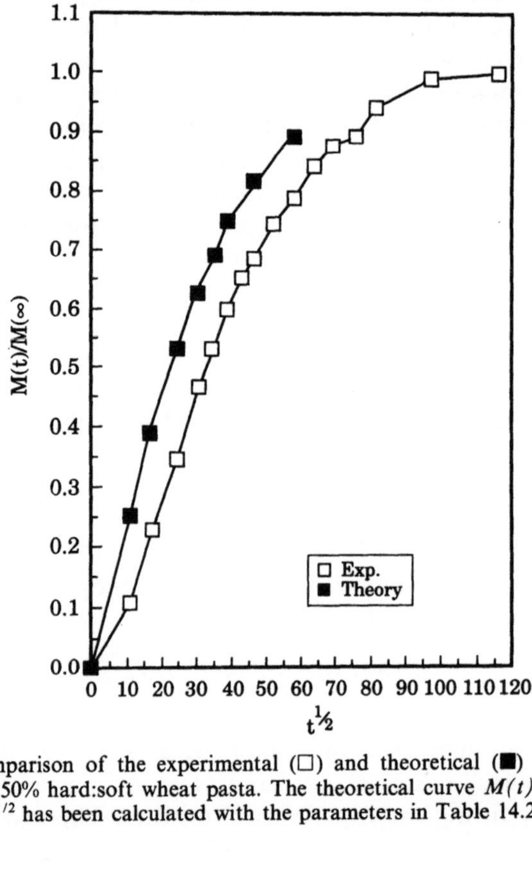

Fig. 14.12 A comparison of the experimental (□) and theoretical (■) bulk rehydration curves for the 50:50% hard:soft wheat pasta. The theoretical curve $M(t)$ as a function of $t^{1/2}$ has been calculated with the parameters in Table 14.2.

Table 14.2 Parameters in the non-Fickian diffusion model of pasta

Wheat	D_0 (10^{-9} cm^2 s^{-1})	A	n (experiment)	n (theory)
Hard	0.3	7.0	0.71	0.59
50:50	3.0	3.7	0.66	0.44
Soft	4.5	5.2	0.59	0.27

pasta samples. Even if these assumptions were valid, an expression of the simple polynomial form Kt^n rarely applies to more than 60% of the fractional solvent uptake, after which additional terms are required. To study properly the nature of the rehydration kinetics, it is necessary to monitor the time course of the moisture profiles.

It is apparent from the pasta imaging experiments that the amount of hard wheat in the pasta increases as the rate of water penetration (D_0) decreases and the diffusion becomes more non-Fickian (n increases). Despite these reasonable deductions, the theoretical fit to the data can still be considerably improved by refining the model. For example, an improved model can be devised by introducing relaxation times associated with the diffusion coefficient and expansion mechanisms. This may also explain the apparent delay in the experimental plot in Fig. 14.12 compared to the theoretical results. However, incorporating these refinements also increases the number of parameters in the model, and so it has not been attempted in these preliminary investigations.

14.7 Conclusions

It is clear that MRI does have the potential of revolutionizing food processing science. Feasibility studies have been reported on many different food systems, but, as yet, there have been relatively few quantitative imaging studies on real food systems. The detailed analysis of the rehydration of extruded pasta has shown the considerable effort needed to quantify and interpret the imaging results. The next decade should see increasing numbers of such studies and improved theoretical models. The pasta rehydration experiment has used only one particular imaging protocol. But there are many further possibilities. For example, radial profiles of the apparent water diffusivity could be measured directly by replacing the CPMG preparation sequence in Fig. 14.1 with a diffusion-weighted pulsed gradient preparation sequence. In this way it should be possible to compare profiles of the water self-diffusion coefficient with the moisture content profiles and to establish directly the dependence of the water diffusivity on water content. This is preferable to postulating some dependence such as equation (14.8) and then deriving parameters by fitting the moisture profiles. Provided either the relaxation times and/or the water self-diffusion coefficient are temperature dependent, MRI can, in principle, be used to image temperature distributions. It should therefore be possible to measure radial temperature profiles with the above radial microimaging protocol. These, in turn, could be modelled in terms of the thermal diffusivity in a similar way to the water diffusivity described in this chapter.

References

Belton, P.S., Colquhoun, I.J. and Hills, B.P. (1993) Applications of NMR to food science. *Annu. Rep. NMR Spectrosc.*, **26**, 1–53.

Callaghan, P.T. (1991) *Principles of Nuclear Magnetic Resonance Microscopy*, Oxford Science Publications, Oxford.

Dickinson, R.J., Hall, A.S, Hind, A.J. and Young, I.R. (1986) Measurement of changes in tissue temperature by MRI. *Comp. Assisted Tomogr.*, **10**, 468–472.

Duce, S.L., Amin, A., Horsfield, M.A., Tyszka, M. and Hall, L.D. (*1995*) *Int. Dairy J.*, in press.

Gassner, G. (1989) in *Nuclear Magnetic Resonance in Agriculture* (eds. P.E. Pfeffer and W.V. Gerasimowicz), CRC Press, Boca Raton, FL, p. 405.

Hills, B.P. (1992) The proton exchange cross relaxation model of water proton relaxation in biopolymer systems. *Mol. Phys.*, **76**, 489–508.

Hills, B.P. and Babonneau, F. (*1995*) Quantitative radial imaging of porous particle beds with varying water contents. *Magn. Reson. Imaging*, **12**, 1065–1074.

Hills, B.P., Belton, P.S. and Quantin, V.M. (1993a) Water proton relaxation in heterogeneous systems I. Saturated randomly packed suspensions of impenetrable particles. *Mol. Phys.*, **78**, 893–908.

Hills, B.P., Smart, S.C., Quantin, V.M. and Belton, P.S. (1993b) NMR microimaging studies of water diffusivity in saturated, microporous systems. *Magn. Reson. Imaging*, **11**, 1175–1184.

Hills, B.P., Wright, K.M., Wright, J. and Hall, L.D. (*1995*) MRI studies of drying in granular beds of non-porous particles. *Magn. Reson. Imaging*, **12**, 1053–1063.

Howe, F.A. (1988) Relaxation times in paramagnetically doped agarose gel as a function of temperature and ion concentration. *Magn. Reson. Imaging*, **6**, 263–270.

Jenner, C.F., Xia, Y., Eccles, C.D. and Callaghan, P.T. (1988) Circulation of water within a wheat grain revealed by NMR microimaging. *Nature (London)*, **336**, 399–407.

Kauten, R.J., Maneval, J.E. and McCarthy, M.J. (1991) Fast determination of spatial localized volume fractions in emulsions. *J. Food Sci.*, **56**, 799–801.

McCarthy, K.L. and Heil, J.R. (1990) in *Food Emulsions and Foams: Theory and Practice* (eds. P.J. Wan, J.L. Cavallo, F.Z. Saleeb and M.J. McCarthy), American Institute of Chemical Engineers, New York, pp. 71–75.

McCarthy, K.L., Kauten, R.J. and Agemura, C.K. (1992) Application of NMR imaging to the study of velocity profiles during extrusion processing. *Trends Food Sci. Technol.*, **3**, 215–219.

McCarthy, M.J., *Magnetic Resonance Imaging in Foods*, Chapman and Hall, London, 1994.

McCarthy, M.J. and Kauten, R.J. (1990) Magnetic resonance imaging applications in food research. *Trends Food Sci. Technol.*, **1**, 134–139.

McCarthy, M.J., Perez, E. and Ozilgen, M. (1991) Model for transient moisture profiles of a drying apple slab using data obtained with MRI. *Biotechnol. Prog.*, **7**, 540–543.

Peppas, N.A. and Brannon-Peppas, L. (1994) Water diffusion and sorption in amorphous macromolecular systems and foods. *J. Food Eng.*, **22**, 189–210.

Ruan, R. Schmidt, S. Schmidt, A.R. and Litchfield, J.B. (1991) Non-destructive measurement of transient moisture profiles and the moisture diffusion coefficient in a potato during drying and absorption by NMR imaging. *J. Food Process. Eng.*, **14**, 297–313.

Rutledge, D.N., Rene, F., Hills, B.P. and Foucat, L. (1994) Magnetic resonance imaging studies of the freeze-drying kinetics of potato. *J. Food Process. Eng..*, **17**, 325–352.

Sun, X., Litchfield, J.B. and Schmidt, S.J. (1993) Temperature mapping in a model food gel using MRI. *J. Food Sci.*, **58**, 168–172.

Stejskal, E.O. and Tanner, J.E. (1965) Spin diffusion measurements: spin echoes in the presence of a time dependent field gradient. *J. Chem. Phys.*, **42**, 288–292.
Wright, J.J. and Hall, L.D. (1994) Image obtained as part of a MAFF-DTI funded LINK programme, at the Herchel-Smith Laboratory, Cambridge.

Subject Index

MIX
Papier aus verantwortungsvollen Quellen
Paper from responsible sources
FSC® C105338

If you have any concerns about our products,
you can contact us on
ProductSafety@springernature.com

In case Publisher is established outside the EU,
the EU authorized representative is:
Springer Nature Customer Service Center GmbH
Europaplatz 3, 69115 Heidelberg, Germany

Printed by Libri Plureos GmbH
in Hamburg, Germany